ANNUAL EDITIONS

World Politics 04/05

Twenty-fifth Edition

EDITOR

Helen E. Purkitt

U.S. Naval Academy

Dr. Helen Purkitt obtained her Ph.D. in International Relations from the University of Southern California. She is Professor of Political Science at the U.S. Naval Academy. Her research and teaching interests include political psychology, African politics, international security, and environment politics. She is currently completing a study of how the United States government can better monitor and deter emerging dual use biotechnology threats. She is also currently collecting data for a basic research project on changing reference points in political problem framing and is writing an online textbook on how people think about politics and make political decisions for Political Psychology classes. Recently completed projects include a study of dual-use biotechnology trends in developing countries for the Defense Threat Reduction Agency (DTRA) and a co-authored book with Stephen Burgess entitled, *South Africa's Weapons of Mass Destruction* (forthcoming, 2005: Indiana University Press).

McGraw-Hill/Dushkin

2460 Kerper Blvd., Dubuque, IA 52001

Visit us on the Internet
http://www.dushkin.com

Credits

1. **New World Order**
   Unit photo—Royalty-Free/CORBIS
2. **World Economy**
   Unit photo—Royalty-Free/CORBIS
3. **Weapons of Mass Destruction**
   Unit photo—StockTrek/Getty Images
4. **North America**
   Unit photo—PhotoLink/Getty Images
5. **Latin America**
   Unit photo—Ryan McVay/Getty Images
6. **Europe**
   Unit photo—Royalty-Free/CORBIS
7. **Former Soviet Union**
   Unit photo—Mel Curtis/Getty Images
8. **The Pacific Basin**
   Unit photo—David Buffington/Getty Images
9. **Middle East and Africa**
   Unit photo—M. Freeman/PhotoLink/Getty Images
10. **International Organizations and Global Issues**
    Unit photo—Royalty-Free/CORBIS

Copyright

Cataloging in Publication Data
Main entry under title: Annual Editions: World Politics. 2004/2005.
1. World Politics—Periodicals. I. Purkitt, Helen E., *comp.* II. Title: World Politics.
ISBN 0–07–286157–6   658'.05   ISSN 1098–0300

© 2005 by McGraw-Hill/Dushkin, Dubuque, IA 52001, A Division of The McGraw-Hill Companies.

Copyright law prohibits the reproduction, storage, or transmission in any form by any means of any portion of this publication without the express written permission of McGraw-Hill/Dushkin, and of the copyright holder (if different) of the part of the publication to be reproduced. The Guidelines for Classroom Copying endorsed by Congress explicitly state that unauthorized copying may not be used to create, to replace, or to substitute for anthologies, compilations, or collective works.

Annual Editions® is a Registered Trademark of McGraw-Hill/Dushkin, A Division of The McGraw-Hill Companies.

Twenty-Fifth Edition

Main cover image © EPA Photo/EPA/Stephen Chernin; Background cover image © PhotoDisk Imaging/Getty Images
Printed in the United States of America   1234567890QPDQPD0987654   Printed on Recycled Paper

Editors/Advisory Board

Members of the Advisory Board are instrumental in the final selection of articles for each edition of ANNUAL EDITIONS. Their review of articles for content, level, currentness, and appropriateness provides critical direction to the editor and staff. We think that you will find their careful consideration well reflected in this volume.

EDITOR
Helen E. Purkitt
U.S. Naval Academy

ADVISORY BOARD

Valerie J. Assetto
Colorado State University

William O. Chittick
University of Georgia

Alistair D. Edgar
Wilfrid Laurier University

Lev S. Gonick
Case Western Reserve University

Dennis R. Gordon
Santa Clara University

Malcolm J. Grieve
Acadia University

Ryan C. Hendrickson
Eastern Illinois University

William D. Jackson
Miami University

Young Whan Kihl
Iowa State University

William H. Meyer
University of Delaware

Suzanne Ogden
Northeastern University

John T. Rourke
University of Connecticut, Storrs

Bhim Sandhu
West Chester University

Bernard Schechterman
University of Miami

Choudhury M. Shamim
California State University-Fullerton

Primo Vannicelli
University of Massachusetts, Boston

Staff

EDITORIAL STAFF
Larry Loeppke, Managing Editor
Susan Brusch, Senior Developmental Editor
Jay Oberbroeckling, Developmental Editor
Lenny J. Behnke, Permissions Coordinator
Lori Church, Permissions Coordinator
Shirley Lanners, Permissions Coordinator
Bonnie Coakley, Editorial Assistant

TECHNOLOGY STAFF
Craig Purcell, eContent Coordinator

PRODUCTION STAFF
Beth Kundert, Production Manager
Trish Mish, Production Assistant
Karl Voss, Lead Typesetter
Jean Smith, Typesetter
Karen Spring, Typesetter
Sandy Wille, Typesetter
Tara McDermott, Designer

To the Reader

In publishing ANNUAL EDITIONS we recognize the enormous role played by the magazines, newspapers, and journals of the public press in providing current, first-rate educational information in a broad spectrum of interest areas. Many of these articles are appropriate for students, researchers, and professionals seeking accurate, current material to help bridge the gap between principles and theories and the real world. These articles, however, become more useful for study when those of lasting value are carefully collected, organized, indexed, and reproduced in a low-cost format, which provides easy and permanent access when the material is needed. That is the role played by ANNUAL EDITIONS.

Annual Editions: World Politics 04/05 is aimed at filling a void in materials for learning about world politics and foreign policy. The articles are chosen for those who are new to the study of world politics. The goal is to help students learn more about international issues that often seem remote but may have profound consequences for a nation's well-being, security, and survival.

International relations can be viewed as a complex and dynamic system of actions and reactions by a diverse set of factors. Such a dynamic process may produce new situations that induce further actions. The articles in this volume convey just how dynamic, interdependent, and complex the relations among different types of international factors are in contemporary international relations.

Interdependence means that events in places as far away as Latin America, Asia, the Middle East, and Africa may effect the United States, just as America's actions, and inaction, have significant repercussions for other states. Interdependence also refers to the increased role of non-state factors such as international corporations, the United Nations, and a rich array of non-governmental factors such as Cable News Network (CNN) and the terrorist network of Bin Ladin's al Qaeda.

The September 11, 2001 terrorist attack on the World Trade Towers and the Pentagon tragically underscored the reality that non-state factors increasingly influence the scope, nature, and pace of events worldwide. However, the U.S.-led military interventions in Afghanistan in 2002, U.S. military operation and occupation of Iraq starting in 2003, escalation of tensions between the U.S. and North Korea, and ongoing conflicts among countries throughout the world confirm that inter-state conflicts will also continue as a key feature of international relations.

Today, international events proceed at such a rapid pace that what is said about international affairs today may be outdated by tomorrow. It is important, therefore, that readers develop a mental framework or theory of the international system as a complex system of loosely connected and diverse sets of factors who interact around an ever-changing agenda of international issues.

This collection of articles about international events provides up-to-date information, commentaries about the current set of issues on the world agenda, and analyses of the significance of the issues and emerging trends for the structure and functioning of the post–cold war international system.

This twenty-fifth edition of Annual Editions: World Politics is divided into 10 units. The end of the cold war means that we can no longer view international relations through the prism of a bipolar system. Instead, national, regional, subregion, and transnational issues are increasingly important aspects of international relations in the emerging multipolar and multidimensional world system.

Articles in unit 1 summarize themes and broad areas of international concerns in a period of high uncertainty about future security threats. Each article in unit 1 offers alternative views about the emerging new world order and the important trends and factors that will shape world politics in the twenty-first century.

Now that globalism is a defining characteristic of the modern international system, it is imperative that students of international relations understand the linkages between economic and political trends. The articles in unit 2 address a wide variety of issues in the modern world economy from the potentially destabilizing consequences of expanding consumer debt worldwide, to the ways that illicit trade and smuggling of natural resources fuel civil wars, as well as how geography—along with limited capital-technology and trade revenues—limits the ability of many countries in the Southern hemisphere to achieve high levels of economic development. However, as the world shifts from a hydrocarbon molecule based economy to a bio-based economy, genes may replace petroleum as the basic unit of commerce. Thus, future conflicts may involve conflicts between gene-rich but technologically-poor states and gene-poor but technologically-rich nation-states over who will control access to increasingly valuable genes that are used in a variety of commercial biotechnology applications.

The articles in unit 3 discuss specific issues and emerging trends related to the spread of weapons of mass destruction. The lack of extensive stockpiles of weapons of mass destruction (WMD) or evidence of ongoing nuclear, chemical, or biological weapons programs in Iraq after the U.S. military intervention raises a host of questions about the adequacy of U.S. and other national intelligence and policy making agencies' capabilities. This also brings up more general issues related to how to monitor WMD weapons proliferation in the future. The public scandal involving Dr. Khan—the father of the Pakistani who was involved in selling nuclear technology and expertise to North Korea, Libya, and Iran—highlights the importance of a growing black market in WMD materials around the world. Today, the world faces nation-state WMD proliferation threats, at the same time new and unprecedented security threats are tied to new ways to manufacture chemicals and the use of genetic modification techniques to build exotic new biological warfare weapons. Several new questions must now be addressed related to how to deter, defend against, and cope with the effects of weapons of mass destruction. The unanswered questions are important ones to address because the probability is greater today than in the past that WMD weapons will be used in the future by either nation-state antagonists, terrorist groups, or even lone individuals. The anthrax attacks in 2001 in the United States serve as precautionary reminder to authorities worldwide that they must now prepare to defend against new threats at the local, national, and international levels of world society.

Articles in units 4 through 9 focus on the implications of radical Islamic fundamentalism, the U.S. War on terrorism, and longer term international and regional trends in seven geographical areas or subsystems: North America, Latin America, Europe, Russia and the other independent nation-states of the former Soviet Union, the Pacific Basin, the Middle East, and Sub-Saharan Africa. A common theme running through these articles is the increased challenges facing local, national, and regional political authorities who must simultaneously cope with the problems generated by economic globalism and sub-national political changes.

The final unit of this reader, unit 10, examines unresolved issues and new trends in the post–cold war era related to the different roles that the United Nations plays for member nation-states and how to cope with a variety of existing and new global problems and trends. The fact that the United States returned to the United Nations for help in managing the transition to civilian rule in Iraq at the end of 2003 reminds us that this highly beleaguered organization remains one of the world's most useful institutions for managing conventional and unconventional security threats in international relations. While the United States, along with many other western countries are focused primarily on increasing cooperation in fighting a global War on Terrorism, most governments around the world are more concerned with other security issues and the need to fight against other networks involved with the illegal movement of drugs, arms, intellectual property, people, and money.

Meanwhile, the rapid spread of the bird flu in 2004 following on the heels of the 2003 SARS epidemic and the continuing spread of the HIV/AIDS pandemic around the world serve as potent reminders of why greater international cooperation will be needed in the future. Even before the HIV/AIDS pandemic peaks in Southern Africa, there are signs indicating that the epicenter of the global pandemic is already shifting to Eurasia. The spread of the disease in Africa is a humanitarian tragedy of epic proportions that has lowered the life expectancy, disrupted the economy, and devastated societies. As the HIV/AIDS pandemic spreads to Russia, India, and China, the global balance of power may be altered. Some experts warn that the effects of the spreading HIV/AIDS pandemic, much like the effects of the black plague in feudal Europe, may help to trigger major changes in power structures within existing nation-states and even the structure of the international system.

Increased numbers of analysts now call the twenty-first Century the biological century in recognition of the dramatic political, economic, and social changes that are likely to occur as the result of major research breakthroughs and bio-tech innovations. The dark side of the biological revolution is that novel processes and procedures can also be used to create new types of weapons of mass destruction that may be difficult to recognize or prevent. As these developments unfold, international politics promise to be a complex mix of old and new trends and factors. Terrorist, mercenaries, and the increased use of "child soldiers" all promise to figure as prominent factors in future years. Longstanding problems including water shortages, global warming, resource pollution, and famine or malnourishment for millions of the world's citizens remain unchecked problems. At the same time conspicuous overconsumption in the world's richer countries means that obesity is now the leading cause of health problems in the developed world.

I would like to thank Larry Loeppke and his associates at McGraw-Hill/Dushkin for their help in putting this volume together. Many users of *Annual Editions: World Politics* took the time to contribute articles or comments on this collection of readings. I greatly appreciate these suggestions and the article evaluations. Please continue to provide feedback to guide the annual revision of this anthology by filling out the postage-paid *article rating form* on the last page of this book.

Helen Purkitt

Helen E. Purkitt
Editor

Contents

To the Reader — iv
Topic Guide — xiii
Selected World Wide Web Sites — xvii

UNIT 1
New World Order

Five articles consider some of the challenges facing the world: the impact of local conflicts on foreign policy, major influences on domestic and international security, and the consequences of globalization.

Unit Overview — xx

1. **Clash of Globalizations,** Stanley Hoffmann, *Foreign Affairs,* July/August 2002

   The events of **September 11, 2001 marked the beginning of a new era**, but how the events translate into global politics is moot. Stanley Hoffman summarizes current approaches used to understand international relations and **concludes that three realities characterize the modern state system:** Great powers' rivalries have not disappeared; interstate war is less common; all states' foreign policies are increasingly shaped by domestic politics in addition to economic and military power. — 3

2. **War,** Lawrence Freedman, *Foreign Policy,* July/August 2003

   Modern wars range from preemptive wars by the United States using high-tech arsenals to swiftly defeat third world regimes to brutal civil wars. Lawrence Freedman examines **several generalizations about the nature of war** and concludes, among other things that future warfare will most likely be **asymmetric conflicts** that will be **difficult ones for the United States to win** and will **involve large numbers of casualties**. — 8

3. **The Protean Enemy,** Jessica Stern, *Foreign Affairs,* July/August 2003

   Al Qaeda survives by constantly evolving, adapting its mission, working with new groups, recruiting new members, and by maintaining virtual networks worldwide through the use of the Internet. To be effective, Western counter-terrorist tactics must also be highly adaptive. Stern recommends appealing to radical Islamic revivalists who oppose violence and working to prevent al Qaeda from acquiring weapons of mass destruction. — 12

4. **A World of Exiles,** *The Economist,* January 4, 2003

   Emigres have long sought to pressure the government in their adopted home. Today **emigres are also making their influence felt in their home countries**. The mingling of homeland interests with concerns back home explains many aspects of modern international relations. Thus, diaspora members helped Croatia win early international recognition and paid for much of Eritrea's war with Ethiopia. — 17

5. **The People's Sovereignty,** George Soros, *Foreign Policy,* January/February 2004

   George Soros, a philanthropist who has given more than $5 billion to populist causes, explains how the priciple of **sovereignty can block efforts to help people** in other nation-states. His foundation's approach for overcoming this barrier is to **give money directly to** support **local governments and nongovernmental organizations (NGOs)**. — 20

The concepts in bold italics are developed in the article. For further expansion, please refer to the Topic Guide and the Index.

UNIT 2
World Economy

Three articles examine the global marketplace as politics redefine the rules of the economic game.

Unit Overview — 22

6. **Charging Ahead,** Joshua Kurlantzick, *The Washington Monthly,* May 2003

   For decades the American consumer market helped drive the world economy. Americans spent and foreigners invested in the U.S. economy. Joshua Kurlantzick explains how America's biggest **new export—credit cards—could** have perverse and unintended negative consequences throughout the world that might **bring down the world economy**. — 24

7. **The Market for Civil War,** Paul Collier, *Foreign Policy,* May/June 2003

   A recent study of civil conflict over the past 40 years finds that **poverty and illegal trade in natural resources**, rather than ancient political feuds and ethnic tensions, are the **main causes of civil wars**. — 28

8. **From Petro to Agro: Seeds of a New Economy,** Robert E. Armstrong, *Defense Horizons,* October 2002

   As the world shifts from a hydrocarbon molecule based economy to a bio-based economy, **genes will replace petroleum as the basic unit of commerce**. National security implications accompanying this shift may include a downgrading of the importance of dealings with oil-rich countries and an increase in conflicts between gene-rich but technologically-poor states and gene-poor but technologically-rich nations over **who will control access to high valued genes** used in commercial applications. — 33

UNIT 3
Weapons of Mass Destruction

Five articles discuss nuclear proliferation and the use of toxic weapons.

Unit Overview — 42

9. **Ex-Inspector Says C.I.A. Missed Disarray in Iraqi Arms Program,** James Risen, *The New York Times,* January 26, 2004

   Former head of the U.S. Iraqi Inspection Team, David A. Kay, reported that **Iraq's WMD programs were in a state of disarray in the years prior to the U.S. invasion** and that the **program has plunged into a "vortex of corruption" in the late 1990s**. According to Kay, U.S. intelligence missed signs of the chaos in the Iraqi WMD program due to an over-reliance on intelligence from spy satellites, intercepted communications, foreign spies, defectors, and exiles. — 45

10. **The Nuclear Crisis on the Korean Peninsula: Avoiding the Road to Perdition,** Selig S. Harrison, *Current History,* April 2003

    If the United States confronts North Korea it is likely to exacerbate the nuclear crisis and undermine United States relations with Northeast Asia. Instead of confrontation, the Task Force on U.S.-Korea Policy **recommends a policy of constructive engagement** with friends and allies in the region to help **exert maximum pressure on North Korea** and pursue a long-term policy of liberalizing the North Korean system. — 48

11. **N. Korea Shops Stealthily for Nuclear Arms Gear,** Joby Warrick, *Washington Post,* August 15, 2003

    An intercepted French cargo ship was found to be carrying aluminum pipes that were ultimately destined for North Korea's secret nuclear bomb program. The captured ship offered a "glimpse into the shadowy world of weapons proliferation, in which missile parts and bomb materials circle the globe undetected, secreted away in cargo containers and suitcases." This "gray zone" of proliferation persists due to weak states, open borders, lack of controls, and a ready market of buyers and sellers for weapons of mass destruction. — 63

The concepts in bold italics are developed in the article. For further expansion, please refer to the Topic Guide and the Index.

12. **Nuclear Nightmares,** Bill Keller, *The New York Times Magazine,* May 26, 2002

    "Experts on terrorism and proliferation agree that ***sooner or later, an attack will happen in the United States***. When and how remain the most challenging questions." Given the difficulties involved in obtaining the amount of fissile materials needed for a full-fledged bomb, ***many experts*** are ***now predicting that terrorists are most likely to use radiation and other nuclear materials*** to cause disruptions, terror, and deaths. 67

13. **Towards an Internet Civil Defence Against Bioterrorism,** Ronald E. LaPorte et al., *The Lancet Infectious Diseases,* September 2001

    There is little evidence that the large resources put into bioterrorism preparedness work. We must face the disturbing fact that it is ***very difficult to predict and guard against bioterrorism*** because there are too many targets, too many means to penetrate the targets, and the bioterrorists are crafty. Instead of building an inflexible Maginot line of defence as we are now, perhaps ***we should consider an ever alert, flexible electronic-matrix of civil defence***. 77

UNIT 4
North America

The five articles in this unit discuss current and future United States and Canadian roles in world policy and international trade.

Unit Overview 80

Part A. The United States

14. **Bush's Revolution,** Ivo H. Daalder and James M. Lindsay, *Current History,* Volume 102, No 667, November 2003

    Bush's foreign policy has been designed to challenge the existing order. Daalder and Lindsay review how his doctrines and actions ***have changed the course of American foreign*** policy and explain how the consequences of his actions will be felt for years to come. 83

15. **Supremacy by Stealth: Ten Rules for Managing the World,** Robert D. Kaplan, *The Atlantic Monthly,* July/August 2003

    Robert Kaplan offers some ***rules and tactics of how the United States should operate*** on a tactical level to manage an unruly world. 91

16. **The Watchful and the Wary,** Robert Dreyfuss, *Mother Jones,* July/August 2003

    The ***War on Terrorism has changed a 30-year prohibition on the FBI and CIA from spying on Americans***. Today, the FBI and CIA are building a massive intelligence network designed to spy on terrorists—and on everyday Americans. Temporary actions after September 11, 2001 are not a permanent routine aspect of law enforcement. 101

Part B. Canada

17. **Economic Crossroads on the Line,** Michael Grunwald, *Washington Post,* December 26, 2001

    After September 11th the United States and Canada moved to increase barriers along their 5,500 mile frontier. The dramatic costs of doing so have helped energize long-term commitments from both nations to use technology to create a "smarter border" to increase security. Nevertheless, the ***changes created delays*** for Canadians, 90 percent of who live within 100 miles of the border, and ***highlighted attitudinal differences between citizens of the two countries*** on several issues related to the War on Terrorism. 104

18. **Canada Links Arrest of 19 to Possible Terrorism Ties,** Clifford Krauss, *The New York Times,* August 24, 2003

    Canadian security officials—for possible ties to terrorism—detained several immigrants from Pakistan after they were ***found taking flying lessons at a school near an Ontario nuclear power plant***. 109

The concepts in bold italics are developed in the article. For further expansion, please refer to the Topic Guide and the Index.

UNIT 5
Latin America

Two selections consider Latin American relations in the Western Hemisphere with regard to politics, economic reform, and trade.

Unit Overview   110

19. **Free Trade on Trial,** The Economist, January 3, 2004

    On the tenth anniversary, the North American Free-Trade Agreement (NAFTA) remains unpopular in all three countries: the United States, Canada, and Mexico. Intra-area trade and foreign investment expanded greatly but economic growth in Mexico has been dismal, manufacturing jobs have declined in the United States, and environmental problems associated with the Maquiladora cluster has worsened. Most now agree that the benefits of the free trade agreement were overstated.   113

20. **Latin America's New Political Leaders: Walking on a Wire,** Michael Shifter, Current History, February 2003

    "Today's underlying political currents in Latin America are less about ideology and more about a public desire to find leaders who can effectively address everyday problems, and who do so honestly. The formulas of the past are questioned—whether 'socialism' in the 1970s or 'neoliberalism' in the 1990s—and largely dismissed. With traditional ideas and structures breaking apart, new leaders are being called on to produce results."   116

UNIT 6
Europe

Five selections review some of the historic events that will alter Western and Central Europe. Topics include the European Union's expansion and Central/Eastern Europe's strivings toward democracy.

Unit Overview   122

Part A. West Europe

21. **Europe Enlarged, America Detached?,** Simon Serfaty, Current History, March 2003

    The single currency in 2002, enlargement to 10 new members in 2002, the constitutional convention initiated in 2003, and the Intergovernmental Conference (IGC) scheduled for 2004 are marks of the final transformation of Europe. While the EU is achieving a new territorial and political synthesis, NATO has increasingly become an afterthought. Serfaty describes **the "new normalcy" and urges more, not less, integration between the EU and NATO**.   125

22. **America as European Hegemon,** Christopher Layne, The National Interest, Summer 2003

    Layne describes U.S. aims in Eurpoe during the post-war period to illustrate that the United States has always sought to assert its hegemony and France and Germany seek to create a European counter balance to U.S. hegemony. **Within a widened Europe, France and Germany—with Russia and sometimes China—are developing new habits of diplomatic cooperation to oppose Washington**, while the United Kingdom and newer members of a widened Europe work closely with the United States.   131

23. **Forget Asylum-Seekers: It's the People Inside who Count,** The Economist, May 10, 2003

    "The real issue for European societies is not how to keep new foreigners out but how to integrate the minorities they already have."   139

The concepts in bold italics are developed in the article. For further expansion, please refer to the Topic Guide and the Index.

24. **How the Armies of Europe Let Their Guard Down,** Philip Shiskin, *Wall Street Journal,* February 18, 2003

The 17 European countries in NATO have about 2.3 million active-duty troops, about a million more than the U.S. does. But many of NATO's European national forces are poorly equipped, in part because so much money is spent on pay and benefits for a substantially older military force than those maintained by the United States. ***While the U.S. spends 36% of its defense budget on pay and benefits, most NATO members earmark an average of 65%.*** Personnel expenditures leave much less for technology, weapons, and other gear a modern force needs. **142**

Part B. Central Europe

25. **A Nervous New Arrival on the European Union's Block,** *The Economist,* August 30, 2003

With nearly 40 million people, ***Poland accounts for roughly half the population and half the GDP of all ten incoming countries joining the European Union in 2004.*** But ***it must also do the most to get into shape.*** Poland sees EU membership as playing a big part in its future security but also hopes that the EU keeps rolling eastward. **145**

UNIT 7
Former Soviet Union

Two articles examine the relationship between the Bush administration and Russia and the recruiting of new terrorist members.

Unit Overview   148

26. **US-Russian Relations: Between Realism and Reality,** Celeste A. Wallander, *Current History,* October 2003

The **United States and Russia** have tried to form a strategic partnership but "***competing interests, divergent domestic views, and mismatched political and economic systems keep getting in the way.***" **151**

27. **The Terrorist Notebooks,** Martha Brill Olcott and Bakhtiyar Babajanov, *Foreign Policy,* March/April 2003

Excerpts from a young man recruited for jihad as one of a group of Central Asians, mostly Uzbek by nationality, describes their training at local terrorist schools in the mid-1990s. While many of these recruits were killed during U.S. bombings in Afghanistan, there remain many ***young people with limited education and diminishing economic prospects who live in communities throughout Central Asia that are likely to be future recruits for radical forms of Islam.*** **156**

UNIT 8
The Pacific Basin

Four articles examine some of the countries who are instrumental in the economic evolution of the Pacific Basin.

Unit Overview   166

28. **Changing Course on China,** Elizabeth Economy, *Current History,* September 2003

Relations between ***China and the United States changed from strategic competitors to partners against terror after September 11, 2001***—but Taiwan, human rights, and trade issues remain sticking points. **169**

The concepts in bold italics are developed in the article. For further expansion, please refer to the Topic Guide and the Index.

29. **How to Deal with North Korea,** James T. Laney and Jason T. Shaplen, *Foreign Affairs,* March/April 2003

"***Pyongyang's*** belligerent behavior should not obscure other ***dramatic conciliatory steps North Korea has taken in recent years***—steps suggesting that, even now, a solution is within reach. The ***trick is to craft a plan that does not reward the North for its misdeeds***. In such a plan, all major outside powers should guarantee the security of the entire Korean Peninsula first. This will remove Pyongyang's excuse for nuclear proliferation—and break the deadlock on the world's last Cold War frontier."   **175**

30. **Can India Overtake China?,** Yasheng Huang and Tarun Khanna, *Foreign Policy,* July/August 2003

While China's export-led manufacturing boom is largely a creation of foreign direct investment (FDI), India has spawned a number of domestic entrepreneurs that now compete internationally with the best that Europe and the United States has to offer. India's stronger infrastructure, and more efficient capital markets and legal system are additional reasons why ***India's homegrown entrepreneurs may have a long-term advantage over China's inefficient banks and capital markets***.   **180**

31. **Dangerous Neighbours,** Ahmed Rashid, *Far Eastern Economic Review,* January 9, 2003

Several of Afghanistan's neighbors are keen to sponsor rival warlords much like they did in the early 1990s. The ***United States fears that the country will split along ethnic lines if U.S. troups leave***. Meanwhile, 90 percent of the attacks they face in the country are coming from groups based in Pakistan. Peace will require a stronger central government and economic development.   **184**

UNIT 9
Middle East and Africa

Four articles review the current state of the Middle East and Africa with regard to conflict, extremism, and democratic trends.

Unit Overview   **186**

Part A. The Middle East

32. **"Why Do They Hate Us?",** Peter Ford, *The Christian Science Monitor,* September 27, 2001

In this report, a variety of ***Muslims from different backgrounds***, including those who are sympathetic toward the United States, explain why the ***carnage of September 11, 2001*** was considered to be ***retribution for 50 years of U.S. policies*** in the region.   **189**

33. **The Reluctant Nation Builders,** Alan Sorensen, *Current History,* December 2003

The Bush administration experiment with nation-building in Iraq was not based on an extensive post-war plan or lessons from other recent nation-building exercises. Sorensen explains ***why America's makeshift attempt to remake Iraq could prove hard to sustain***. "If Iraq descends into chaos or comes to be regarded as a costly distraction from the war on terror instead of its central front, American resolve could falter."   **199**

34. **The Fall of the House of Saud,** Robert Baer, *The Atlantic Monthly,* May 2003

Americans long considered Saudi Arabia a politically stable and friendly government that provided lucrative business relationships. The country, however, is run by an increasingly dysfunctional royal family that has been funding militant Islamic movements abroad in an attempt to protect itself from them at home. Robert Baer, a former CIA operative, explains why today's ***Saudi Arabia can't last much longer*** and why the social and economic fallout of its demise could be calamitous.   **203**

The concepts in bold italics are developed in the article. For further expansion, please refer to the Topic Guide and the Index.

Part B. Sub-Saharan Africa

35. **America and Africa,** Salih Booker, William Minter, and Ann-Louise Colgan, *Current History,* May 2003

    "Africa's issues are global issues—HIV/AIDS, human development, new models for economic growth, peace, and democracy. Worldwide consciousness of HIV/AIDS pandemic has even forced its way into the pages of a United States president's State of the Union address. In practice, however, priorities are being set by another agenda—a war agenda."   **212**

UNIT 10
International Organizations and Global Issues

Two articles discuss international organizations and world peace, UN reform, and the ever changing world which goverments face.

Unit Overview   **216**

36. **United Nations,** Madeleine K. Albright, *Foreign Policy,* September/October 2003

    The former Secretary of State explains why despite all its problems, the **United Nations remains the world's best hope against disease, poverty, global crime, and war**—and all at a reasonable price.   **219**

37. **The Five Wars of Globalization,** Moises Naim, *Foreign Policy,* January/February 2003

    In addition to terrorism, **governments are fighting wars against other networks involved in drugs, arms, intellectual property, people, and money**. Governments will lose these wars until they adopt **new strategies to deal with a larger, unprecedented struggle that now shapes the world** as much as confrontations between nation-states once did.   **224**

Index   **230**
Test Your Knowledge Form   **234**
Article Rating Form   **235**

The concepts in bold italics are developed in the article. For further expansion, please refer to the Topic Guide and the Index.

Topic Guide

This topic guide suggests how the selections in this book relate to the subjects covered in your course. You may want to use the topics listed on these pages to search the Web more easily.

On the following pages a number of Web sites have been gathered specifically for this book. They are arranged to reflect the units of this *Annual Edition*. You can link to these sites by going to the DUSHKIN ONLINE support site at *http://www.dushkin.com/online/*.

ALL THE ARTICLES THAT RELATE TO EACH TOPIC ARE LISTED BELOW THE BOLD-FACED TERM.

Afghanistan
31. Dangerous Neighbours

Africa
7. The Market for Civil War
35. America and Africa

Al Qaeda
3. The Protean Enemy

Alternative visions of world politics
1. Clash of Globalizations
2. War
4. A World of Exiles
5. The People's Sovereignty
8. From Petro to Agro: Seeds of a New Economy
15. Supremacy by Stealth: Ten Rules for Managing the World
37. The Five Wars of Globalization

Arms control
9. Ex-Inspector Says C.I.A. Missed Disarray in Iraqi Arms Program
10. The Nuclear Crisis on the Korean Peninsula: Avoiding the Road to Perdition
12. Nuclear Nightmares
37. The Five Wars of Globalization

Arms trade
11. N. Korea Shops Stealthily for Nuclear Arms Gear

Asia
10. The Nuclear Crisis on the Korean Peninsula: Avoiding the Road to Perdition
11. N. Korea Shops Stealthily for Nuclear Arms Gear
27. The Terrorist Notebooks
28. Changing Course on China
30. Can India Overtake China?

Bush, George W.
9. Ex-Inspector Says C.I.A. Missed Disarray in Iraqi Arms Program
14. Bush's Revolution
35. America and Africa

Canada
17. Economic Crossroads on the Line
18. Canada Links Arrest of 19 to Possible Terrorism Ties
19. Free Trade on Trial

Central Intelligence Agency
9. Ex-Inspector Says C.I.A. Missed Disarray in Iraqi Arms Program
16. The Watchful and the Wary

Civil War
7. The Market for Civil War

Communication
1. Clash of Globalizations

Consumer debt
6. Charging Ahead

Cultural customs and values
1. Clash of Globalizations
32. "Why Do They Hate Us?"
34. The Fall of the House of Saud
37. The Five Wars of Globalization

Cultural values
1. Clash of Globalizations
32. "Why Do They Hate Us?"
34. The Fall of the House of Saud

Democracy
1. Clash of Globalizations
16. The Watchful and the Wary

Democratic politics
20. Latin America's New Political Leaders: Walking on a Wire

Democratization
35. America and Africa

Dependencies, international
1. Clash of Globalizations

Developing world
4. A World of Exiles
5. The People's Sovereignty
7. The Market for Civil War
8. From Petro to Agro: Seeds of a New Economy
19. Free Trade on Trial
20. Latin America's New Political Leaders: Walking on a Wire
37. The Five Wars of Globalization

Development, economic
1. Clash of Globalizations

Development, social
1. Clash of Globalizations

xiii

Domestic and foreign policy linkages
37. The Five Wars of Globalization

Drugs
37. The Five Wars of Globalization

Economics
1. Clash of Globalizations
34. The Fall of the House of Saud
37. The Five Wars of Globalization

Economics and politics
20. Latin America's New Political Leaders: Walking on a Wire

Elections
20. Latin America's New Political Leaders: Walking on a Wire

Energy
34. The Fall of the House of Saud

Environmental issues
7. The Market for Civil War
8. From Petro to Agro: Seeds of a New Economy

Ethnic conflict
4. A World of Exiles
7. The Market for Civil War
23. Forget Asylum-Seekers: It's the People Inside who Count
31. Dangerous Neighbours
32. "Why Do They Hate Us?"
33. The Reluctant Nation Builders

Europe, central
19. Free Trade on Trial
21. Europe Enlarged, America Detached?
25. A Nervous New Arrival on the European Union's Block

Europe, western
21. Europe Enlarged, America Detached?
22. America as European Hegemon
23. Forget Asylum-Seekers: It's the People Inside who Count
24. How the Armies of Europe Let Their Guard Down

Financial management
6. Charging Ahead

Foreign policy, U.S.
9. Ex-Inspector Says C.I.A. Missed Disarray in Iraqi Arms Program
12. Nuclear Nightmares
13. Towards an Internet Civil Defence Against Bioterrorism
14. Bush's Revolution
15. Supremacy by Stealth: Ten Rules for Managing the World
16. The Watchful and the Wary
17. Economic Crossroads on the Line
19. Free Trade on Trial
22. America as European Hegemon
26. US-Russian Relations: Between Realism and Reality
28. Changing Course on China
31. Dangerous Neighbours
32. "Why Do They Hate Us?"
33. The Reluctant Nation Builders
34. The Fall of the House of Saud
35. America and Africa

Future
12. Nuclear Nightmares
34. The Fall of the House of Saud

Future terrorist threats
12. Nuclear Nightmares

Global business practices
6. Charging Ahead

Globalism
1. Clash of Globalizations
2. War
5. The People's Sovereignty
6. Charging Ahead
8. From Petro to Agro: Seeds of a New Economy
15. Supremacy by Stealth: Ten Rules for Managing the World
19. Free Trade on Trial
30. Can India Overtake China?
37. The Five Wars of Globalization

Globalization
37. The Five Wars of Globalization

Homeland defense
12. Nuclear Nightmares
13. Towards an Internet Civil Defence Against Bioterrorism
16. The Watchful and the Wary
17. Economic Crossroads on the Line
18. Canada Links Arrest of 19 to Possible Terrorism Ties

Human rights
16. The Watchful and the Wary
35. America and Africa
37. The Five Wars of Globalization

Human security
5. The People's Sovereignty
7. The Market for Civil War
36. United Nations
37. The Five Wars of Globalization

Immigration
37. The Five Wars of Globalization

International trade
7. The Market for Civil War
37. The Five Wars of Globalization

Internet
13. Towards an Internet Civil Defence Against Bioterrorism

Iraq
9. Ex-Inspector Says C.I.A. Missed Disarray in Iraqi Arms Program
33. The Reluctant Nation Builders

Islamic extremism
32. "Why Do They Hate Us?"
34. The Fall of the House of Saud

Latin America
19. Free Trade on Trial

Latin America, Mexico
20. Latin America's New Political Leaders: Walking on a Wire

Latin American economies
19. Free Trade on Trial

Lending policies
6. Charging Ahead

Low-intensity conflict
1. Clash of Globalizations
2. War
7. The Market for Civil War
15. Supremacy by Stealth: Ten Rules for Managing the World
32. "Why Do They Hate Us?"
33. The Reluctant Nation Builders
37. The Five Wars of Globalization

Middle East
32. "Why Do They Hate Us?"
33. The Reluctant Nation Builders
34. The Fall of the House of Saud

Military: Warfare and Terrorism
12. Nuclear Nightmares
28. Changing Course on China
32. "Why Do They Hate Us?"
37. The Five Wars of Globalization

Military power
2. War
12. Nuclear Nightmares
15. Supremacy by Stealth: Ten Rules for Managing the World
22. America as European Hegemon
24. How the Armies of Europe Let Their Guard Down
33. The Reluctant Nation Builders

NATO
22. America as European Hegemon
24. How the Armies of Europe Let Their Guard Down

Natural resources
7. The Market for Civil War

Nuclear terrorism
12. Nuclear Nightmares

Nuclear weapons
9. Ex-Inspector Says C.I.A. Missed Disarray in Iraqi Arms Program
10. The Nuclear Crisis on the Korean Peninsula: Avoiding the Road to Perdition
11. N. Korea Shops Stealthily for Nuclear Arms Gear
12. Nuclear Nightmares
18. Canada Links Arrest of 19 to Possible Terrorism Ties

Osama bin Laden
3. The Protean Enemy

Political economy
4. A World of Exiles
6. Charging Ahead
7. The Market for Civil War
8. From Petro to Agro: Seeds of a New Economy
19. Free Trade on Trial
30. Can India Overtake China?

Population
4. A World of Exiles
23. Forget Asylum-Seekers: It's the People Inside who Count

Poverty
4. A World of Exiles
7. The Market for Civil War

Russia
26. US-Russian Relations: Between Realism and Reality

Science
12. Nuclear Nightmares

September 11th attacks, U.S.
32. "Why Do They Hate Us?"

Socioeconomic forces
6. Charging Ahead

Technology
12. Nuclear Nightmares

Terrorism
2. War
12. Nuclear Nightmares
13. Towards an Internet Civil Defence Against Bioterrorism
15. Supremacy by Stealth: Ten Rules for Managing the World
17. Economic Crossroads on the Line
18. Canada Links Arrest of 19 to Possible Terrorism Ties
32. "Why Do They Hate Us?"
34. The Fall of the House of Saud

Terrorist tactics
3. The Protean Enemy
27. The Terrorist Notebooks

Trade
6. Charging Ahead
7. The Market for Civil War
8. From Petro to Agro: Seeds of a New Economy
11. N. Korea Shops Stealthily for Nuclear Arms Gear
19. Free Trade on Trial
34. The Fall of the House of Saud

Transition economies
4. A World of Exiles
5. The People's Sovereignty
30. Can India Overtake China?
33. The Reluctant Nation Builders
37. The Five Wars of Globalization

Underdeveloped countries
1. Clash of Globalizations

United Nations
36. United Nations

War on terrorism
3. The Protean Enemy

Weapons of mass destruction
9. Ex-Inspector Says C.I.A. Missed Disarray in Iraqi Arms Program
10. The Nuclear Crisis on the Korean Peninsula: Avoiding the Road to Perdition
11. N. Korea Shops Stealthily for Nuclear Arms Gear
12. Nuclear Nightmares
13. Towards an Internet Civil Defence Against Bioterrorism
18. Canada Links Arrest of 19 to Possible Terrorism Ties

World Wide Web Sites

The following World Wide Web sites have been carefully researched and selected to support the articles found in this reader. The easiest way to access these selected sites is to go to our DUSHKIN ONLINE support site at *http://www.dushkin.com/online/*.

AE: World Politics 04/05

The following sites were available at the time of publication. Visit our Web site—we update DUSHKIN ONLINE regularly to reflect any changes.

General Sources

CIA Factbook
http://www.cia.gov/cia/publications/factbook/
This site provides information on various countries.

FACTs
http://www.ploughshares.ca
Useful site for research on inter-state conflicts.

Country Indicators for Foreign Policy
http://www.carleton.ca/cifp/
Statistical data on nation-states compiled by Carlton University, Canada.

Belfer Center for Science and International Affairs (BCSIA)
http://www.ksg.harvard.edu/bcsia/
BCSIA is a center for research, teaching, and training in international affairs.

Carnegie Endowment for International Peace
http://www.ceip.org
One of the goals of this organization is to stimulate discussion and learning among experts and the public on a wide range of international issues. The site provides links to the journal *Foreign Policy* and to the Moscow Center.

Central Intelligence Agency
http://www.odci.gov
Use this official home page to learn about many facets of the CIA and to get connections to other sites and resources, such as *The CIA Factbook,* which provides extensive statistical information about every country in the world.

The Heritage Foundation
http://www.heritage.org
This page offers discussion about and links to many sites of the Heritage Foundation and other organizations having to do with foreign policy and foreign affairs.

World Wide Web Virtual Library: International Affairs Resources
http://www.etown.edu/vl/
Surf this site and its links to learn about specific countries and regions, to research think tanks and organizations, and to study such vital topics as international law, development, the international economy, human rights, and peacekeeping.

UNIT 1: New World Order

Globalization Index
http://www.atkearney.com/main.taf?p=5,4,1,64
This site provides an explanation of how the Globalization Index is contructed and also provides country rankings.

The Globalization Website
http://www.emory.edu/SOC/globalization/
Discusses globalization and is a guide to available sources on globalization.

Women in International Politics
http://www.guide2womenleaders.com
This site contains data on women who have served as political leaders.

Avalon Project at Yale Law School
http://www.yale.edu/lawweb/avalon/terrorism/terror.htm
The Avalon Project Web site features documents in the fields of law, history, economics, diplomacy, politics, government, and terrorism.

Human Rights Web
http://www.hrweb.org
This useful site offers ideas on how individuals can get involved in helping to protect human rights around the world.

U.S. Air Force Institute for National Security Studies
http://www.usafa.af.mil/inss/occasion.htm
The full-text commissioned peer review reports on a variety of security topics affecting the United States and the world can be found at this Web site, sponsored by the Department of Defense.

UNIT 2: World Economy

International Monetary Fund
http://www.imf.org
This link brings you to the homepage for the International Monetary Fund.

Transparency International
http://www.transparency.org/pressreleases_archive/2002/2002.08.28.cpi.en.html
This site contains a Corruption Perceptions Index.

TI Bribe Payers Index
http://www.transparency.org/surveys/index.html#bpi
This site gives a breakdown of power capability comparisons amoung various countries as well as economic analyses.

Graphs Comparing Countries
http://humandevelopment.bu.edu/use_exsisting_index/start_comp_graph.cfm
This site allows you to compare various countries and nation-states with statistics using a visual tool.

International Political Economy Network
http://csf.colorado.edu/ipe/
This premier site for research and scholarship includes electronic archives.

Organization for Economic Cooperation and Development/FDI Statistics
http://www.oecd.org/daf/investment/
Explore world trade and investment trends and statistics on this site. It provides links to many related topics and addresses global economic issues on a country-by-country basis.

xvii

www.dushkin.com/online/

Virtual Seminar in Global Political Economy/Global Cities & Social Movements
http://csf.colorado.edu/gpe/gpe95b/resources.html

This site of Internet resources is rich in links to subjects of interest in regional environmental studies, covering topics such as sustainable cities, megacities, and urban planning.

World Bank
http://www.worldbank.org

News (press releases, summaries of new projects, speeches) and coverage of numerous topics regarding development, countries, and regions are provided at this site. Go to the research and growth section of this site to access specific research and data regarding the world economy.

UNIT 3: Weapons of Mass Destruction

National Defense University Website
http://www.ndu.edu

This contains information on current studies. This site also provides a look at the school where many senior marine, naval officers, and senior civilians attend prior to assuming top-level positions.

U.S.-Russia Developments
http://www.acronym.org.uk/start

This is a site maintained by Acronym Institute for Disarmament Diplomacy which provides information on U.S. and Russian disarmament activity.

The Bulletin of the Atomic Scientists
http://www.bullatomsci.org

This site allows you to read more about the Doomsday Clock and other issues as well as topics related to nuclear weaponry, arms control, and disarmament.

Federation of American Scientists
http://www.fas.org

This site provides useful information about and links to a variety of topics related to chemical and biological warfare, missiles, conventional arms, and terrorism.

ISN International Relations and Security Network
http://www.isn.ethz.ch

This site, maintained by the Center for Security Studies and Conflict Research, is a clearinghouse for extensive information on international relations and security policy.

The RMA Debate: Terrorism and Counter-terrorism
http://www.comw.org/rma/fulltext/terrorism.html

The RMA Debate is a gateway to full-text online resources about the Revolution in Military Affairs, information war, and asymmetrical warfare.

Terrorism Research Center
http://www.terrorism.com

The Terrorism Research Center features definitions and research on terrorism, counterterrorism documents, a comprehensive list of Web links, and profiles of terrorist and counterterrorist groups.

UNIT 4: North America

U.S. Department of State
http://www.state.gov/index.cfm

The site provides information organized by categories as well as "background notes" on specific countries and regions.

The Henry L. Stimson Center—Peace Operations and Europe
http://www.stimson.org/fopo/?SN=FP20020610372

The Future of Peace Operations has begun to address specific areas concerning Europe and operations. The site links to useful UN, NATO, and EU documents, research pieces, and news sites.

The North American Institute
http://www.northamericaninstitute.org

NAMI, a trinational public-affairs organization, is concerned with the emerging "regional space" of Canada, the United States, and Mexico and the development of a North American community. It provides links for study of trade, the environment, and institutional developments.

UNIT 5: Latin America

Inter-American Dialogue
http://www.iadialog.org

This is the Web site for IAD, a premier U.S. center for policy analysis, communication, and exchange in Western Hemisphere affairs. The 100-member organization has helped to shape the agenda of issues and choices in hemispheric relations.

UNIT 6: Europe

Central Europe Online
http://www.centraleurope.com

This site contains daily updated information under headings such as news on the Web today, economics, trade, and currency.

Europa: European Union
http://europa.eu.int

This server site of the European Union will lead you to the history of the EU (and its predecessors), descriptions of EU policies, institutions, and goals, and documentation of treaties and other materials.

NATO Integrated Data Service
http://www.nato.int/structur/nids/nids.htm

Check out this Web site to review North Atlantic Treaty Organization documentation, to read *NATO Review,* and to explore key issues in the field of European security and transatlantic cooperation.

Social Science Information Gateway
http://sosig.esrc.bris.ac.uk

A project of the Economic and Social Research Council (ESRC), this is an online catalog of thousands of Internet resources relevant to political education and research.

UNIT 7: Former Soviet Union

Russia Today
http://www.russiatoday.com

This site includes headline news, resources, government, politics, election results, and pressing issues.

Russian and East European Network Information Center, University of Texas at Austin
http://reenic.utexas.edu/reenic/index.html

This is *the* Web site for information on the former Soviet Union.

www.dushkin.com/online/

UNIT 8: The Pacific Basin

Crisisweb: The International Crisis Group (ICG)
http://www.crisisweb.org/home/index.cfm

ICG is an organization "committed to strengthening the capacity of the international community to anticipate, understand, and act to prevent and contain conflict." Go to this site to view the latest reports and research concerning conflicts around the world.

ASEAN Web
http://www.asean.or.id

This site of the Association of South East Asian Nations provides an overview of Asia: Web resources, summits, economic affairs, political foundations, and regional cooperation.

Inside China Today
http://www.insidechina.com

Part of the European Internet Network, this site leads you to information on all of China, including recent news, government, and related sites.

Japan Ministry of Foreign Affairs
http://www.mofa.go.jp

Visit this official site for Japanese foreign policy statements and press releases, archives, and discussions of regional and global relations.

UNIT 9: Middle East and Africa

Africa News Online
http://www.africanews.org

Open this site for up-to-date information on all of Africa, with reports from Africa's leading newspapers, magazines, and news agencies. Coverage is country-by-country and regional. Internet links are among the resource pages.

ArabNet
http://www.arab.net

This page of ArabNet, the online resource for the Arab world in the Middle East and North Africa, presents links to 22 Arab countries. Each country page classfies information using a standardized system.

Columbia International Affairs Online
http://www.ciaonet.org/cbr/cbr00/video/cbr_v/cbr_v.html

At this site find excerpts from al Qaeda's 2-hour videotape used to recruit young Muslims to fight in a holy war. The tape demonstates al Qaeda's use of the Internet and media outlets for propaganda and persuasion purposes.

ei: Electronic Intifada
http://electronicintifada.net/new.shtml

EI is a major Palestinian portal for information about the Palestinian-Israeli conflict from a Palestinian perspective.

IslamiCity
http://islamicity.com

This is one of the largest Islamic sites on the Web, reaching 50 million people a month. Based in California, it includes public opinion polls, links to television and radio broadcasts, and religious guidance.

MEMRI: The Middle East Research Institute
http://www.memri.org/video

Arab satellite channels air recent video clips on topics related to Islamic culture, fundamentalism, and terrorism from this site. For translations of what Arab leaders are telling their followers, go to http://www.memri.org.

UNIT 10: International Organizations and Global Issues

United Nations
http://untreaty.un.org

This site contains text on over 30,000 UN treaties.

HIV/AIDS
http://www.unaids.org

This is a site giving information on the rising toll of HIV/AIDS.

Commission on Global Governance
http://www.sovereignty.net/p/gov/gganalysis.htm

This site provides access to *The Report of the Commission on Global Governance,* produced by an international group of leaders who want to find ways in which the global community can better manage its affairs.

Global Trends 2005 Project
http://www.csis.org/gt2005/sumreport.html

The Center for Strategic and International Studies explores the coming global trends and challenges of the new millenium. Read their summary report at this Web site. Also access Enterprises for the Environment, Global Information Infrastructure Commission, and Americas at this site.

InterAction
http://www.interaction.org

InterAction encourages grassroots action, engages policy makers on advocacy issues, and uses this site to inform people on its initiatives to expand international humanitarian relief and development assistance programs.

IRIN
http://www.irinnews.org

The UN Office for the Coordination of Humanitarian Affairs provides free analytical reports, fact sheets, interviews, daily country updates, and weekly summaries through this site and e-mail distribution service. The site is a good source of news for crisis situations as they occur.

The North-South Institute
http://www.nsi-ins.ca/ensi/index.html

Searching this site of the North-South Institute—which works to strengthen international development cooperation and enhance gender and social equity—will help you find information and debates on a variety of global issues.

Uniited Nations Home Page
http://www.un.org

Here is the gateway to information about the United Nations. Also see *http:/www.undp.org/missions/usa/usna/htm* for the U.S. Mission at the UN.

We highly recommend that you review our Web site for expanded information and our other product lines. We are continually updating and adding links to our Web site in order to offer you the most usable and useful information that will support and expand the value of your Annual Editions. You can reach us at: *http://www.dushkin.com/annualeditions/*.

UNIT 1
New World Order

Unit Selections

1. **Clash of Globalizations**, Stanley Hoffmann
2. **War**, Lawrence Freedman
3. **The Protean Enemy**, Jessica Stern
4. **A World of Exiles**, The Economist
5. **The People's Sovereignty**, George Soros

Key Points to Consider

- Do you agree or disagree with Lawrence Freedman's thesis that the United States will experience difficulties and large numbers of casualties in future efforts to enemies in asymmetric conflicts?

- Jessica Stern recommends appealing to radical Islamic revivalists who oppose violence and working to prevent al Qaeda members from acquiring weapons of mass destruction as the best way to defeat al Qaeda's terrorist networks. Do you agree or disagree? Why?

- Are independent nation-states likely to remain the most important factor in international relations or will national sovereignty be subverted or even replaced by other factors?

- What types of non-state factors are likely to increase in influence in the future?

 **Links: www.dushkin.com/online/**
These sites are annotated in the World Wide Web pages.

Globalization Index
http://www.atkearney.com/main.taf?p=5,4,1,64

The Globalization Website
http://www.emory.edu/SOC/globalization/

Women in International Politics
http://www.guide2womenleaders.com

Avalon Project at Yale Law School
http://www.yale.edu/lawweb/avalon/terrorism/terror.htm

Human Rights Web
http://www.hrweb.org

U.S. Air Force Institute for National Security Studies
http://www.usafa.af.mil/inss/occasion.htm

At the beginning of the twenty-first century, there is a noticeable increase in efforts to predict important changes and to understand new patterns of relationships shaping international relations. Anyone can engage in the sport as there is little agreement among "futurists" about what to expect regarding causes of tension, types of conflict, or the patterns of interaction that may characterize international relations.

With the demise of the cold war, analysts focus more on the political and economic ramifications of the emerging international system. Many cite "globalization" as the dominant characteristic of the new international order. Globalization refers to the increased global interdependence of economic, communication, and transport systems. Globalization also refers to innovations in computer and other high-tech capabilities that are increasingly being used by people in all parts of world society. While many futurists stress the novel features of the new world order, others emphasize that many of the major changes in trade, communication, and transportation taking place today are similar to the profound remaking of the world during earlier centuries. During the 15th and 16th centuries, European nation-states and societies in Africa, the Americas, the Middle East, and to a lesser extent Asia were tied together in the first modern world economy.

Economic and cultural trends in the modern era of globalization may differ from earlier eras of intense globalization because modern trends are triggering integrative and disintegrative changes that are global in scope and at an unprecedented pace of change. The most significant disintegrative global trends include the rise of cultural extremism in Islamic, Judaic, and Christian cultures; increased economic inequality between the developed and developing sectors of world society; and the diffusion of high-technology weaponry. These trends often spark violent responses that trigger or aggravate international conflicts. In "Clash of Globalizations," Stanley Hoffmann concludes that the events of September 11 marked the beginning of a new era, but one where Great Powers' rivalries will continue to shape international relations. This changing vision of the world also emphasizes the increasingly important role that domestic politics now plays in the formulation of foreign policies for states. In another analysis entitled "War," Lawrence Freeman describes how the world is currently experiencing several different types of wars and concludes that future wars will be asymmetric conflicts that will be difficult ones for the United States to win and ones that will claim the lives of large numbers of people.

For some, the terrorist attacks against the World Trade Center in New York and the Pentagon in Washington D.C. on September 11, 2001, and the anthrax letter attacks the following month highlighted the vulnerabilities of economically developed societies to attacks by disaffected radicals who now pursue their political goals by killing large numbers of civilians. In "The Protean Enemy," Jessica Stern outlines how al Qaeda survives by constantly adapting its mission, working with new groups, recruiting new members, and by maintaining virtual Internet networks. Stern argues that Western counter-terrorist tactics must be equally adaptive. She recommends appealing to Islamic revivalists who both oppose violence and believe in doing what is necessary to prevent al Qaeda members from acquiring weapons of mass destruction.

The juxtaposition of contradictory trends in the international system also allows futurologists to speculate on the structure of the emerging world order. While some analysts emphasize the multipolar and multidimensional aspects of the international system, other analysts predict that the future world order will continue to be a unipolar world dominated by the United States. Many analysts now focus more on the role of non-state factors such as al Qaeda while other analysts go so far as to predict the demise of democratic capitalism and free-market integration and the rise of new factors that will replace the nation-state system before the end of the twenty-first Century.

Perhaps the most distinctive characteristic of the current era in international relations is the extensive degree of complexity that exists among a diverse set of factors. Today, large numbers of émigrés who migrated from their home country to new nation-states in an effort to make a better life or to escape political, religious, or other forms of persecution are making their presence felt by influencing the policies of their host country government. The result is a mingling of homeland interests with concerns back home that is complicating international relations. These transnational loyalties among diasporas members helped Croatia win early international recognition. Remittances from émigrés back home helped finance an inter-state war between Eritrea and Ethiopia that resulted in the deaths of thousands in trench warfare on a scale not seen since World War I.

Wealthy individuals such as Bill and Melinda Gates can also impact the way many governments in Africa and throughout the developing world now combat the HIV/AIDS pandemic and spread of other infectious diseases. In a similar fashion, another wealthy philanthropist, George Soros, through his private foundation grants has given more than $5 billion to populist causes that often makes a substantial difference on the ground. Soros has found that it is possible to have an effect by giving money directly to support local governments and nongovernmental organizations. Thus, thousands of daily transactions and exchanges throughout the world may be undermining the sovereignty of nation-states to such an extent that the future system may be very different than the one we know today.

Lasting peace and security requires local, national, regional, and international factors to manage shared resources such as water, while effectively coping with new transnational problems such as the spread of infectious diseases. At the same time, local and national authorities must meet the basic needs of growing populations and promote economic growth. How well political regimes cope with environmental and related challenges is receiving more attention today because it is now recognized that such factors serve as root causes or triggers fueling local conflicts. Conflicts in Somalia and Rwanda were not spontaneous outbreaks of clan warfare or ethnic violence. Rather, the underlying strains of hunger, drought, the longer-term lack of cultivable land, and the breakdown of traditional clan structures were the result. Along with those factors, population increases also fueled recent conflicts as did the residue of past political and economic relationships.

Occasionally, problems caused by environmental factors facilitate peace. A devastating drought in Mozambique, in addition to damage caused by prolonged civil war, led warring factions to agree to a negotiated peace settlement in the early 1990s. In a similar fashion, the worsening problems created by drought throughout the Middle East, along with the desires of President Assad of Syria and former Prime Minister Barak of Israel, prompted the leaders of the two countries to resume peace talks at the end of 1999. However, the breakdown of the peace process in the Middle East, the resumption of the intifada in 2000, the increased number of suicide bombings, and statements by Prime Minister Sharon that Palestinian leader Arafat was "irrelevant to a negotiated settlement" underscore how difficult it will be to construct lasting peace agreements among parties engaged in protracted conflicts. Some observers now predict that future conflicts about how to share increasingly scarce water may do more to aggravate and prolong the Arab-Israeli conflict than the wall that the Sharon government of Israel is currently completing despite pleas by the heads of nation-states and other factors in the international system.

Article 1

Clash of Globalizations

Stanley Hoffmann

A NEW PARADIGM?

WHAT IS THE STATE OF international relations today? In the 1990s, specialists concentrated on the partial disintegration of the global order's traditional foundations: states. During that decade, many countries, often those born of decolonization, revealed themselves to be no more than pseudostates, without solid institutions, internal cohesion, or national consciousness. The end of communist coercion in the former Soviet Union and in the former Yugoslavia also revealed long-hidden ethnic tensions. Minorities that were or considered themselves oppressed demanded independence. In Iraq, Sudan, Afghanistan, and Haiti, rulers waged open warfare against their subjects. These wars increased the importance of humanitarian interventions, which came at the expense of the hallowed principles of national sovereignty and nonintervention. Thus the dominant tension of the decade was the clash between the fragmentation of states (and the state system) and the progress of economic, cultural, and political integration—in other words, globalization.

Everybody has understood the events of September 11 as the beginning of a new era. But what does this break mean? In the conventional approach to international relations, war took place among states. But in September, poorly armed individuals suddenly challenged, surprised, and wounded the world's dominant superpower. The attacks also showed that, for all its accomplishments, globalization makes an awful form of violence easily accessible to hopeless fanatics. Terrorism is the bloody link between interstate relations and global society. As countless individuals and groups are becoming global actors along with states, insecurity and vulnerability are rising. To assess today's bleak state of affairs, therefore, several questions are necessary. What concepts help explain the new global order? What is the condition of the interstate part of international relations? And what does the emerging global civil society contribute to world order?

SOUND AND FURY

TWO MODELS made a great deal of noise in the 1990s. The first one—Francis Fukuyama's "End of History" thesis—was not vindicated by events. To be sure, his argument predicted the end of ideological conflicts, not history itself, and the triumph of political and economic liberalism. That point is correct in a narrow sense: the "secular religions" that fought each other so bloodily in the last century are now dead. But Fukuyama failed to note that nationalism remains very much alive. Moreover, he ignored the explosive potential of religious wars that has extended to a large part of the Islamic world.

Fukuyama's academic mentor, the political scientist Samuel Huntington, provided a few years later a gloomier account that saw a very different world. Huntington predicted that violence resulting from international anarchy and the absence of common values and institutions would erupt among civilizations rather than among states or ideologies. But Huntington's conception of what constitutes a civilization was hazy. He failed to take into account sufficiently conflicts within each so-called civilization, and he overestimated the importance of religion in the behavior of non-Western elites, who are often secularized and Westernized. Hence he could not clearly define the link between a civilization and the foreign policies of its member states.

Other, less sensational models still have adherents. The "realist" orthodoxy insists that nothing has changed in international relations since Thucydides and Machiavelli: a state's military and economic power determines its fate; interdependence and international institutions are secondary and fragile phenomena; and states' objectives are imposed by the threats to their survival or security. Such is the world described by Henry Kissinger. Unfortunately, this venerable model has trouble integrating change, especially globalization and the rise of nonstate actors. Moreover, it overlooks the need for international cooperation that results from such new threats as the proliferation of weapons of mass destruction (WMD). And it ignores what the scholar Raymond Aron called the "germ of a universal consciousness": the liberal, promarket norms that developed states have come to hold in common.

Taking Aron's point, many scholars today interpret the world in terms of a triumphant globalization that submerges borders through new means of information and communication. In this universe, a state choosing to stay closed invariably faces decline and growing discontent among its subjects, who are eager for material progress. But if it opens up, it must accept a reduced role that is mainly limited to social protection, physical protection against aggression or civil war, and maintaining national identity. The champion of this epic without heroes is *The New York Times* columnist Thomas Friedman. He contrasts barriers with open vistas, obsolescence with modernity, state control with free markets. He sees in globalization the light of dawn, the "golden straitjacket" that will force contentious publics to understand that the logic of globalization is that of peace (since war would interrupt globalization and therefore progress) and democracy (because new technologies increase individual autonomy and encourage initiative).

BACK TO REALITY

THESE MODELS come up hard against three realities. First, rivalries among great powers (and the capacity of smaller states to exploit such tensions) have most certainly not disappeared. For a while now, however, the existence of nuclear weapons has produced a certain degree of prudence among the powers that have them. The risk of destruction that these weapons hold has moderated the game and turned nuclear arms into instruments of last resort. But the game could heat up as more states seek other WMD as a way of narrowing the gap between the nuclear club and the other powers. The sale of such weapons thus becomes a hugely contentious issue, and efforts to slow down the spread of all WMD, especially to dangerous "rogue" states, can paradoxically become new causes of violence.

Second, if wars between states are becoming less common, wars within them are on the rise—as seen in the former Yugoslavia, Iraq, much of Africa, and Sri Lanka. Uninvolved states first tend to hesitate to get engaged in these complex conflicts, but they then (sometimes) intervene to prevent these conflicts from turning into regional catastrophes. The interveners, in turn, seek the help of the United Nations or regional organizations to rebuild these states, promote stability, and prevent future fragmentation and misery.

Third, states' foreign policies are shaped not only by realist geopolitical factors such as economics and military power but by domestic politics. Even in undemocratic regimes, forces such as xenophobic passions, economic grievances, and transnational ethnic solidarity can make policymaking far more complex and less predictable. Many states—especially the United States—have to grapple with the frequent interplay of competing government branches. And the importance of individual leaders and their personalities is often underestimated in the study of international affairs.

For realists, then, transnational terrorism creates a formidable dilemma. If a state is the victim of private actors such as terrorists, it will try to eliminate these groups by depriving them of sanctuaries and punishing the states that harbor them. The national interest of the attacked state will therefore require either armed interventions against governments supporting terrorists or a course of prudence and discreet pressure on other governments to bring these terrorists to justice. Either option requires a questioning of sovereignty—the holy concept of realist theories. The classical realist universe of Hans Morgenthau and Aron may therefore still be very much alive in a world of states, but it has increasingly hazy contours and offers only difficult choices when it faces the threat of terrorism.

At the same time, the real universe of globalization does not resemble the one that Friedman celebrates. In fact, globalization has three forms, each with its own problems. First is economic globalization, which results from recent revolutions in technology, information, trade, foreign investment, and international business. The main actors are companies, investors, banks, and private services industries, as well as states and international organizations. This present form of capitalism, ironically foreseen by Karl Marx and Friedrich Engels, poses a central dilemma between efficiency and fairness. The specialization and integration of firms make it possible to increase aggregate wealth, but the logic of pure capitalism does not favor social justice. Economic globalization has thus become a formidable cause of inequality among and within states, and the concern for global competitiveness limits the aptitude of states and other actors to address this problem.

Optimism regarding globalization rests on very fragile foundations.

Next comes cultural globalization. It stems from the technological revolution and economic globalization, which together foster the flow of cultural goods. Here the key choice is between uniformization (often termed "Americanization") and diversity. The result is both a "disenchantment of the world" (in Max Weber's words) and a reaction against uniformity. The latter takes form in a renaissance of local cultures and languages as well as assaults against Western culture, which is denounced as an arrogant bearer of a secular, revolutionary ideology and a mask for U.S. hegemony.

Finally there is political globalization, a product of the other two. It is characterized by the preponderance of the United States and its political institutions and by a vast array of international and regional organizations and transgovernmental networks (specializing in areas such as policing or migration or justice). It is also marked by private institutions that are neither governmental nor purely national—say, Doctors Without Borders or Amnesty International. But many of these agencies lack democratic accountability and are weak in scope, power, and authority. Furthermore, much uncertainty hangs over the fate of American hegemony, which faces significant resistance abroad and is affected by America's own oscillation between the temptations of domination and isolation.

The benefits of globalization are undeniable. But Friedman-like optimism rests on very fragile foundations. For one thing, globalization is neither inevitable nor irresistible. Rather, it is largely an American creation, rooted in the period after World War II and based on U.S. economic might. By extension, then, a deep and protracted economic crisis in the United States could have as devastating an effect on globalization as did the Great Depression.

Second, globalization's reach remains limited because it excludes many poor countries, and the states that it does transform react in different ways. This fact stems from the diversity of economic and social conditions at home as well as from partisan politics. The world is far away from a perfect integration of markets, services, and factors of production. Sometimes the simple existence of borders slows down and can even paralyze this integration; at other times it gives integration the flavors and colors of the dominant state (as in the case of the Internet).

Third, international civil society remains embryonic. Many nongovernmental organizations reflect only a tiny segment of the populations of their members' states. They largely represent only modernized countries, or those in which the weight of the state is not too heavy. Often, NGOs have little independence from governments.

Fourth, the individual emancipation so dear to Friedman does not quickly succeed in democratizing regimes, as one can see today in China. Nor does emancipation prevent public institutions such as the International Monetary Fund, the World Bank, or the World Trade Organization from remaining opaque in their activities and often arbitrary and unfair in their rulings.

Fifth, the attractive idea of improving the human condition through the abolition of barriers is dubious. Globalization is in fact only a sum of techniques (audio and videocassettes, the Internet, instantaneous communications) that are at the disposal of states or private actors. Self-interest and ideology, not humanitarian reasons, are what drive these actors. Their behavior is quite different from the vision of globalization as an Enlightenment-based utopia that is simultaneously scientific, rational, and universal. For many reasons—misery, injustice, humiliation, attachment to traditions, aspiration to more than just a better standard of living—this "Enlightenment" stereotype of globalization thus provokes revolt and dissatisfaction.

Another contradiction is also at work. On the one hand, international and transnational cooperation is necessary to ensure that globalization will not be undermined by the inequalities resulting from market fluctuations, weak state-sponsored protections, and the incapacity of many states to improve their fates by themselves. On the other hand, cooperation presupposes that many states and rich private players operate altruistically—which is certainly not the essence of international relations—or practice a remarkably generous conception of their long-term interests. But the fact remains that most rich states still refuse to provide sufficient development aid or to intervene in crisis situations such as the genocide in Rwanda. That reluctance compares poorly with the American enthusiasm to pursue the fight against al Qaeda and the Taliban. What is wrong here is not patriotic enthusiasm as such, but the weakness of the humanitarian impulse when the national interest in saving non-American victims is not self-evident.

IMAGINED COMMUNITIES

AMONG the many effects of globalization on international politics, three hold particular importance. The first concerns institutions. Contrary to realist predictions, most states are not perpetually at war with each other. Many regions and countries live in peace; in other cases, violence is internal rather than state-to-state. And since no government can do everything by itself, interstate organisms have emerged. The result, which can be termed "global society," seeks to reduce the potentially destructive effects of national regulations on the forces of integration. But it also seeks to ensure fairness in the world market and create international regulatory regimes in such areas as trade, communications, human rights, migration, and refugees. The main obstacle to this effort is the reluctance of states to accept global directives that might constrain the market or further reduce their sovereignty. Thus the UN's powers remain limited and sometimes only purely theoretical. International criminal justice is still only a spotty and contested last resort. In the world economy—where the market, not global governance, has been the main beneficiary of the state's retreat—the network of global institutions is fragmented and incomplete. Foreign investment remains ruled by bilateral agreements. Environmental protection is badly ensured, and issues such as migration and population growth are largely ignored. Institutional networks are not powerful enough to address unfettered short-term capital movements, the lack of international regulation on bankruptcy and competition, and primitive coordination among rich countries. In turn, the global "governance" that does exist is partial and weak at a time when economic globalization deprives many states of independent monetary and fiscal policies, or it obliges them to make cruel choices between economic competitiveness and the preservation of social safety nets. All the while, the United States displays an increasing impatience toward institutions that weigh on American freedom of action. Movement toward a world state looks increasingly unlikely. The more state sovereignty crumbles under the blows of globalization or such recent developments as humanitarian intervention and the fight against terrorism, the more states cling to what is left to them.

Second, globalization has not profoundly challenged the enduring national nature of citizenship. Economic life takes place on a global scale, but human identity remains national—hence the strong resistance to cultural homogenization. Over the centuries, increasingly centralized states have expanded their functions and tried to forge a sense of common identity for their subjects. But no central power in the world can do the same thing today, even in the European Union. There, a single currency and advanced economic coordination have not yet produced a unified economy or strong central institutions endowed with legal autonomy, nor have they resulted in a sense of postnational citizenship. The march from national identity to one that would be both national and European has only just begun. A world very partially unified by technology still has no collective consciousness or collective solidarity. What states are unwilling to do the world market cannot do all by itself, especially in engendering a sense of world citizenship.

Third, there is the relationship between globalization and violence. The traditional state of war, even if it is limited in scope, still persists. There are high risks of regional explosions in the Middle East and in East Asia, and these could seriously affect relations between the major powers. Because of this threat, and because modern arms are increasingly costly, the "anarchical society" of states lacks the resources to correct some of globalization's most flagrant flaws. These very costs, combined with the classic distrust among international actors who prefer to try to preserve their security alone or through traditional alliances, prevent a more satisfactory institutionalization of world politics—for example, an increase of the UN's powers. This step could happen if global society were provided with sufficient forces to prevent a conflict or restore peace—but it is not.

Globalization, far from spreading peace, thus seems to foster conflicts and resentments. The lowering of various barriers celebrated by Friedman, especially the spread of global media, makes it possible for the most deprived or oppressed to compare their fate with that of the free and well-off. These dispossessed then ask for help from others with common resentments, ethnic origin, or religious faith. Insofar as globalization enriches some and uproots many, those who are both poor and uprooted may seek revenge and self-esteem in terrorism.

GLOBALIZATION AND TERROR

TERRORISM is the poisoned fruit of several forces. It can be the weapon of the weak in a classic conflict among states or within a state, as in Kashmir or the Palestinian territories. But it can also be seen as a product of globalization. Transnational terrorism is made possible by the vast array of communication tools. Islamic terrorism, for example, is not only based on support for the Palestinian struggle and opposition to an invasive American presence. It is also fueled by a resistance to "unjust" economic globalization and to a Western culture deemed threatening to local religions and cultures.

If globalization often facilitates terrorist violence, the fight against this war without borders is potentially disastrous for both economic development and globalization. Antiterrorist measures restrict mobility and financial flows, while new terrorist attacks could lead the way for an antiglobalist reaction comparable to the chauvinistic paroxysms of the 1930s. Global terrorism is not the simple extension of war among states to nonstates. It is the subversion of traditional ways of war because it does not care about the sovereignty of either its enemies or the allies who shelter them. It provokes its victims to take measures that, in the name of legitimate defense, violate knowingly the sovereignty of those states accused of encouraging terror. (After all, it was not the Taliban's infamous domestic violations of human rights that led the United States into Afghanistan; it was the Taliban's support of Osama bin Laden.)

But all those trespasses against the sacred principles of sovereignty do not constitute progress toward global society, which has yet to agree on a common definition of terrorism or on a common policy against it. Indeed, the beneficiaries of the antiterrorist "war" have been the illiberal, poorer states that have lost so much of their sovereignty of late. Now the crackdown on terror allows them to tighten their controls on their own people, products, and money. They can give themselves new reasons to violate individual rights in the name of common defense against insecurity—and thus stop the slow, hesitant march toward international criminal justice.

Another main beneficiary will be the United States, the only actor capable of carrying the war against terrorism into all corners of the world. Despite its power, however, America cannot fully protect itself against future terrorist acts, nor can it fully overcome its ambivalence toward forms of interstate cooperation that might restrict U.S. freedom of action. Thus terrorism is a global phenomenon that ultimately reinforces the enemy—the state—at the same time as it tries to destroy it. The states that are its targets have no interest in applying the laws of war to their fight against terrorists; they have every interest in treating terrorists as outlaws and pariahs. The champions of globalization have sometimes glimpsed the "jungle" aspects of economic globalization, but few observers foresaw similar aspects in global terrorist and antiterrorist violence.

Finally, the unique position of the United States raises a serious question over the future of world affairs. In the realm of interstate problems, American behavior will determine whether the nonsuperpowers and weak states will continue to look at the United States as a friendly power (or at least a tolerable hegemon), or whether they are provoked by Washington's hubris into coalescing against American preponderance. America may be a hegemon, but combining rhetorical overkill and ill-defined designs is full of risks. Washington has yet to understand that nothing is more dangerous for a "hyperpower" than the temptation of unilateralism. It may well believe that the constraints of international agreements and organizations are not necessary, since U.S. values and power are all that is needed for world order. But in reality, those same international constraints provide far better opportunities for leadership than arrogant demonstrations of contempt for others' views, and they offer useful ways of restraining unilateralist behavior in other states. A hegemon concerned with prolonging its rule should be especially interested in using internationalist methods and institutions, for the gain in influence far exceeds the loss in freedom of action.

In the realm of global society, much will depend on whether the United States will overcome its frequent indifference to the costs that globalization imposes on poorer countries. For now, Washington is too reluctant to make resources available for economic development, and it remains hostile to agencies that monitor and regulate the global market. All too often, the right-leaning tendencies of the American political system push U.S. diplomacy toward an excessive reliance on America's greatest asset—military strength—as well as an excessive reliance on market capitalism and a "sovereigntism" that offends and alienates. That the mighty United States is so afraid of the world's imposing its "inferior" values on Americans is often a source of ridicule and indignation abroad.

ODD MAN OUT

FOR ALL THESE TENSIONS, it is still possible that the American war on terrorism will be contained by prudence, and that other governments will give priority to the many internal problems created by interstate rivalries and the flaws of globalization. But the world risks being squeezed between a new Scylla and Charybdis. The Charybdis is universal intervention, unilaterally decided by American leaders who are convinced that they have found a global mission provided by a colossal threat. Presentable as an epic contest between good and evil, this struggle offers the best way of rallying the population and overcoming domestic divisions. The Scylla is resignation to universal chaos in the form of new attacks by future bin Ladens, fresh humanitarian disasters, or regional wars that risk escalation. Only through wise judgment can the path between them be charted.

We can analyze the present, but we cannot predict the future. We live in a world where a society of uneven and often virtual states overlaps with a global society burdened by weak public institutions and underdeveloped civil society. A single power dominates, but its economy could become unmanageable or disrupted by future terrorist attacks. Thus to predict the future confidently would be highly incautious or naive. To be sure, the world has survived many crises, but it has done so at a very high price, even in times when WMD were not available.

Precisely because the future is neither decipherable nor determined, students of international relations face two missions. They must try to understand what goes on by taking an inventory of current goods and disentangling the threads of present networks. But the fear of confusing the empirical with the normative should not prevent them from writing as political philosophers at a time when many philosophers are extending their conceptions of just society to international relations. How can one make the global house more livable? The answer presupposes a political philosophy that would be both just and acceptable even to those whose values have other foundations. As the late philosopher Judith Shklar did, we can take as a point of departure and as a guiding thread the fate of the victims of violence, oppression, and misery; as a goal, we should seek material and moral emancipation. While taking into account the formidable constraints of the world as it is, it is possible to loosen them.

STANLEY HOFFMANN is Buttenwieser University Professor at Harvard University and a regular book reviewer for *Foreign Affairs*.

Reprinted by permission of *Foreign Affairs*, July/August 2002, pp. 104-115. © 2002 by the Council of Foreign Relations, Inc.

WAR

Which is more representative of modern war: The United States unleashing high-tech arsenals to defeat dubious Third World regimes swiftly or machete-wielding insurgents fighting brutal civil wars in Africa? The short answer: both. Yet neither of these scenarios conforms to the classic model of warfare as a titanic struggle between rival great powers. It's time to update the textbooks and reappraise the nature of war.

By Lawrence Freedman

"War Is the Continuation of Politics by Other Means"

Yes. After more than 170 years, the thesis of Prussian military theorist Karl von Clausewitz still applies. War is violence with a purpose. What has changed is whose purposes are being served and their nature. Clausewitz was most interested in great powers struggling for dominance, drawing upon the whole resources of their states, and throwing vast armies against each other. Today, with the United States as the dominant superpower, the also-rans in the international hierarchy know there is little point in trying to gain ascendancy through arms races and alliance formation. And in a postcolonial world characterized by economic interdependence, there are fewer reasons to pursue the old mercantilist agenda of conquering and occupying productive territory, protecting trade routes, and gaining influence by planting the national flag on foreign shores.

Traditional power struggles still prevail in some regions of the world, such as Africa, where rival factions vie for dominance, countries remain marginalized from the global economy, and violence is endemic. By and large, these regions produce civil wars—the most common type of modern warfare. Although there is no novelty in conflicts caused by groups seeking secession or insurrection, global communications have internationalized civil wars by drawing attention to humanitarian distress. As such, when major states intervene abroad, they normally claim to do so in the name of universal values rather than selfish national interests.

"Wars Are Never Formally Declared Anymore"

Right. Back in the days when interstate conflicts were the norm, governments used formal declarations to endow themselves with extraordinary wartime powers, such as rooting out "enemy aliens" residing on their soil, controlling economic activity, or suppressing domestic dissent. Formal declarations also provided a basis for regulating and containing war: Combatants acquired a distinct legal status, as did noncombatants, who were classified as neutrals.

However, since the Second World War, governments have avoided formal declarations. Citizens now view extra wartime powers as superfluous and alarming. And neutral countries would be reluctant to undertake any actions—not least those potentially helpful to the government declaring war—that might compromise their impartiality. What's more, adequate legal basis for war can normally be found in Article 51 of the Charter of the United Nations, which acknowledges the "inherent right of individual or collective self-defense." Even preemptive wars can be justified as self-defense, if a state can show that an attack is imminent.

The main consequence of the contemporary reluctance to declare war has been a search for euphemisms, such as "enforcement actions" and "use of force," on the grounds that as soon as a conflict is officially called a war, then all the most inconvenient legal consequences kick in. Just ask the unfortunate inmates of Camp X-Ray on the U.S. Naval Base Guantanamo Bay, Cuba, some of whom the United States has denied formal legal protections under the Geneva Convention by labeling them "unlawful combatants."

"Democracies Do Not Go To War With Each Other"

Irrelevant. The dichotomy between bellicose authoritarian states and peace-loving democratic states (until roused by the former's aggression) does not do justice to the complexity of modern international affairs. In colonial times, countries that now make up the majority of the international system were part of nondemocratic empires, largely run by European democracies. Decolonization in the 20th century led to the creation of many new states, each with a distinctive political system, sometimes apparently democratic and often not at all. These systems have had to cope with tremendous economic and social problems, which is why conflict in the modern world takes on so many different forms. As could be seen in the Balkans, conflicts can erupt between countries that are notionally democratic, even if in a bowdlerized form.

Democracy comes wrapped in many packages, not always particularly liberal or stable. And stable, liberal democracies have shown themselves to be quite warlike when convinced of a just cause. Sometimes, popular opinion compels a democratically elected government to go to war against its own better judgment—as when moral outrage over atrocities in Cuba prompted former U.S. President William McKinley reluctantly to ask Congress to declare war against Spain in 1898. Likewise, even authoritarian states periodically find it necessary to appeal to popular sentiment when making the case for war. Soviet premier Joseph Stalin did not encourage his people to fight the Nazis in the name of communism. Rather, he declared the struggle to be a Great Patriotic War to defend Mother Russia.

Such a great disparity of power now exists between democratic countries and the rest of the world that Western governments feel they can risk fighting wars that would have had little domestic support in the past. The promise of swift, decisive victory against a weaker foe sometimes mitigates concerns over high casualties or military quagmire.

"A Just War Is Backed by the United Nations"

No. The United Nations was founded to prevent wars, not authorize them. Although Chapter 7 of the U.N. Charter contains a mechanism to enforce the views of the Security Council if "international peace and security" are at risk, the United Nations has rarely granted permission to go to war. The U.N.-authorized "police action" to defend South Korea following the North's invasion in June 1950 was made possible by the Soviet Union's boycott of the Security Council. The Security Council endorsed war against Iraq in 1990 when the Cold War was effectively over and relations between the West and the Soviet Union were remarkably good. In between, the U.N. Security Council debated many wars, often with "concern" or even "regret," but East-West antagonism, coupled with rival veto powers, precluded a shared view among U.N. members on what constituted a legitimate war.

And while the Security Council is responsible for preserving international peace and security, this objective is not necessarily the same as justice. Threats to the rights of states may stir the United Nations to action, but threats against civilian populations and minority groups are more problematic. Russia and China, mindful of their own secessionist movements, vigorously dislike any interference in the internal affairs of states. Hence, they opposed U.N. military action to defend the Kosovar Albanians against the Serbs in 1999. Similarly, even though Iraq had violated U.N. resolutions and brutally repressed its own people, the United Nations was reluctant to sanction a war to overthrow a member state.

For Western governments, the international legitimacy that comes with the United Nations' good-housekeeping seal of approval can make it easier to acquire the greater prize of domestic legitimacy. In times of war, however, votes in the U.S. Congress or British Parliament still count more than votes in the Security Council.

"The War on Terrorism Is Not a War"

Generally true. However, al Qaeda is a unique case. Wars do not have to involve states. Even before the destruction of the World Trade Center, the United States and al Qaeda had engaged in a low-level war of attrition. Al Qaeda's 1998 attacks against U.S. embassies in East Africa prompted the United States to respond with cruise missile strikes against suspected terrorist targets in Sudan and Afghanistan. The sheer scale of the terrorist attacks on September 11, 2001, provoked a more drastic response, with President George W. Bush standing before Congress to declare a war on terrorism that would begin with al Qaeda but "not end until every terrorist group of global reach has been found, stopped, and defeated."

Proclaiming a war against terrorism might be justified, but this characterization exaggerates the importance of the military dimension. Terrorist groups wish their struggles to be perceived as war, since that focuses attention on their political demands. Most governments prefer to describe terrorists as criminals, since that description focuses attention on their brutal tactics and delegitimizes their political agenda. But the U.S. campaign against Osama bin Laden in Afghanistan was not a typical counterterrorist operation. Despite all the talk about al Qaeda being a stateless group, the terrorist organization's close association with the Taliban regime led to a conflict that soon took on the characteristics of a regular interstate war.

Still, more often than not, the foot soldiers in the "war" against terrorism are law enforcement officials and intelligence operatives. In the end, terrorism is best defeated through isolating militants from their claimed constituency, demonstrating the shameful and counterproductive nature of their methods, and, if possible, addressing the grievances upon which they feed. Terrorist campaigns will peter out if they are prevented from gaining funds or are continually thwarted in their efforts to mount attacks.

"Information Technology Has Cleared the Fog of War"

Wrong. The U.S. military benefited just as much as the U.S. economy from the revolution in information technology during the 1990s. Information can flow instantaneously across all dimensions of the battle space (the three-dimensional version of the old battlefield, encompassing land, oceans, air, and space), picked up from numerous, intrusive sensors and disseminated amongst all forces. Because such high-quality information flows make an enormous difference in contemporary conflicts, there are new incentives to disrupt or confuse these flows—including those supporting critical infrastructure, such as energy, transportation, and banking.

Another more serious problem is that with so much information around, on the Internet and through the media, assessing its reliability and judging the appropriate response is difficult. Political and military leaders must still decide how to prioritize the many messages they receive, how to interpret raw data picked up from numerous sources, and how to put that data into context. Critical information, including plans hatched by the enemy, may still not be available. In the battles for hearts and minds, great damage can be done by rumors and callous images that may circulate rapidly without any checks on their validity.

The mistakes and confusion resulting from an age of information surplus are different from those in an age of scarcity, but they can still cause problems. It is always important to remember that information is not the same as intelligence, which in turn is not the same as wisdom.

"All Future Warfare Will Be Asymmetric"

True. Lt. Col. Bryan McCoy, commanding officer of the 3rd Battalion, 4th Marine Regiment in Iraq, was reported to have complained: "The enemy has gone asymmetric on us. There's treachery. There are ambushes. It's not straight-up conventional fighting." Though they would not swap places with their enemies, Western military planners worry over the variety of ways a weaker opponent might use unconventional tactics to offset their high-tech advantage.

In one sense, all wars are asymmetric, as no two countries ever have the same force structures and capabilities, let alone geography, political systems, and strategic cultures. In the past, these differences evened out as both sides sought to turn a conflict to their comparative advantage. Often, in the end, the key asymmetry lay in resources: the ability of the economic and political systems to cope with the attritional costs of war and to replenish their armed forces. By the nuclear age, the ultimate symmetry was reached—as two superpowers developed the capacity to annihilate each other completely—and this threat of mutual assured destruction made war appear the ultimate folly.

With the end of the Cold War, the military superiority of the United States and its allies allows them to win any conventional war fought between regular forces on an open battle space. The extent of this dominance did not become apparent until the 1991 Gulf War, when Iraqi forces were overwhelmed first by U.S. airpower and then by the United States' highly mobile and lethal ground forces. For any prospective enemies of the United States, the take-away lesson from this war was that they were bound to lose if they fought on U.S. terms.

Opponents of the West might try to remedy this imbalance through weapons of mass destruction or superterrorism, to be directed against a nation's homeland. Likewise, drawing Western forces into apparently unwinnable wars and deploying guerrilla tactics of ambush and harassment, until casualties accumulate to an intolerable level, worked well for the North Vietnamese. Such tactics were far less successful for former Iraqi President Saddam Hussein.

Critical to such conflicts is what strategic theorist Alex George describes as asymmetrical motivation: For one side, the stakes may be so high that losing a war is unthinkable; for the other, the stakes may be limited, leading to impatience with conflicts that drag out and cause cumulative casualties. The recent war in Iraq provided a test of the proposition that the attacks of September 11, 2001, made Americans more resolute and less risk-averse, by turning what might otherwise have been a war of choice into a war of necessity. The risks, however, turned out not to be that great, so it remains uncertain how much asymmetrical motivation can compensate for a drastic asymmetry in capabilities.

"Airpower Is Decisive in Modern War"

Only up to a point. The 1991 Gulf War demonstrated what air superiority could achieve. Iraqi aircraft had no chance in dogfights and were caught in their bunkers by precision-guided weapons. Air defenses were soon degraded to the point where allied aircraft could operate relatively freely. The war did not, however, fully resolve the long-standing dispute between what used to be called the strategic and tactical use of airpower.

The theory that strategic airpower alone could be decisive in war by destroying the very will of a government and its people to fight was popular before the Second World War. Yet the terrible air raids during that war reinforced social solidarity, if anything. The greatest value of airpower lay in destroying the infrastructure of war, from production to supply lines, and supporting troops on the ground in a tactical role.

These days, bombings directed at a civilian population and its infrastructure are deemed unacceptable, but some military planners believe precision targeting of enemy leadership can provide a strategic substitute. The evidence of both the 1991 and 2003 wars in Iraq suggests that short of a direct, "decapitating" hit, spectacular demolitions of largely empty buildings with some symbolic political significance bring few benefits. Stray weapons invariably hit marketplaces and residential buildings, thereby undermining the claim of sparing civilians.

Where airpower really makes the difference is in its classic "tactical" role—battering unprotected regular forces, destroying their equipment, and providing lethal firepower during engagements. There are limits to what can be done against guerrilla tactics from the air, especially when targets are integrated within the civilian population. However, with modern unmanned aerial vehicles, such as the U.S. Predator drone, excellent intelligence can now be obtained on quite small groups.

"Modern War Means Fewer Casualties"

No. These days, Western countries seek to limit their casualties and play to their high-tech advantage by fighting capital-intensive rather than labor-intensive wars. Indeed, all their recent wars, including the war to overthrow Saddam Hussein in Iraq, have seen casualties in tens or hundreds (on the Western side), rather than thousands, which would have been considered extraordinarily low in the past. These days, friendly-fire incidents and other accidents are just as likely as combat to cause casualties.

This low death toll, however, is distorted by the West's à la carte approach toward military intervention. During the 1990s, the United States and Europe often avoided getting too involved in messy civil wars that didn't fit their idealized scenarios, with massive casualties (overwhelmingly civilian) caused by low technology such as small arms and machetes. The millions of victims of the vicious African wars of the last decade—including in Angola, Democratic Republic of Congo, Rwanda, Sierra Leone, and Somalia—would have good reason to question the proposition that contemporary war inflicts fewer casualties.

It didn't take much violence to deter the United States, the champion of the revolution in military affairs, from testing its theories of modern war more widely. The United States withdrew from Somalia in 1993 after a nasty incident left 18 American soldiers and as many as 1,000 Somalis dead. After that experience, the United States was loath to intervene to prevent the Rwandan genocide, in which an estimated 800,000 people died.

"The United States Can Win a War at Any Time in Any Place"

No. The United States has the power to prevail in most conflicts, but whether it has the necessary determination in the face of its own casualties and those it inflicts on others is another matter. Americans cannot always assume conflicts will be sharp and quick—a dangerously beguiling vision, recently reinforced by toppling regimes in Afghanistan and Iraq, whose people were unwilling to fight. And the demands of fighting in many theaters of war can perilously overstretch resources, to the point where the economic strength upon which U.S. power is based may dissipate.

The war in Iraq demonstrated that the United States has pulled even further ahead of the crowd in terms of military capabilities since Operation Desert Storm. But the war also confirmed that the political situations left behind may be more complex and unsatisfactory than those that followed the sharp, interstate wars of the past. Defeated countries require substantial amounts of U.S. economic assistance and political attention, as well as residual military commitments. The lesson of history is that it is easier to acquire empires (even benevolent ones) than to sustain them.

The unintended consequences of any war, for good as well as bad, tend to be as important as the intended. Sometimes wars are necessary and must be fought. But despite all the promise of the information age, airpower, and precision technology, war remains irredeemably violent and should always be approached with care.

Lawrence Freedman is professor of war studies at the University of London's King's College and a member of the FOREIGN POLICY *editorial board.*

From *Foreign Policy*, July/August 2003, pp. 16-24. © 2003 by the Carnegie Endowment for International Peace (www.foreignpolicy.com). Permission conveyed through Copyright Clearance Center, Inc.

The Protean Enemy

Jessica Stern

WHAT'S NEXT FROM AL QAEDA?

HAVING SUFFERED the destruction of its sanctuary in Afghanistan two years ago, al Qaeda's already decentralized organization has become more decentralized still. The group's leaders have largely dispersed to Pakistan, Iran, Iraq, and elsewhere around the world (only a few still remain in Afghanistan's lawless border regions). And with many of the planet's intelligence agencies now focusing on destroying its network, al Qaeda's ability to carry out large-scale attacks has been degraded.

Yet despite these setbacks, al Qaeda and its affiliates remain among the most significant threats to U.S. national security today. In fact, according to George Tenet, the CIA's director, they will continue to be this dangerous for the next two to five years. An alleged al Qaeda spokesperson has warned that the group is planning another strike similar to those of September 11. On May 12, simultaneous bombings of three housing complexes in Riyadh, Saudi Arabia, killed at least 29 people and injured over 200, many of them Westerners. Intelligence officials in the United States, Europe, and Africa report that al Qaeda has stepped up its recruitment drive in response to the war in Iraq. And the target audience for its recruitment has also changed. They are now younger, with an even more "menacing attitude," as France's top investigative judge on terrorism-related cases, Jean-Louis Bruguière, describes them. More of them are converts to Islam. And more of them are women.

What accounts for al Qaeda's ongoing effectiveness in the face of an unprecedented onslaught? The answer lies in the organization's remarkably protean nature. Over its life span, al Qaeda has constantly evolved and shown a surprising willingness to adapt its mission. This capacity for change has consistently made the group more appealing to recruits, attracted surprising new allies, and—most worrisome from a Western perspective—made it harder to detect and destroy. Unless Washington and its allies show a similar adaptability, the war on terrorism won't be won anytime soon, and the death toll is likely to mount.

MALLEABLE MISSIONS

WHY DO religious terrorists kill? In interviews over the last five years, many terrorists and their supporters have suggested to me that people first join such groups to make the world a better place—at least for the particular populations they aim to serve. Over time, however, militants have told me, terrorism can become a career as much as a passion. Leaders harness humiliation and anomie and turn them into weapons. Jihad becomes addictive, militants report, and with some individuals or groups—the "professional" terrorists—grievances can evolve into greed: for money, political power, status, or attention.

In such "professional" terrorist groups, simply perpetuating their cadres becomes a central goal, and what started out as a moral crusade becomes a sophisticated organization. Ensuring the survival of the group demands flexibility in many areas, but especially in terms of mission. Objectives thus evolve in a variety of ways. Some groups find a new cause once their first one is achieved—much as the March of Dimes broadened its mission from finding a cure for polio to fighting birth defects after the Salk vaccine was developed. Other groups broaden their goals in order to attract a wider variety of recruits. Still other organizations transform themselves into profit-driven organized criminals, or form alliances with groups that have ideologies different from their own, forcing both to adapt. Some terrorist groups hold fast to their original missions. But only the spry survive.

Consider, for example, Egyptian Islamic Jihad (EIJ). EIJ's original objective was to fight the oppressive, secular rulers of Egypt and turn the country into an Islamic state. But the group fell on hard times after its leader, Sheikh Omar Abdel Rahman, was imprisoned in the United States and other EIJ leaders were killed or forced into exile. Thus in the early 1990s, Ayman al-Zawahiri decided to shift the group's sights from its "near enemy"—the secular rulers of Egypt—to the "far enemy," namely the United States and other Western countries. Switching goals in this way allowed the group to align itself with another terrorist aiming to attack the West and able to provide a significant influx of cash: Osama bin Laden. In return for bin Laden's financial assistance, Zawahiri provided some 200 loyal, disciplined, and well-trained followers, who became the core of al Qaeda's leadership.

A second group that has changed its mission over time to secure a more reliable source of funding is the Islamic Movement of Uzbekistan (IMU), which, like EIJ, eventually joined forces with the Taliban and al Qaeda. The IMU's original mission was to topple Uzbekistan's corrupt and repressive post-Soviet dictator, Islam Karimov. Once the IMU formed an alliance with the Taliban's leader, Mullah Omar, however, it began promoting the Taliban's anti-American and anti-Western agenda, also condemning music, cigarettes, sex, and alcohol. This new puritanism reduced its appeal among its original, less-ideological supporters in Uzbekistan—one downside to switching missions.

Even Osama bin Laden himself has changed his objectives over time. The Saudi terrorist inherited an organization devoted to fighting Soviet forces in Afghanistan. But he turned it into a flexible group of ruthless warriors ready to fight on behalf of multiple causes. His first call to holy war, issued in 1992, urged believers to kill American soldiers in Saudi Arabia and the Horn of Africa but barely mentioned Palestine. The second, issued in 1996, was a 40-page document listing atrocities and injustices committed against Muslims, mainly by Western powers. With the release of his third manifesto in February 1998, however, bin Laden began urging his followers to start deliberately targeting American civilians, rather than soldiers. (Some al Qaeda members were reportedly distressed by this shift to civilian targets and left the group.) Although this third declaration mentioned the Palestinian struggle, it was still only one among a litany of grievances. Only in bin Laden's fourth call to arms—issued to the al Jazeera network on October 7, 2001, to coincide with the U.S. aerial bombardment of Afghanistan—did he emphasize Israel's occupation of Palestinian lands and the suffering of Iraqi children under UN sanctions, concerns broadly shared in the Islamic world. By extending his appeal, bin Laden sought to turn the war on terrorism into a war between all of Islam and the West. The events of September 11, he charged, split the world into two camps—believers and infidels—and the time had come for "every Muslim to defend his religion."

One of the masterminds of the September 11 attacks, Ramzi bin al-Shibh, later described violence as "the tax" that Muslims must pay "for gaining authority on earth." This comment points to yet another way that al Qaeda's ends have mutated over the years. In his putative autobiography, Zawahiri calls the "New World Order" a source of humiliation for Muslims. It is better, he says, for the youth of Islam to carry arms and defend their religion with pride and dignity than to submit to this humiliation. One of al Qaeda's aims in fighting the West, in other words, has become to restore the dignity of humiliated young Muslims. This idea is similar to the anticolonialist theoretician Frantz Fanon's notion that violence is a "cleansing force" that frees oppressed youth from "inferiority complexes," "despair," and "inaction," making them fearless and restoring their self-respect. The real target audience of violent attacks is therefore not necessarily the victims and their sympathizers, but the perpetrators and their sympathizers. Violence becomes a way to bolster support for the organization and the movement it represents. Hence, among the justifications for "special operations" listed in al Qaeda's terrorist manual are "bringing new members to the organization's ranks" and "boosting Islamic morale and lowering that of the enemy." The United States may have become al Qaeda's principal enemy, but raising the morale of Islamist fighters and their sympathizers is now one of its principal goals.

FRIENDS OF CONVENIENCE

APART FROM the flexibility of its mission, another explanation for al Qaeda's remarkable staying power is its willingness to forge broad—and sometimes unlikely—alliances. In an effort to expand his network, bin Laden created the International Islamic Front for Jihad Against the Jews and Crusaders (IIF) in February 1998. In addition to bin Laden and EIJ's Zawahiri, members included the head of Egypt's Gama'a al Islamiya, the secretary-general of the Pakistani religious party known as the Jamiat-ul-Ulema-e-Islam (JUI), and the head of Bangladesh's Jihad Movement. Later, the IIF was expanded to include the Pakistani *jihadi* organizations Lashkar-e-Taiba, Harkat-ul-Mujahideen, and Sipah-e-Sahaba Pakistan, the last an anti-Shi'a sectarian party. Senior al Qaeda lieutenant Abu Zubaydah was captured at a Lashkar-e-Taiba safe house in Faisalabad in March 2002, suggesting that some of Lashkar-e-Taiba's members are facilitating and assisting the movement of al Qaeda members in Pakistan. And Indian sources claim that Lashkar-e-Taiba is now trying to play a role similar to that once played by al Qaeda itself, coordinating and in some cases funding pro-bin Laden networks, especially in Southeast Asia and the Persian Gulf.

In addition to its formal alliances through the IIF, bin Laden's organization has also nurtured ties and now works closely with a variety of still other groups around the world, including Ansar al Islam, based mainly in Iraq and Europe; Jemaah Islamiah in Southeast Asia; Abu Sayyaf and the Moro Islamic Liberation Front in the Philippines; and many Pakistani *jihadi* groups. In some cases, al Qaeda has provided these allies with funding and direction. In others, the groups have shared camps, operatives, and logistics. Some "franchise groups," such as Jemaah Islamiah, have raised money for joint operations with al Qaeda.

Perhaps most surprising (and alarming) is the increasing evidence that al Qaeda, a Sunni organization, is now cooperating with the Shi'a group Hezbollah, considered to be the most sophisticated terrorist group in the world. Hezbollah, which enjoys backing from Syria and Iran, is based in southern Lebanon and in the lawless "triborder" region of South America, where Paraguay, Brazil, and Argentina meet. The group has also maintained a fundraising presence in the United States since the 1980s. According to the CIA's Tenet, however, the group has lately stepped up its U.S. activities and was recently spotted "casing and surveilling American facilities." Although low-level cooperation between al Qaeda and Hezbollah has been evident for some time—their logistical cooperation was revealed in the trial of al Qaeda operatives involved in the 1998 embassy bombing attacks in east Africa—the two groups have formed a much closer relationship since al Qaeda was evicted from its base in Afghanistan. Representatives of the two groups have lately met up in Lebanon, Paraguay, and an unidentified African country. According to a report in Israel's *Ha'aretz* newspaper, Imad Mughniyah, who directs Hezbollah in the triborder area, has also been appointed by Iran to coordinate the group's activities with Hamas and Palestinian Islamic Jihad.

The triborder region of South America has become the world's new Libya, a place where terrorists with widely disparate ideologies—Marxist Colombian rebels, American white supremacists, Hamas, Hezbollah, and others—meet to swap tradecraft. Authorities now worry that the more sophisticated groups will invite the American radicals to help them. Moneys raised for terrorist organizations in the United States are often funneled through Latin America, which has also become an important stopover point for operatives entering the United States. Reports that Venezuela's President Hugo Chávez is allowing Colombian rebels and militant Islamist groups to operate in his country are meanwhile becoming more credible, as are claims that Venezuela's Margarita Island has become a terrorist haven.

As these developments suggest and Tenet confirms, "mixing and matching of capabilities, swapping of training, and the use of common facilities" have become the hallmark of professional terrorists today. This fact has been borne out by the leader of a Pakistani *jihadi* group affiliated with al Qaeda, who recently told me that informal contacts between his group and Hezbollah, Hamas, and others have become common. Operatives with particular skills loan themselves out to different groups, with expenses being covered by the charities that formed to fund the fight against the Soviet Union in Afghanistan.

Meanwhile, the Bush administration's claims that al Qaeda cooperated with the "infidel" (read: secular) Saddam Hussein while he was still in office are now also gaining support, and from a surprising source. Hamid Mir, bin Laden's "official biographer" and an analyst for al Jazeera, spent two weeks filming in Iraq during the war. Unlike most reporters, Mir wandered the country freely and was not embedded with U.S. troops. He reports that he has "personal knowledge" that one of Saddam's intelligence operatives, Farooq Hijazi, tried to contact bin Laden in Afghanistan as early as 1998. At that time, bin Laden was publicly still quite critical of the Iraqi leader, but he had become far more circumspect by November 2001, when Mir interviewed him for the third time. Mir also reports that he met a number of Hezbollah operatives while in Iraq and was taken to a recruitment center there.

NEW-STYLE NETWORKS

AL QAEDA SEEMS to have learned that in order to evade detection in the West, it must adopt some of the qualities of a "virtual network": a style of organization used by American right-wing extremists for operating in environments (such as the United States) that have effective law enforcement agencies. American antigovernment groups refer to this style as "leaderless resistance." The idea was popularized by Louis Beam, the self-described ambassador-at-large, staff propagandist, and "computer terrorist to the Chosen" for Aryan Nations, an American neo-Nazi group. Beam writes that hierarchical organization is extremely dangerous for insurgents, especially in "technologically advanced societies where electronic surveillance can often penetrate the structure, revealing its chain of command." In leaderless organizations, however, "individuals and groups operate independently of each other, and never report to a central headquarters or single leader for direction or instruction, as would those who belong to a typical pyramid organization." Leaders do not issue orders or pay operatives; instead, they inspire small cells or individuals to take action on their own initiative.

Lone-wolf terrorists typically act out of a mixture of ideology and personal grievances. For example, Mir Aimal Kansi, the Pakistani national who shot several CIA employees in 1993, described his actions as "between jihad and tribal revenge"—jihad against America for its support of Israel and revenge against the CIA, which he apparently felt had mistreated his father during Afghanistan's war against the Soviets. Meanwhile, John Allen Muhammad, one of the alleged "Washington snipers," reportedly told a friend that he endorsed the September 11 attacks and disapproved of U.S. policy toward Muslim states, but he appears to have been principally motivated by anger at his ex-wife for keeping him from seeing their children, and some of his victims seem to have been personal enemies. As increasingly powerful weapons become more and more available, lone wolves, who face few political constraints, will become more of a threat, whatever their primary motivation.

The Internet has also greatly facilitated the spread of "virtual" subcultures and has substantially increased the capacity of loosely networked terrorist organizations. For example, Beam's essay on the virtues of "leaderless resistance" has long been available on the Web and, according to researcher Michael Reynolds, has been highlighted by radical Muslim sites. Islamist Web sites also offer on-line training courses in the production of explosives and urge visitors to take action on their own. The "encyclopedia of jihad," parts of which are available on-line, provides instructions for creating "clandestine activity cells," with units for intelligence, supply, planning and preparation, and implementation.

The obstacles these Web sites pose for Western law enforcement are obvious. In one article on the "culture of jihad" available on-line, a Saudi Islamist urges bin Laden's sympathizers to take action without waiting for instructions. "I do not need to meet the Sheikh and ask his permission to carry out some operation," he writes, "the same as I do not need permission to pray, or to think about killing the Jews and the Crusaders that gather on our lands." Nor does it make any difference whether bin Laden is alive or dead: "There are a thousand bin Ladens in this nation. We should not abandon our way, which the Sheikh has paved for you, regardless of the existence of the Sheikh or his absence." And according to U.S. government officials, al Qaeda now uses chat rooms to recruit Latino Muslims with

U.S. passports, in the belief that they will arouse less suspicion as operatives than would Arab-Americans. Finally, as the late neo-Nazi William Pierce once told me, using the Web to recruit "leaderless resisters" offers still another advantage: it attracts better-educated young people than do more traditional methods, such as radio programs.

Already the effects of these leaderless cells have been felt. In February 2002, Ahmed Omar Saeed Sheikh, the British national who was recently sentenced to death for his involvement in the abduction and murder of *Wall Street Journal* reporter Daniel Pearl, warned his Pakistani interrogators that they would soon confront the threat of small cells, working independently of the known organizations that Pakistani President Pervez Musharraf had vowed to shut down. Sure enough, soon after Omar Sheikh made this threat, unidentified terrorists killed 5 people in an Islamabad church known to be frequented by U.S. embassy personnel, and another group killed 11 French military personnel in Karachi in May. And in July, still other unidentified terrorists detonated a truck bomb at the entrance of the U.S. consulate in Karachi, killing 12 Pakistanis.

JOINING THE FAMILY

VIRTUAL LINKS are only part of the problem; terrorists, including members of bin Laden's IIF, have also started to forge ties with traditional organized crime groups, especially in India. One particularly troubling example is the relationship established between Omar Sheikh and an ambitious Indian gangster named Aftab Ansari. Asif Reza Khan, the "chief executive" for Ansari's Indian operations, told interrogators that he received military training at a camp in Khost, Afghanistan, belonging to Lashkar-e-Taiba, and that "leaders of different militant outfits in Pakistan were trying to use his network for the purpose of jihad, whereas [Ansari] was trying to use the militants' networks for underworld operations."

Khan told his interrogators that the don provided money and hideouts to his new partners, in one case transferring $100,000 to Omar Sheikh—money that Omar Sheikh, in turn, wired to Muhammad Atta, the lead hijacker in the September 11 attacks. According to Khan, Ansari viewed the $100,000 gift as an "investment" in a valuable relationship.

Still another set of unlikely links has sprung up in American prisons, where Saudi charities now fund organizations that preach radical Islam. According to Warith Deen Umar, who hired most of the Muslim chaplains currently active in New York State prisons, prisoners who are recent Muslim converts are natural recruits for Islamist organizations. Umar, incidentally, told *The Wall Street Journal* that the September 11 hijackers should be honored as martyrs, and he traveled to Saudi Arabia twice as part of an outreach program designed to spread Salafism (a radical Muslim movement) in U.S. prisons.

Another organization now active in U.S. prisons is Jamaat ul-Fuqra, a terrorist group committed to purifying Islam through violence. (Daniel Pearl was abducted and murdered in Pakistan while attempting to interview the group's leader, Sheikh Gilani, to investigate the claim that Richard Reid—who attempted to blow up an international flight with explosives hidden in his shoes—was acting under Gilani's orders.) The group functions much like a cult in the United States; members live in poverty in compounds, some of which are heavily armed. Its members have been convicted of fraud, murder, and several bombings, but so far, most of their crimes have been relatively small scale. Clement Rodney Hampton-El, however, convicted of participating with Omar Abdel Rahman in a 1993 plot to blow up New York City landmarks, was linked to the group, and U.S. law enforcement authorities worry that the Fuqra has since come under the influence of al Qaeda.

Still another surprising source of al Qaeda recruits is Tablighi Jamaat (TJ), a revivalist organization that aims at creating better Muslims through "spiritual jihad": good deeds, contemplation, and proselytizing. According to the historian Barbara Metcalf, TJ has traditionally functioned as a self-help group, much like Alcoholics Anonymous, and most specialists claim that it is no more prone to violence than are the Seventh-Day Adventists, with whom TJ is frequently compared. But several Americans known to have trained in al Qaeda camps were brought to Southwest Asia by TJ and appear to have been recruited into *jihadi* organizations while traveling under TJ auspices. For example, Jose Padilla (an American now being held as an "enemy combatant" for planning to set off a "dirty" radiological bomb in the United States) was a member of TJ, as were Richard Reid and John Walker Lindh (the so-called American Taliban). According to prosecutors, the "Lackawanna Six" group (an alleged al Qaeda sleeper cell from a Buffalo, New York, suburb) similarly first went to Pakistan to receive TJ religious training before proceeding to the al Farooq training camp in Afghanistan. A Pakistani TJ member told me that *jihadi* groups openly recruit at the organization's central headquarters in Raiwind, Pakistan, including at the mosque. And TJ members in Boston say that a lot of Muslims end up treating the group, which is now active in American inner cities and prisons, as a gateway to *jihadi* organizations.

As such evidence suggests, although it may have been founded to create better individuals, TJ has produced offshoots that have evolved into more militant outfits. In October 1995, Pakistani authorities uncovered a military plot to assassinate Prime Minister Benazir Bhutto and establish a theocracy. Most of the officers involved in the attempted coup were members of TJ. The group is said to have been strongly influenced by retired Lieutenant General Javed Nasir, who served as Pakistan's intelligence chief from 1990 to 1993 but was sacked under pressure from the United States for his support of militant Islamists around the world.

Totalitarian Islamist revivalism has become the ideology of the dystopian new world order. In an earlier era, radicals might have described their grievances through other ideological lenses, perhaps anarchism, Marxism, or Nazism. Today they choose extreme Islamism.

Radical transnational Islam, divorced from its countries of origin, appeals to some jobless youths in depressed parts of Europe and the United States. As the French scholar Olivier Roy points out, leaders of radical Islamic groups often come from the middle classes, many of them having trained in technical fields, but their followers tend be working-class dropouts.

Focusing on economic and social alienation may help explain why such a surprising array of groups has proved willing to join forces with al Qaeda. Some white supremacists and extremist Christians applaud al Qaeda's rejectionist goals and may eventually contribute to al Qaeda missions. Already a Swiss neo-Nazi named Albert Huber has called for his followers to join forces with Islamists. Indeed, Huber sat on the board of directors of the Bank al Taqwa, which the U.S. government accuses of being a major donor to al Qaeda. Meanwhile, Matt Hale, leader of the white-supremacist World Church of the Creator, has published a book indicting Jews and Israelis as the real culprits behind the attacks of September 11. These groups, along with Horst Mahler (a founder of the radical leftist German group the Red Army Faction), view the September 11 attacks as the first shot in a war against globalization, a phenomenon that they fear will exterminate national cultures. Leaderless resisters drawn from the ranks of white supremacists or other groups are not currently capable of carrying out massive attacks on their own, but they may be if they join forces with al Qaeda.

MODERN METHODOLOGY

AL QAEDA HAS lately adopted innovative tactics as well as new alliances. Two new approaches are particularly alarming to intelligence officials: efforts to use surface-to-air missiles to shoot down aircraft and attempts to acquire chemical, nuclear, or biological weapons.

In November 2002, terrorists launched two shoulder-fired SA-7 missiles at an Israeli passenger jet taking off from Mombasa, Kenya, with 271 passengers on board. Investigators say that the missiles came from the same batch as those used in an earlier, also unsuccessful attack on a U.S. military jet in Saudi Arabia. And intelligence officials believe that Hezbollah contacts were used to smuggle the missiles into Kenya from Somalia.

Meanwhile, according to Barton Gellman of *The Washington Post*, documents seized in Pakistan in March 2003 reveal that al Qaeda has acquired the necessary materials for producing botulinum and salmonella toxins and the chemical agent cyanide—and is close to developing a workable plan for producing anthrax, a far more lethal agent. Even more worrisome is the possibility that al Qaeda, perhaps working with Hezbollah or other terrorist groups, will recruit scientists with access to sophisticated nuclear or biological weapons programs, possibly, but not necessarily, ones that are state-run.

To fight such dangerous tactics, Western governments will also need to adapt. In addition to military, intelligence, and law enforcement responses, Washington should start thinking about how U.S. policies are perceived by potential recruits to terrorist organizations. The United States too often ignores the unintended consequences of its actions, disregarding, for example, the negative message sent by Washington's ongoing neglect of Afghanistan and of the chaos in postwar Iraq. If the United States allows Iraq to become another failed state, groups both inside and outside the country that support al Qaeda's goals will benefit.

Terrorists, after all, depend on the broader population for support, and the right U.S. policies could do much to diminish the appeal of rejectionist groups. It does not make sense in such an atmosphere to keep U.S. markets closed to Pakistani textiles or to insist on protecting intellectual property with regard to drugs that needy populations in developing countries cannot hope to afford.

In countries where extremist religious schools promote terrorism, Washington should help develop alternative schools rather than attempt to persuade the local government to shut down radical *madrasahs*. In Pakistan, many children end up at extremist schools because their parents cannot afford the alternatives; better funding for secular education could therefore make a positive difference.

The appeal of radical Islam to alienated youth living in the West is perhaps an even more difficult problem to address. Uneasiness with liberal values, discomfort with uncertain identities, and resentment of the privileged are perennial problems in modern societies. What is new today is that radical leaders are using the tools of globalization to construct new, transnational identities based on death cults, turning grievances and alienation into powerful weapons. To fight these tactics will require getting the input not just of moderate Muslims, but of radical Islamist revivalists who oppose violence.

To prevent terrorists from acquiring new weapons, meanwhile, Western governments must make it harder for radicals to get their hands on them. Especially important is the need to continue upgrading security at vulnerable nuclear sites, many of which, in Russia and other former Soviet states, are still vulnerable to theft. The global system of disease monitoring—a system sorely tested during the SARS epidemic—should also be upgraded, since biological attacks may be difficult to distinguish from natural outbreaks. Only by matching the radical innovation shown by professional terrorists such as al Qaeda—and by showing a similar willingness to adapt and adopt new methods and new ways of thinking—can the United States and its allies make themselves safe from the ongoing threat of terrorist attack.

JESSICA STERN is Lecturer in Public Policy at Harvard's John F. Kennedy School of Government and the author of *The Ultimate Terrorists* and the forthcoming *Terror in the Name of God: Why Religious Militants Kill*.

A World of Exiles

Emigrés have long sought to bring pressure to bear on governments in their adopted countries. Now their influence is being felt at home too

WHY does Macedonia have no embassy in Australia? Why might a mountain in northern Greece soon be disfigured by an image of Alexander the Great 73 metres (nearly 240 feet) high? Who paid for the bloody war between Ethiopia and Eritrea? How did Croatia succeed in winning early international recognition as an independent country? And why do Mexican candidates for political office campaign in the United States?

The short answer to each of these questions is a diaspora—a community of people living outside their country of origin. Macedonia has no embassy in Australia because Greeks think the former Yugoslav republic that calls itself Macedonia has purloined the name from them, and the Greek vote counts for a lot in Australia. So, as a sop to local Greeks outraged by its decision to recognise the upstart Macedonia, the Australian government has not yet allowed it to open an embassy in Canberra.

The case of the missing embassy is an extreme, but typical, example of how diasporas have long exerted their influence: they have lobbied in their adopted countries for policies favourable to the homeland. But now something new is taking place: diasporas are increasingly exerting influence on the politics of the countries they have physically, but not emotionally, abandoned. An example of this trend is the case of the monumental Alexander. The Greek diaspora is so proud of Alexander the Great, whose Macedonian kingdom encompassed what are now parts of northern Greece, and so keen to establish him as Greek, that it wants to carve his effigy on a cliff face on Mount Kerdyllion. The Greek authorities in Athens are horrified, but the Alexander the Great Foundation, based in Chicago, is eager to get chipping, and says its members will cover the $45m cost. Grotesque as it may consider the scheme—the monument would be four times the size of the American presidents carved on Mount Rushmore—the Greek government may yield. It is to rich Greek-Americans that it turns when it wants to promote its interests in America.

Similarly, it was to its citizens abroad that Eritrea looked when it decided to wage a pointless border war between 1998 and 2000. Small, poor and just six years old, the country was in no position to fight its much bigger neighbour, Ethiopia. But of Eritrea's 3.8m people, about 333,000 were émigrés and, astonishingly, the government was able to tax their personal income at 2% a year. This helped to finance, and thus to perpetuate, a terrible war.

Croats abroad also did their bit for their country, both before and after independence in 1991. In the early 1990s, not long after European communism had collapsed but before the Yugoslav federation had begun to disintegrate, the cry went up in Croatia for Croats of the diaspora to come home. Some did, returning to fight in the war that broke out in 1991. Other Croats abroad raised money: as much as $30m had been mustered by 1991. Meanwhile, Croat exiles were lobbying hard in Germany, which in turn bounced the European Union into early recognition of the new state. Fiercely nationalist exiles forked out at least $4m for the 1990 election campaign of Franjo Tudjman, Croatia's arch-nationalist president, and in return were awarded representation in parliament in 1992, by which time the country had won its independence. Twelve out of the 120 seats were allotted to diaspora Croats, who cast their votes in consulates abroad, or in community centres, clubs and churches designated by the authorities in Zagreb. By contrast, only seven seats were set aside for Croatia's ethnic minorities.

Since 1996, Mexicans abroad have also had the right to vote in national elections, although the legislation to allow them to do so without coming home has yet to be passed. Still, with 10.8m citizens of voting age out of the country, many political candidates reckon it is worth their while to campaign in the United States, where 99% of the absentees live. Even if just a small proportion comes back on polling day, that may be enough to tip an election.

The diaspora's new right to vote, however theoretical it remains, was a right reluctantly given. The Institutional Revolutionary Party (PRI), which ruled Mexico uninterruptedly for over 60 years, thought expatriate Mexicans were unlikely to support it. It was probably right. Although the diaspora was not composed of political exiles implacably hostile to the government of the country they had left, the émigrés were probably sophisticated enough to dislike the PRI's self-serving policies. Moreover, nationalism, which so often makes exiles sympathetic to the government back home, was hardly an issue in Mexican elections.

As a rule of thumb, though, émigrés are nationalists, even though they may at the same time be loyal citizens of their adopted country (96% of Australia's Croats are naturalised, and not known as lukewarm in their Ozziness). Perhaps the strength of nationalist feeling has something to do with feelings of guilt among those fortunate enough to live abroad, especially when the home country is under some kind of threat. Perhaps it has something to do with not having to live with the consequences of nationalism *pur et dur*. Perhaps it is because exile sharpens the sense of the country left behind. Issues may simply seem clearer from afar. In any event, absence certainly seems to make the heart grow fonder—and fiercer.

If you doubt that, just imagine what would happen in Ireland, north and south of the border, if Irish-Americans were allowed to vote in Ireland's or Ulster's elections. Irish-Americans tend to be strong supporters of republicanism, and in the 1970s and 1980s they raised huge amounts of money for republican terrorist groups such as the IRA. Or imagine what would happen to Cuba's politics if the exiles of southern Florida could vote as well as rant.

Wired and wonderful

In the past, absence also made the power of the diaspora grow weaker. Scatter a few million émigrés across the globe, and, being everywhere in a minority, they are weak. They are influential only where they are concentrated, as, say, Swedes are in southern Finland; or where they are especially well-organised, as Jews are in the United States and Armenians are in France. That is certainly the way it used to be. But nowadays jet planes, rapid communication and in particular the internet have enabled dispersed exiles to come together cheaply and effectively for the first time in history.

This change is most evident not among the best known, older-established diasporas but among the younger ones. Thus the influence on China of the huge community of overseas Chinese is, so far at least, chiefly felt through commerce and investment; it is not yet directly political. Similarly, émigré Indians have yet to exert much out-of-body-politic influence on the homeland. Expatriate Scots count for little in Scotland. And, though passionately interested in Israel and ready to support it financially, the Jewish diaspora—the first to be given the name, after the Babylonian captivity—is probably more influential outside than within the Jewish state.

Look instead at the Tamils, and in particular at the long war fought by the Tamil Tigers against the government of Sri Lanka. Superficially, this looks like a classic struggle between an oppressed group trying to win the right to secede and an intransigent government unwilling to let it. The Tamils have indeed been hard done by in the past, and Sri Lanka's government forces have committed their share of atrocities. The guerrillas have a skilful leader, some useful exiles abroad and foreign friends who lend support. Such has been the pattern often enough in struggles elsewhere—as, for instance, when the United States wanted to break away from Britain.

But the Sri Lankan civil war has not been a standard affair. For a start, it has been unusually brutal: 65,000 lives have been lost. The Tigers have ruthlessly exploited not just child soldiers but also suicide bombers. And they have done so for most of the past 19 years with the support of a generous community of exiles. Some 60m Tamils live in India, and Sri Lanka's politics have on at least one occasion fatefully affected India's: Rajiv Gandhi, a former prime minister of India, was assassinated by a Tamil suicide bomber in 1991 in retribution for India's involvement in Sri Lanka's civil war. But the truly Sri Lankan Tamil diaspora lies not in India but in Europe and North America. In 2001 the United Nations put the number of Tamil refugees abroad at 817,000. Taking into account those with citizenship of other countries, the total may be even bigger.

Many are poor, but a glance at the *Tamil Guardian*—"a weekly update for the global Tamil community"—suggests that the diaspora includes many others who are educated, prosperous and committed to the homeland. A broadsheet, it reprints serious articles from such papers as the *New York Times*, as well as covering the politics of Sri Lanka and the sporting and cultural activities of the diaspora; the Tigers receive uncritically fulsome coverage. Nor does it appear to want for advertisements.

Expatriate Tamils do not rely on their newspaper alone. Anything judged to be of interest to the community—a critical article in *The Economist*, for instance—may be circulated on the internet among Tamil émigrés, particularly academics, and a flood of rather similar protests may ensue. A demonstration can be similarly conjured up if necessary, as when, in 2001, Sri Lanka appointed as high commissioner to Australia a general accused of brutality.

Both through voluntary contributions and through extortion, the diaspora has been used to help pay for the Tigers' long military campaign, which involves boats and naval forces as well as rockets, missiles and the usual paraphernalia required by soldiers—plus the cyanide pills that all Tigers are sworn to swallow if captured. At the same time, the diaspora has given succour to the exiled leadership, whose main base was for decades in London.

Second thoughts from abroad

If for many years the diaspora helped to sustain its side of Sri Lanka's vicious war, all the while condoning rather than condemning Tamil atrocities, so it has recently started to exercise a more benign influence. A few years ago, America, Britain and India all decided that the Tigers were not an entirely wholesome liberation group, and took steps to declare them terrorists instead; Britain proscribed them in February 2001. The disconcerted diaspora began to close its wallet.

Then came September 11th 2001, followed by a string of hideous suicide bombings in Israel, and the diaspora realised that the Tigers' record of 150 or more such bombings did not put them in good company. It was time to call a halt. A ceasefire was signed in December 2001 and talks have since been held in Thailand and Norway. Unexpectedly, the Tigers' leaders have even dropped their insistence on a completely independent homeland, saying they will settle for "substantial autonomy" instead. The outlook is now hopeful. For this, the diaspora can take some credit.

Arrivederci, Buenos Aires

With luck, the Tamil diaspora will soon be called upon to perform a more traditional role, that of helping to pay for the development of their homeland. Some diasporas are poorer than the people they left behind: so impoverished are the 537,000 Argentines of Italian origin that regions of Italy like the Veneto have organised "emigrant re-entry projects" to try to find jobs for those who want to come home. Most exile groups, however, are relatively well off. Even refugees who have fled their country with little or nothing tend either to go home eventually or to make good in a new country. It is largely the prosperity of emigrants, combined with their levels of education, that gives them their influence back home. All in all, émigrés of one kind or another send about $100 billion home each year through official channels, 60% of it to poor countries, which may receive another $15 billion unofficially.

Much of this money is sent by underpaid Filipinas in Asia or exploited Bangladeshis in the Gulf. Yet exiles' earnings may well be higher than those at home, sometimes much higher, thanks not only to wage differentials but also to their qualifications: perhaps a third of highly educated Ghanaians live abroad, and three-quarters of Jamaica's population with higher education can be found in the United States alone. El Salvador values its emigrants' remittances ($1.75 billion in 2000) so much that it has made provision for legal aid in the United States to those of its citizens who want to claim or prolong political asylum.

Most poor countries are now resigned to losing their exiles physically, but that does not mean they cannot get hold of some of their money, or their expertise. Increasingly, this is an organised endeavour. So, for example, if you had been flying into Accra on July 22nd 2001, you might well have been going to a Ghana Homecoming Summit, organised by the country's Investment Promotion Centre to harness the skills of the Ghanaian diaspora and get it to cough up even more than the $300m-400m that it already sends home each year.

Why should it? Leaving aside sentimental considerations, one inducement is increasingly frequently offered: in return for money, voting rights. Apart from Mexico, Croatia and Eritrea, Armenia and India have also promised them. Filipinos would like them. Turkey, like Eritrea and Mexico, has amended its constitution to give them, but Germany, where the Turkish diaspora is heavily concentrated, is not keen to have foreign elections held on its soil.

Other countries have found other ways of exploiting their expatriates' political energies. Eritrea is one of the most advanced, perhaps because about 90% of eligible Eritreans abroad voted in the 1993 referendum on independence. Diaspora Eritreans then helped to draft the constitution, which guarantees them voting rights in future elections.

An alternative is to bring the exiles home in person. Turkey's Islamist party, now in government, has parachuted diaspora leaders into safe electoral seats in Turkey to reward them for fund-raising abroad. Afghanistan has pondered putting its ex-king back on the throne. Bulgaria has turned its ex-monarch into a prime minister. The Balts have been émigré importers on an almost industrial scale. Since becoming independent in 1991, Estonia has recruited from the diaspora two foreign ministers and a defence minister, plus lots of civil servants, especially in the foreign ministry. Latvia's popular president was brought back from Canada. It has also had the services of an American-Latvian defence minister, a bunch of members of parliament and a handful of diplomats, all mustered from the ranks of its émigrés. Lithuania's huge diaspora has supplied it with a president, the current chief of general staff (both Lithuanian-Americans) and several historians, novelists and poets. Unlike most expatriate Balts, Lithuania's are not all fierce nationalists.

The diaspora's influence is not always welcomed. Some exiles are simply mistrusted, especially if they were abroad when the going was hard. Thus Thabo Mbeki and those of his colleagues who struggled against apartheid from outside South Africa do not aways command the respect of those who stayed and fought it from within. These exiles, though, were more like political exiles of the conventional kind—the enemies of English monarchs who plotted from France, the Cubans who launched the Bay of Pigs invasion from the United States, Russian revolutionaries who wandered Europe to escape the tsar, and so on. Diaspora exiles are different: their motives for leaving are often economic as much as political, and many have no intention of going home.

Home and away, all at once

Still, they can be a pain in the domestic neck. Kosovo's ethnic-Albanian émigrés helped to pay for and arm the guerrillas who proved crucial in NATO's war to rid Kosovo of its Serb oppressors. But those same émigrés, many of them left-wing in ideology and criminal in their connections, were less welcome when the time came to build the peace. Similarly, Albania's diaspora, smaller and on the whole right-wing, disastrously backed Sali Berisha, until the pyramid schemes that briefly beguiled his countrymen came crashing down. The Turks of northern Cyprus are not always convinced that the Turkish-Cypriot community in London, almost as numerous, is lobbying for the objectives they really seek.

Since September 11th, diasporas have come under new scrutiny to see whether they harbour or breed terrorists. They were already the object of growing interest among academics; and think-tanks such as the Rand Corporation in California had issued warnings that some diasporas might become fifth columns for hostile governments. Yet the influence of émigrés can be exaggerated. Eva Ostergaard-Nielsen, of the London School of Economics, points out that diasporas seldom make a government adopt a policy unless that policy is also in the country's national interest. But they do undoubtedly, and increasingly, mingle homeland interests with those of their adopted country, and carry their own concerns back home. As Tip O'Neill, an American politician of the old school, once said, "All politics is local." Now more than ever.

From *The Economist*, January 4, 2003, pp. 41-43. © 2003 by the Economist Newspaper, Ltd. Reprinted by permission. Further reproduction prohibited. www.economist.com

The People's Sovereignty

How a new twist on an old idea can protect the world's most vulnerable populations

By George Soros

Sovereignty is an anachronistic concept originating in bygone times when society consisted of rulers and subjects, not citizens. It became the cornerstone of international relations with the Treaty of Westphalia in 1648. During the French Revolution, the king was overthrown and the people assumed sovereignty. But a nationalist concept of sovereignty soon superseded the dynastic version. Today, though not all nation-states are democratically accountable to their citizens, the principle of sovereignty stands in the way of outside intervention in the internal affairs of nation-states.

But true sovereignty belongs to the people, who in turn delegate it to their governments. If governments abuse the authority entrusted to them and citizens have no opportunity to correct such abuses, outside interference is justified. By specifying that sovereignty is based on the people, the international community can penetrate nation-states' borders to protect the rights of citizens. In particular, the principle of the people's sovereignty can help solve two modern challenges: the obstacles to delivering aid effectively to sovereign states, and the obstacles to global collective action dealing with states experiencing internal conflict.

External aid does not necessarily interfere with the sovereignty of states; governments can take aid or leave it. Having spent nearly $5 billion in such assistance over the years, I have experienced all the pitfalls that beset foreign aid. In 1984, I established the first national foundation inside Communist Hungary, followed by national foundations in some 32 additional nations. These foundations have been operating with total annual budgets averaging around $450 million for the last decade.

Although offers of external assistance do not undermine state sovereignty, foreign aid should not flow through national governments alone but should also support local governments and nongovernmental organizations (NGOS). Democratic governments should not object to aid directed at such groups. But precisely those governments that do not qualify for official assistance tend to oppose these nongovernmental channels. Such objections make a prima facie case that those regimes are violating the people's sovereignty; thus, the case for supporting civil society grows even stronger.

That principle has guided my network of foundations. In every country, we create a local board of citizens and channel our support through it. These boards work with the government when possible; where they cannot, they confine their support to civil society and resist state interference. So far, the foundations have successfully fought repression because governments are loath to publicly crack down on organizations that serve the interests of the people. Consider what happened in Yugoslavia toward the end of the Slobodan Milosevic era: Despite outlawing my foundation, Belgrade never enforced the decision, which allowed the foundation to continue operating.

Outside governments and international aid organizations are in a much stronger position than private foundations to resist governmental meddling in assistance directed to NGOs. Even the most repressive regimes seek to maintain the fiction that they have the people's interest at heart, leaving themselves susceptible to diplomatic disapproval. Although external pressure can be counterproductive—the land issue in Zimbabwe touched a nerve with the African public, and Zimbabwean President Robert Mugabe deflected widespread international disapproval by posing as a warrior against colonial oppression—a suitable pressure point can often be found. For instance, when Egyptian authorities jailed democracy activist Saad Eddin Ibrahim in 2000 for, among other charges, accepting unauthorized foreign financial support, the United States retaliated by freezing a supplemental aid package to the country. The Egyptian Court of Cessation eventually acquitted Ibrahim in March 2003, reaffirming freedom of speech and freedom to receive funds from abroad.

Even the most repressive regimes seek to maintain the fiction that they have the people's interest at heart, leaving themselves susceptible to diplomatic disapproval.

Since armed conflicts and repressive regimes may pose dangers beyond the borders of the countries concerned, all democratic nations have an interest in overcoming collective action problems and promoting open societies all over the world. The earlier preventive action begins, the less costly and more effective it is likely to be. For example, in the former Yugoslavia, early outside pressure on Milosevic—either when he abolished that autonomy of Kosovo in 1990 or when the Yugoslav navy bombarded Dubrovnik a year later—could have averted the tragedies that befell the region over the next decade.

The Baltic states, particularly Latvia and Estonia, provide a positive example of conflict prevention. These states were forcibly integrated into the Soviet Union in 1940; much of the local population was deported and other nationalities brought in. When the Baltics regained their independence in 1991, they struggled with a strong impulse to deny citizenship rights to members of these other nationalities. Such mistreatment of the sizeable Russian populations within these countries could have provided Russia with a compelling excuse for armed intervention, but the Organization for Security and Co-operation in Europe (OSCE) and the European Union pressured the Baltic states to guarantee minorities legal rights and protections. My foundations (among others) provided language instruction and supported other forms of ethnic reconciliation. A potential crisis was defused.

Unfortunately, though, a non-crisis makes no headlines. As things stand now, conditions must deteriorate significantly before foreign governments are willing to take a firm stand. But by the time gruesome television images provoke outrage in Western audiences, it will be too late to prevent a crisis. And as crises multiply, the public becomes less responsive, allowing dangerous situations to fester. The tardy U.S. intervention in Liberia is typical.

Of course, predicting which grievances will develop into bloodshed is impossible; the most effective preventive action reduces the potential for crises to develop in the first place. The best way to accomplish this goal is by fostering open, democratic societies. That has been the objective of my foundations since before the disintegration of the Soviet empire. It must be pursued on a larger scale.

That pursuit brings us back to a reconsideration of the principle of sovereignty. As U.N. Secretary-General Kofi Annan has stated, "State sovereignty, in its most basic sense, is being redefined—not least by the forces of globalisation and international co-operation. States are…instruments at the service of their peoples, and not vice versa." Indeed, the rulers of a sovereign state have a responsibility to protect the state's citizens. When they fail to do so, the responsibility is transferred to the international community. Global attention is often the only lifeline available to the oppressed.

George Soros is a financier, philanthropist, and the author of The Bubble of American Supremacy (*New York: Public Affairs, 2004*).

UNIT 2
World Economy

Unit Selections

6. **Charging Ahead**, Joshua Kurlantzick
7. **The Market for Civil War**, Paul Collier
8. **From Petro to Agro: Seeds of a New Economy**, Robert E. Armstrong

Key Points to Consider

- Can you think of some additional ways to disrupt illegal transnational capital and resource transfers throughout the world by terrorists and criminals?

- Be sure you can explain some of the potential unintended negative consequences that could occur throughout the world from the rising use of U.S. credit cards.

- According to the study discussed by Paul Collier, how does poverty and illegal trade in natural resources cause civil wars?

- If Robert Armstrong is correct that countries with a genetically diverse plant and animal gene pool are likely to benefit from the ongoing biological revolution, which countries in the world are likely to increase in wealth and power in the future? Which countries are likely to decline?

 **Links: www.dushkin.com/online/**
These sites are annotated in the World Wide Web pages.

International Monetary Fund
http://www.imf.org

Transparency International
http://www.transparency.org/pressreleases_archive/2002/2002.08.28.cpi.en.html

TI Bribe Payers Index
http://www.transparency.org/surveys/index.html#bpi

Graphs Comparing Countries
http://humandevelopment.bu.edu/use_exsisting_index/start_comp_graph.cfm

International Political Economy Network
http://csf.colorado.edu/ipe/

Organization for Economic Cooperation and Development/FDI Statistics
http://www.oecd.org/daf/investment/

Virtual Seminar in Global Political Economy/Global Cities & Social Movements
http://csf.colorado.edu/gpe/gpe95b/resources.html

World Bank
http://www.worldbank.org

The United States government's preparations for the military intervention in Iraq and the subsequent civil unrest and political violence in Iraq after the U.S. military captured Baghdad in 2003 tended to capture most people's attention in 2003 and 2004. Several economic trends that may lead to serious problems and widespread economic dislocations throughout the world went largely unnoticed as media attention focused more on the aftershocks of the war in Iraq.

The Iraqi conflict heightened economic conflicts and cleavages among Western allies, with rising powers such as China, and many countries in the developing world. Tensions among these states are likely to worsen in future years. Media attention focused on the rise of consumer boycotts as the "freedom fries" movement and other types of proposed boycotts after France and Germany refused to back the United States in efforts to obtain a UN resolution supporting the Iraqi war. Less attention was paid to ongoing trade disputes between the U.S. and the European Community (EC). In 2003, the U.S. petitioned to the World Trade Organization (WTO) to move against EC governments for violating WTO rules regarding subsidizing domestic farmers while Europeans charged that the U.S. government used preferential tariffs, tax, and agricultural subsidy policies to illegally aid U.S. farmers. This longstanding trade dispute is aggravated by protests by Green Party supporters and others throughout Europe who want to block U.S. agricultural products that have been genetically modified from being imported anywhere in Europe. The trade disputes spill over to efforts to revise international economic rules on trade that are designed to help developing countries to compete more effectively in world trade. During 2003 and early 2004 ongoing trade negotiators missed several Doha deadlines in efforts to reach agreement on a new multinational trade agreement that would be fairer to developing states.

Escalating tensions between the United States and China were temporarily halted during 2003 after the United States succeeded in pressuring China to agree to purchase more U.S. Treasury bonds and loosened controls on foreign currency holdings. China used some of its large pools of foreign currency reserves, which now total $356 billion, to buy a record $41 billion of Treasuries during the first six months of 2003. While this amount was less than is held by Japan, it was far more than any other foreign country. Both of the policy actions were designed to slow down China's growing trade surplus with the United States. To date, China has refused to agree to a longstanding United States' demand to let the Chinese currency rate float against foreign currencies and thus make U.S. exports to China more competitive.

The Chinese trade surplus is only one reason for a large and growing trade deficit that the United States has with the rest of the world. Many economic analysts in the U.S. and abroad are starting to be concerned by the United States' growing trade deficit. While the U.S. has long run a huge trade deficit with most of the world, the current trade deficit of over 489 billion dollars looms larger now that the U.S. is facing a large and growing budget deficit. Most observers of trends in the world economic system have focused recently on the increased number of service jobs, in addition to manufacturing jobs, that are being contracted out by businesses in the developed world to workers in the developing world. Few analysts have paid much attention to the potential unintended effects of America's biggest new export—credit cards. However, Joshua Kurlantzick in "Charging Ahead" describes how the American consumer market, that historically helped drive the world economy, could have unintended consequences, including possibly bringing down the world economy.

After the September 11, 2001 terrorist attack in the United States, the U.S. government redoubled efforts to disrupt capital flows among terrorists and criminal groups living in different nation-states. Recovered documents captured by agents working for the U.S. War on terrorism underscored the importance of unreported international economic exchanges throughout the world. However, the United States is unlikely to ever know all the details of how Bin Laden and other terrorists, in addition to unknown numbers of other individuals, moved money across national borders. This is because the ancient system for transferring money between countries, the hawala that is widespread in the Middle East and Africa, leaves no electronic or physical transfers of funds. Instead, informal mechanisms for moving money across national borders rely on trust among members of interpersonal networks in different countries. Few observers seriously believe that it is possible to eliminate illegal transnational capital and resource transfers, as these changes are now an integral feature of the modern world economy.

Recent economic trends also highlight the fact that the gap between the rich and poor in the world continues to grow. Many analysts today are proposing policies that are designed to stop the growing disparity. The agenda includes debt cancellation for nation-states with per capita incomes of less than $1,000, increased development aid, improved market access for developing countries, and an anti-poverty coalition to represent the interests of the developing countries. The proposals reflect a shared consensus of critics of neo-classical policy prescriptions supported by the G-8, the IMF, and the World Bank on what needs to be done. However, little progress has yet been made in reaching an agreement about who will pay to implement these proposals.

There is a growing sense that some international actions are necessary in order to alleviate the conditions fueling poverty, recruiting for terrorist groups, and civil wars. An additional rationale for implementing such an agenda is offered in "The Market for Civil War." In this article, Paul Collier discusses the results of a recent study of civil conflicts over the past 40 years that found that poverty and illegal trade in natural resources—rather than ancient political feuds and ethnic tensions—are the main causes of civil war.

Charging Ahead

*America's biggest new export—credit cards—
could bring down the world economy.*

By Joshua Kurlantzick

For most of the last two generations, the global economic system operated under a paradoxical division of labor. Americans consumed, while Asians—and much of the rest of the world—saved. The vast American consumer market helped drive the whole world economy. But since Americans were spending at such a prodigious clip, they weren't able to save much. That created two problems: There wasn't a lot of capital for business investment and even less to make up for the country's huge current account deficit—a function of our buying more from foreigners than we sell them. That's where the saving of the rest of the world came in. Foreigners saved so much that they had plenty of capital to invest—and where better to invest it than in the galloping, consumer-driven American economy?

Foreign investors, banks, and other companies purchased American equities, treasuries, and greenbacks, and invested in the United States. (According to a recent Merrill Lynch report, the United States absorbed nearly three-quarters of the savings of the world's major industrial countries in 2002.) This inflow of foreign capital has kept America's current account deficit stable and U.S. inflation low, making it easier for American consumers to keep on buying. Asians, meanwhile, needed our consumption-driven economy because their export-driven economies thrived on Americans who spent every dollar they earned, and then some. This division of labor may have been morally dysfunctional. But as a global economic order, it worked like a charm.

Of course, economists long warned that the system was inherently unstable. If foreigners suddenly lost faith in the U.S. economy and pulled out their billions, the market would bid the value of the dollar down dramatically. Indeed, since the stock market bubble burst in 2000 that's already begun to happen. In the last two years, foreign investment in the U.S. economy has plummeted to levels last seen in the early 1990s. With America at war against terrorism, anxious economists now worry that rising anti-Americanism or just the war-induced strains on the American economy could prolong the foreign investment drought or dry it up even more, leading to a sharp devaluation of the dollar, and perhaps even a cycle of worldwide recession.

There's no way to predict if any of this will come to pass. But the crux of the problem is that these possibilities remain outside America's control. The only way to truly solve the problem is for Americans to save more at home or sell more goods and services overseas. Ironically, though, what may bring the whole system crashing down once and for all is one of America's own most rapidly growing exports: credit cards.

Sticky Rice, Stickier Debt

Until the mid-1990s, consumers outside North America and Western Europe rarely ran up large amounts of personal debt. With the credit card market in the developed world still growing, big credit card companies did not focus on the developing world (which includes most of Asia). Countries like Thailand still hadn't developed large populations of middle class consumers. But traditional mores also played a role. In Asia, where people historically considered saving an important virtue and conducted nearly all transactions in cash, personal debt was less a fact of life than a source of shame. As recently as the mid-1990s, South Koreans saved more than 30 percent of their GDP, while Americans struggled to save a measly 1 percent (though U.S. home values and ownership rates, a form of savings, did rise).

For a variety of reasons, however, the situation has recently begun to change. As some developing countries boomed during the 1990s, they germinated millions of new middle- and upper-class consumers who had more capital and more desire to purchase consumer goods and their own homes. Many of these new consumers were under 40—men and women less tied to traditional mores of saving. Many had traveled in North America and seen how Americans easily obtain loans, sign mortgages, and whip out credit cards. After the Asian financial crisis of

1997, which depleted much of the savings of this new middle class, credit was often the only way to keep buying.

At the same time, the late-1990s financial crises prompted Asian, Latin American, and Eastern European governments to try to stimulate domestic consumption to boost national growth rates. Two key ways to do that were to lower rates of interest and ease the regulation of credit. Over the last three years in Thailand, Prime Minister Thaksin Shinawatra has toured the country to encourage Thais to spend more. In South Korea, the government has called on Koreans to borrow as much as possible. In China, the state has gone so far as to create whole new national holidays to give citizens more time to shop.

Meanwhile, intensive lobbying by banks, finance companies, and credit card firms has prompted governments to lower the minimum-income bar for credit cards and other consumer loans. For the lenders, the motivation was obvious. Credit-card operations can generate returns of more than 50 percent, since card issuers often charge interest rates of more than 20 percent. As Noam Neusner noted in *U.S. News and World Report*, "for banks, personal, unsecured loans—[i.e. issuing credit cards or personal loans to people who are a high risk not to pay their balance] represent one of the most profitable niches." And as markets in the developed world became saturated, debt issuers turned to the developing world to keep their profits rolling in.

This crucial deregulation has allowed banks and card companies to pursue a wider range of clients, many of whom have minimal savings and no credit history. In South Korea, for example, it has become so easy for lenders to issue credit cards that the local media reported last year that banks had mistakenly given household pets—which could hardly have a credit rating—their own credit cards. In golf-mad Thailand, card issuers have offered free links lessons. And credit-card issuers have pursued similar tactics across Latin America and Eastern Europe.

Foreign governments and global lenders have both tried to encourage developing world consumers to open their wallets and global consumers have happily obliged. Until recently, credit cards were unheard of in China. Today, however, Chinese use plastic for more than $200 billion worth of transactions annually. India and Indonesia reported major increases in mortgage applications, consumer loans, and credit card purchases in 2002. Across Asia, Visa International increased the number of cards issued by 25 percent last year alone, and cash withdrawals using Visas rose by 44 percent. According to *The Economist*, households now account for nearly 40 percent of East Asian banks' total lending, up from 27 percent in 1997.

Thailand and South Korea have become even more obvious examples of American-style spending. Visa says that total spending on its cards in Thailand is rising by more than 40 percent annually. In South Korea the story is the same. As one South Korean told *The New York Times* in December, the average Korean worker now carries four credit cards (more than 10 million South Koreans carry four or more, while the average bank worker carries between 10 and 15). Meanwhile, consumers in Eastern Europe, Latin America, and even Africa have actively followed suit. Schroder Salomon Smith Barney, an investment bank, estimates that between 1997 and 2001, household lending in Hungary, Poland, and the Czech Republic rose by more than 25 percent. As *The New York Times* recently reported, Russians also are beginning to utilize debt to pay for mortgages and large appliances. Household spending in South Africa, the biggest economy in Africa, has soared since 2000. And while Latin America has been hit by a series of financial crises from which, unlike Asia, it has not completely recovered, consumer spending—especially spending on credit—is rising there as well, in the region's more developed economies.

Extra Credit

As in the United States, the expansion of household spending in the developing world has had many positive impacts. It has not only helped buoy domestic economies hit hard by the economic shocks of the late 1990s but also provided new markets for foreign companies. Chinese acceptance of credit and auto financing, for instance, has helped Volkswagen make China its largest market; other automakers plan to follow suit. "The Chinese are becoming more used to financing. Once we establish the type of comprehensive GM financing systems we have in the U.S., we expect to see a huge jump in purchases," General Motors China head Philip Murtaugh told me last fall in Shanghai. Leading automakers are also developing plans to expand their product lines in Mexico, Brazil, South Africa, and Chile, in part because of the explosion of borrowing on credit—a development which mirrors the pattern in America during the early 20th century. Rising consumer spending has also filled many countries' tax coffers, since in most developing countries sales taxes are much easier to collect than income taxes—income-tax avoidance is very high.

> Thanks to easy credit and widespread fraud, many lenders in the developing world have found themselves overwhelmed with bad consumer loans and huge numbers of personal bankruptcies.

Yet the hard truth is that no one really knows what the impact of this consumer spending will be. These benefits could be outweighed by the potentially destructive impact of unleashing easy credit and American-style personal spending in developing nations with nascent systems of credit checks, unsustainable trade imbalances, and weak and opaque banking systems. In the United States, though financial companies have more leeway to lend to lower-income consumers than they used to, they are still subject to a range of state and federal regulations. Banks, credit card issuers, and other consumer lenders can also utilize America's well-developed system of credit checks to discover which potential customers are bad credit risks. Consequently, U.S. lenders generally are able to avoid lending or offering credit cards to high-risk clients.

Most developing countries lack the regulations and credit-checking services to deal with their rising numbers of consumer loans, credit cards, and mortgages. In Mexico, for instance, few banks require card recipients to show any credit history, minimum income, or even knowledge of how to use a card. As one Mexico City taxi driver told *The Financial Times*, "There's always someone offering me free credit." In Hong Kong, the lack of regulations has allowed credit card companies to aggressively target university students, helping the students obtain a credit card in minutes with nothing more than a photo identification. (Last May, Hong Kong's Consumer Council, a research and advocacy organization, chastised the government for allowing banks and other card issuers to use misleading tactics in their campaigns for students.)

Widespread corrupt practices, meanwhile, have led merchants in other countries to shun plastic altogether. When I traveled to Argentina in the winter of 2001-2002, I often found it difficult to convince retailers to take my American Express, the supposed gold standard of credit cards. Retailers had become so used to seeing fraudulent cards and other forms of questionable scrip issued by Buenos Aires banks and even provincial officials that they no longer trusted anything but cash. Meanwhile, the national government was taking virtually no steps to investigate any of the cards. Most merchants simply adopted a policy summed up by a sign in one cafe: "Se Acepta Dolares."

As a result of so much fraud and so much easy credit, many lenders in the developing world have found themselves overwhelmed with bad consumer loans and huge numbers of personal bankruptcies. To take just one example, 2.5 million South Koreans have fallen into arrears on their credit-card payments, in a nation of only 48 million people. In February, South Korean banks estimated that nearly 8 percent of credit card bills in the country were outstanding for a month or more, roughly double the percentage in the United States, and last year South Korea suffered its largest number of personal bankruptcies ever. According to banking industry statistics, personal bankruptcies are also rising in China, Mexico, Argentina, Chile, Brazil, and Thailand. In Hong Kong, personal bankruptcies are soaring by more than 100 percent per year. Even small, isolated countries like Bolivia have been affected by the rolling wave of easy credit and bankruptcy.

The rising tide of personal bankruptcies not only bears a social cost of broken families, domestic violence, and suicides, it also cuts into global lenders' ability to finance business loans. This isn't the case in the United States, because our capital base is vastly larger and deeper. But in Asian countries with smaller capital bases the problem is immediate and acute. In South Korea, several large banks, including Kookmin, the country's largest lender, have seen their net losses rise over the past four quarters—Kookmin lost $173 million in the fourth quarter of 2002 alone—because of what analysts told *The Financial Times* was a pattern of "reckless lending to consumers [without] properly assessing the creditworthiness of consumers." As a result, Kookmin and several other large South Korean banks have had to increase their cash reserves to cover non-performing consumer loans; in so doing, they have sharply reduced their loans to businesses. In fact, this spring South Korea's credit card companies and banks have had to issue more than $1 billion in new shares in an effort to address a looming cash crunch, a move which has rattled South Korea's stock markets. A similar pattern has been observed in other countries in Asia, Latin America, and even Russia. Even major financial players can be affected. Last year, HSBC had to drastically increase its provisions for potential defaults, in part because of personal bankruptcies in Hong Kong, which thereby reduced the amount of capital HSBC had to lend.

Drawing funds away from business investment can be a serious drag on economic growth. Banks reduce their capital bases, interest rates are forced up, capital becomes harder for businesses to raise, and companies slip into bankruptcy. Indeed, corporate bankruptcies in South Korea, Hong Kong, Thailand, and elsewhere in Asia have risen over the past two years, as many businesses have been unable to find new sources of financing from banks weighed down by portfolios heavy with non-performing loans. The decline in investment is often compounded by the binge-and-purge effects of high consumer debt, in which consumers run up huge credit card bills and then spend virtually nothing for months. This cycle can make it hard for retailers to make long-term business plans.

Prominent economists have begun to realize the potential dangers excessive consumer lending poses to the developing world. Last fall, the International Monetary Fund's chief representative in Seoul, Paul Gruenwald, expressed serious concern that South Korea's economy could be undermined by the "booming" number of bad loans to households and high credit card delinquency rates. Gruenwald has advised the South Korean government to take measures to reduce Koreans' consumer spending binge, and the IMF has since issued several other warnings about excessive, unregulated consumer lending in the developing world. Taking the advice, the South Korean government recently has announced government measures designed to bail out cash-short Korean credit card companies and restore banks' liquidity. Kim Gwang-Lim, South Korea's vice minister of finance, warned that if such a bailout does not succeed, "the potential impact from credit card companies' problems on the financial markets will be significant. Such problems could trigger chaos… and a possible collapse of the entire financial system."

Unsustainable Development

As 1997's Asian financial crisis demonstrated, crises in one developing nation's economy can rapidly spread to others, sparking a wave of destabilization throughout the world economy. Indeed, many of the countries experiencing rapid run-ups of consumer credit are those which went through snaps of boom and bust in the middle 1990s. Beside the growth of easy personal credit in Asia, economic weakness in Mexico—another country where minimal regulations on consumer loans have led banks to lend recklessly—could ripple throughout Latin America, of which Mexico's is the second-largest economy.

For decades America has relied upon the rest of the world to finance its consumption. But rising amounts of personal debt in foreign countries, combined with a dimming view of the American economy, could put a damper on this trend. As foreign banks and other lending institutions—which do not enjoy the

American luxury of being able to sustain their problems by tapping into pools of foreign investments—slash their capital base due to consumer debt, they may reduce their holdings in the United States. Of course, foreign institutions are not going to just stop investing in America, but even marginal reductions in their investments could have a serious impact on the American economy. Indeed, the pullout has already begun. Foreign direct investment into the United States fell from $301 billion in 2000 to $124 billion in 2001, the most recent year for which comprehensive data are available. And foreign investors have slashed their net purchases of U.S. treasuries, greenbacks, and corporate bonds over the past year—corporate bond purchases fell by a whopping 60 percent in September 2002 from a year before.

If foreign money continues pulling out of the United States, leading economists warn, America's debt and current-account deficits might no longer be sustainable, leading to a dollar crash and deep damage to the American economy. "At some point, probably relatively soon, as foreigners pull out their capital, the current-account deficit just won't be able to be maintained anymore," says Robert E. Scott, head of research at the Economic Policy Institute, a leading liberal D.C. think-tank. In a comprehensive research report, Scott's EPI colleagues conclude that the current-account deficit will soon become unsustainable, the dollar will begin to weaken, and the United States will be forced to address this imbalance in trade by moving away from the strong dollar policy which has been a hallmark of both the Clinton and Bush administrations and allowing the dollar to weaken, thereby accepting a deflation in Americans' real incomes. The EPI economists conclude that bringing the current account deficit into balance would require a 10-percent drop in America's GDP. And EPI is hardly the only organization making these claims. Last fall, Stephen Roach, chief economist at Morgan Stanley and one of the most thoughtful long-term predictors of the world economy, warned clients in a research note that "America's ever-widening current account deficit is on an inherently unstable path."

A long-term slowdown of economic growth in the United States could have pervasive effects around the globe. First, American consumer consumption would almost certainly slacken, as Americans began to save more to pay off higher domestic financing. That drop in consumer spending might then further depress global economic growth. Without the uber-consumer nation running at full speed, years might go by before the world economy escaped the cycle of weak consumption and slow growth. Despite increased consumer spending throughout the developing world, Asian countries like Thailand, Malaysia, and South Korea still depend on exports for as much as 50 percent of their gross domestic product. For these states, as well as for most developing countries in Latin America, the United States is by far the largest market. Similarly, China has built up the largest trade deficit with the United States of any country in the world. A falling dollar, which would make exporting to the United States less profitable, would only compound foreign companies' problems.

Cutting Cards

The consumer-credit problem has not yet developed into a full-blown crisis. Most Asian, Latin American, and European nations have not yet reached American levels of consumer indebtedness and low savings rates. Foreign banks and other lending institutions still possess pools of liquid capital to invest in the United States, keeping America's current account deficit afloat.

There is still time for governments, international financial institutions, credit card companies, banks, and other lenders to work together to prevent a global consumer debt crisis. But, in order to give the developing world the benefits of the consumer credit revolution America has enjoyed and protect developing nations from the potentially destructive impact of unleashing easy credit, countries, banks, and credit card issuers around the globe need to be subjected to stricter rules on lending. In short, they need the type of laws we have here in the United States. Nations could impose age limits and minimum income standards for applying for credit cards—a policy Singapore, among some other countries, has already adopted. They could also force credit card companies to charge reasonable interest rates and to use transparent advertising. The United States and other developed countries could work with developing countries to implement internationally recognized credit-evaluation procedures and crack down on banks that continue offering easy credit while defaults and delinquencies soar.

Unfortunately, such changes don't appear likely. Neither the United States nor the European Union has shown much willingness to police credit card companies and other lending institutions, let alone teach developing nations to police consumer-lending markets in their own countries. The world's financial media have largely ignored the consumer debt problem as they have focused primarily on the admittedly sexier topic of corporate fraud and bankruptcy. Meanwhile, consumers in the developing world just keep charging and charging and charging…

JOSHUA KURLANTZICK *is the foreign editor of* The New Republic.

THE MARKET FOR CIVIL WAR

Ethnic tensions and ancient political feuds are not starting civil wars around the world. A groundbreaking new study of civil conflict over the last 40 years reveals that economic forces—such as entrenched poverty and the trade in natural resources—are the true culprits. The solution? Curb rebel financing, jump-start economic growth in vulnerable regions, and provide a robust military presence in nations emerging from conflict.

By Paul Collier

Every time a civil war breaks out, some historian traces its origin to the 14th century and some anthropologist expounds on its ethnic roots. Don't buy into such explanations too quickly. Certain countries are more prone to civil war than others, but distant history and ethnic tensions are rarely the best explanations for a conflict. Look instead at a nation's recent past and, most important, its economic conditions.

Once a country has reached a per capita income rivaling that of the world's richest nations, its risk of civil war is negligible. Today, about 900 million people live in such societies. Four billion more live in countries that are either already middle income or on track to becoming so, thanks to rapidly growing and diversifying economies. This group, which includes the economic success stories of the post–World War II era, faces fairly low risk of civil war. The potential for conflict is concentrated among the countries inhabited by the world's remaining 1.1 billion people. These countries typically have poor and declining economies and rely on natural resources—such as diamonds or oil—for a large proportion of national income. As the British, French, Portuguese, and Soviet empires successively dissolved during the last century, the number of such countries increased in waves.

Such at-risk countries are engaged in a sort of Russian roulette. Every year that their dismal economic conditions persist increases the odds that their societies will fall into armed conflict. Whether by luck or prudence, many such nations have so far escaped civil war. Others have not. And once civil war has started, the decline in income and the accumulation of arms, fighting skills, and military capabilities greatly increase the risks of further conflict.

To date, academics and policymakers alike have misdiagnosed the nature of the problem; little surprise, then, that their efforts to prevent civil wars have been ineffective. When the world's leaders can identify the real factors most likely to drive such conflicts, they will have a better chance of preventing future wars.

THE MYTH OF ETHNIC STRIFE

Between 1960 and 1999, there were 52 major civil wars for which comprehensive data is available on social, political, historical, economic, and geographic circumstances. Such wars spanned the developing regions, with the typical conflict lasting around seven years and leaving a legacy of persistent poverty and disease in its wake. To understand the causes of these conflicts, economist Anke Hoeffler and I studied each five-year-period from 1960 to 1999 and identified preexisting conditions that helped predict the outbreak of war.

> If a country is mountainous and has a large, lightly populated hinterland, it faces an enhanced risk of rebellion… Nepal is therefore more at risk of civil war, geographically speaking, than Singapore.

For example, income inequality and ethnic-religious diversity are frequently cited as causes for conflict. Yet surprisingly, inequality—either of household incomes or of land ownership—does not appear to increase systematically the risk of civil war. Brazil got away with its high inequality; Colombia didn't. And, in fact, ethnic and religious diversity actually reduces the risk of civil conflict. One important exception: Where the largest ethnic group constitutes a majority but lives alongside a substantial minority, such as in Sri Lanka and Rwanda, the risk of civil war roughly doubles. Once wars start, they also tend to last much longer if the nation in question displays two or three dominant ethnic groups.

Conflicts in ethnically diverse countries may be ethnically patterned without being ethnically caused. International media coverage of civil wars often focuses on history and ethnicity because rebel leaders adopt this sort of discourse. Grievances are to a rebel organization what image is to a business. The rebel group needs to stimulate a sense of collective grievance to build cohesion in its army and to attract funding from its diaspora living in rich countries.

Much to the dismay of democratization activists, democracy fails to reduce the risk of civil war, at least in low-income countries. Indeed, politically repressive societies have no greater risk of civil war than full-fledged democracies. Countries falling between the extremes of autocracy and full democracy—where citizens enjoy some limited political rights—are at a greater risk of war. Low-income societies with new democratic institutions are often at enhanced risk: Just consider the current catastrophe in Ivory Coast, where uncertainty over who could stand for the presidential election in 2000 triggered violent clashes and ongoing political instability.

Wherever a civil war occurs, observers will invariably find some deep history of conflict. But overwhelmingly, conflicts in the distant past are not generating civil wars in the present. The history that matters is recent history, not that of the 14th century. If a country recently experienced a civil war, it is much more likely to have another one. The risk fades the longer peace endures.

Civil war is self-perpetuating, partly because it changes the balance of interests within countries. Groups engaged in conflict invest in armaments, skills, and infrastructure that are only good for violence. These groups' leaders, and indeed all those who gain from lawlessness, prosper during war, even though society as a whole suffers. The part of the elite that prefers peace will have shifted much of its wealth outside the country. Hence, as a result of the conflict, the balance of elite interests shifts toward further conflict.

Geography matters, too. If a country is mountainous and has a large, lightly populated hinterland, it faces an enhanced risk of rebellion. Presumably, rebels are harder to find and defeat in such terrain. Nepal is therefore more at risk of civil war, geographically speaking, than Singapore.

DIAMONDS, A REBEL'S BEST FRIEND

All these factors notwithstanding, economic conditions remain paramount in explaining civil wars. For the average country in our study, the risk of a civil war in each five-year period was around 6 percent, but the risks increased alarmingly if the economy was poor, declining, and dependent on natural resource exports. For a country with conditions like those in the Democratic Republic of the Congo (formerly Zaire) in the late 1990s—with deep poverty, a collapsing economy, and huge miner exploitation—the risk reaches nearly 80 percent.

> When valuable natural resources are discovered in a particular region of a country, the people living in such localities suddenly have an economic incentive to secede, violently if necessary.

Once started, wars last longer in low-income countries, which are prone to rebellion for many reasons: Recruits have less of a stake in the status quo, and central governments are typically weak. Each additional percentage point in the growth rate of per capita income shaves off about 1 percentage point of conflict risk; conversely, wars are more likely to follow periods of economic collapse—such as the conflicts that have surged in Indonesia since the East Asian economic crisis of the late 1990s. If a country's per capita income doubles, its risk of conflict drops by roughly half. Simply put, economic growth matters because opportunities for youth depend upon a robust economy.

Conflict is also more likely in countries that depend heavily on natural resources for their export earnings, in part because rebel groups can extort the gains from this trade to finance their operations. Diamonds funded the National Union for the Total Independence of Angola (UNITA) rebel group during Angola's long civil war, as well as the Revolutionary United Front (RUF) in Sierra Leone; timber funded the Khmer Rouge in Cambodia. Indeed, methods of extortion abound. For example, multinational corporations that extract natural resources must often pay huge sums to ransom kidnapped workers and to protect infrastructure from sabotage at the hands of rebel groups. Laughably, such payments are sometimes charged to the companies' "corporate social responsibility" budgets.

Natural resources also fuel war because they make secession more likely. When valuable natural resources are discovered in a particular region of a country, the people living in such localities suddenly have an economic incentive to secede, violently if necessary. Since most countries are ethnically diverse, the lucky, resource-rich locality is likely to be ethnically distinct as well. Often, the weak political force of ethnic romanticism latches on to the stronger force of economic self-interest so that secessionist movements voice ethnic grievances—Biafra (Nigeria), Cabinda (Angola), and Aceh (Indonesia) come to mind. The incentive to secede is probably compounded by the corrupt, incompetent way to which governments commonly use natural resource wealth: The greed of a resource-rich locality can seem ethically less ugly if a corrupt national elite is already hijacking the resources.

THE WAR DIVIDEND

One striking lesson from these patterns is that the motivations for rebellion generally matter less than the conditions that make a rebellion financially and militarily viable. Civil wars only occur if a rebel organization can build and sustain a private army. These organizations are unlike traditional opposition groups such as political parties or protest movements. They are hierarchical, authoritarian, expensive, and usually small. Where such organizations are financially militarily feasible, rebellions are likely to emerge, promoting whatever political agenda their leaders happen to support.

Global efforts to curb civil war should therefore focus on reducing the viability—rather than just the rationale—of rebellion. Of course, policies should address legitimate grievances, not because addressing them paves a royal road to peace but because they are legitimate.

> Botswana and Sierra Leone were similarly poor countries, both sitting on vast diamond deposits... Botswana harnessed this opportunity, becoming the fastest-growing economy in the world. Sierra Leone used the same resources to impoverish itself...

Those nations currently at war and those that have recently emerged from civil war constitute the core of the problem. Many countries have fallen into a conflict trap: a damaging war that sharply increases the risk of further conflict, followed by a fragile peace, and then back to war. The expected duration of a civil war is currently about eight years—double what it was before the 1980s. Wars therefore do more damage now and thus more powerfully provoke further conflict.

No one knows why wars last longer now. Perhaps global markets in both natural resources and arms make rebellion easier to finance and equip. Rebel groups can now sell the future rights to mineral extraction (conditional on rebel victory) to raise funds for weapons purchases. A similar arrangement reputedly helped French oil Giant Elf Aquitaine (now TotalFinaElf) gain its current access to oil in the Republic of the Congo.

So, what specific measures can countries take to reduce the occurrence and likelihood of civil war?

For developing countries already growing rapidly, the most significant risk may be episodes of economic crisis, such as that experienced by Indonesia in the late 1990s. The opportunity to prevent war in such cases only strengthens the justification for international efforts to avert economic crises. In this light, initiatives to reform and rethink the workings of the global financial system are not merely an academic debate or an effort to ease investors' concerns, but rather a much more serious matter with immediate life-and-death consequences.

Poor countries that are not developing but have so far escaped civil war, such as Zambia and Malawi, are also racing against time. If they do not find ways to accelerate their economic growth and development, they will likely stumble into conflict. Recent casualties include Ivory Coast and Nepal. Nations in these conditions should get the message that change is urgent. Often the remedy should go beyond the standard package of market access, debt relief, and aid programs from the developed countries to include credible policy reform and honest governance within vulnerable countries.

Due to their heavy dependence on natural resource revenues, the governments in many at-risk nations face acute problems of corruption and exposure to international price shocks. But natural resources need not be a curse. Twenty-five years ago, Botswana and Sierra Leone were similarly poor countries, both sitting on vast diamond deposits. Over the ensuing quarter century, Botswana harnessed this opportunity, becoming the fastest-growing economy in the world. Sierra Leone used the same resources to impoverish itself, experiencing the most rapid sustained decline of any country; it now ranks at the bottom of the Human Development Index put out by the United Nations Development Programme. These contrasting examples show that good policy and governance are especially vital where natural resources are discovered.

So far, the record has been dismal: There are many more Sierra Leones than Botswanas. But some encouraging signs are emerging. The "Fowler Report" to the U.N. Security Council in 2000 detailed how UNITA evaded U.N. sanctions against arms smuggling and diamond-based financing; as a result, scrutiny of the international diamond trade increased. Such attention may well have contributed to the demise of UNITA in Angola and the RUF in Sierra Leone, two highly durable, diamond-dependent rebel organizations. Moreover, diamonds are now being tracked through the new Kimberley Process certification scheme, making it harder for rebel groups to obtain financing from these goods. Transparency is the first step toward effective national scrutiny.

And if such strategies work with diamonds, why not replicate the process for timber? The Group of Eight industrialized nations flagged the problems posed by natural resources in its 2002 meeting and may take up the issue again in June 2003 at the Evian summit. Specific political leaders have also begun taking action. British Prime Minister Tony Blair has launched an initiative for greater transparency in the reporting of resource revenues by multinational companies in extractive industries; by nationally owned oil companies such as Angola's Sonangol, which are often a state within a state; and by recipient governments. And in a complementary effort, French President Jacques Chirac has recently called for better mechanisms to cushion low-income countries when the prices of their commodity exports plummet. A new compact could emerge: Rich nations take action to cut rebel financial and cushion adverse shocks, while low-income nations adopt better governance of their revenues from natural resources.

ESCAPING THE CONFLICT TRAP

From 1960 to 1999, international interventions—whether economic or miliary—intended to shorten civil wars were disappointing. Some strategies may have succeeded in individual cases (as in the recent destruction of the Taliban regime in Afghanistan), but no type of intervention has worked regularly.

More effective interventions could target the systems that finance and equip rebel organizations, beyond solely focusing on the trade in diamonds and other commodities. Many rebel movements also receive illicit support from neighboring governments. Such support can be exposed and penalized, and the penalties should outweigh the benefits of rebel alliances. Moreover, governments can discourage huge ransom payments by corporations—such as the $20 million reputedly paid in 1984 by the German engineering company Mannesmann-Anlagenbau AG to the Colombian rebel group ELN, or National Liberation Army, for the release of three of the company's staff. Should such payments be tax deductible and so, in effect, subsidized? Governments could ban the insurance arrangements that facilitate and inevitably inflate such payments. National authorities have started to improve scrutiny of national and international banking systems; recent evidence that al Qaeda is shifting its assets into diamonds suggests that this strategy is becoming effective. National and multilateral policymakers should also look again at drug policy. Wouldn't it be easier and more effective to curb the demand for criminally supplied drugs? And certainly, the flow of arms can be curtailed, especially if efforts are made earlier to catch the big operators, like suspected gunrunner Victor Bout, who is though to have supplied arms to rebel groups in several African nations, including Angola.

The best way to break out of the conflict trap is to ensure that countries that have just ended one conflict do not quickly become enmeshed in another. In some nations, the risks of renewed conflicts are so high that an external military peacekeeping force is normally necessary. The operative word is "external" because high military spending by a post-conflict government actually increases the risk of another war. That external military presence must be credible. In Sierra Leone, the RUF took hostage a large U.N. force that it sensed would not fight, yet when confronted by a smaller British force, the rebel group collapsed.

Unfortunately, peacekeeping missions normally do not last long enough to allow economic recovery to take hold and help keep the peace. The peak time for economic recovery is usually during the middle of the first post-conflict decade. That is also when aid is most effective in promoting economic growth. Unfortunately, international aid is frequently mistimed. It pours in during the first year of peace, when the country's institutions are too weak for the money to be used effectively, then tapers out just when it would be most useful.

Governments in countries recovering from civil war also must give greater priority to economic reform: The post-conflict period is a good time to reform because vested interests are loosened up. For example, after the end of civil conflict in Uganda in 1986, the country's economic policies moved from among the worst in Africa to among the best in the following decade.

Finally, diasporas in rich countries pose a particular danger in post-conflict situations. They tend to be more extreme than the populations they leave behind, and they finance extremist and violent organizations. For example, the Tamil and Irish diasporas in North America have both been gullible financiers of murder in the past. Diaspora organizations can play an important role in economic recovery; their networks of skills and businesses are potentially valuable. Afghanistan is now trying constructively to deploy its diaspora, for example. But governments of rich nations should help keep the behavior of diaspora organizations in their borders within legitimate bounds.

Want to Know More?

The arguments in this article draw on **Breaking the Conflict Trap: Civil War and Development Policy** (Washington: World Bank and Oxford University Press, 2003) by Paul Collier. See also the studies and data from **The Economics of Civil War, Crime and Violence** project, available on the World Banks' Web site. Additional research from the project is available in a special issue of the *Journal of Conflict Resolution* (Vol. 46, No. 1, February 2002), edited by Collier and Nicholas Sambanis. See also the International Peace Academy's (IPA) program on **Economic Agendas in Civil War,** available on IPA's Web site.

On the links between natural resources and conflict, see Michael T. Klare's *Natural Resource Wars: The New Landscape of Global Conflict* (New York: Henry Holt, 2001), Michael Renner's *The Anatomy of Resource Wars* (Washington: Worldwatch Institute, 2002), and the Web site of the nongovernmental organization **Global Witness.** Yahya Sadowski contends that ethnic conflict is not as widespread or as "ethnic" as it seems in **"Think Again: Ethnic Conflict"** (FOREIGN POLICY, Summer 1998). For information on the British initiative on transparency in extractive industries, see the Web site of the United Kingdom's **Department of International Development.** And for views on how rebel groups and local movements promote their causes around the globe, see Clifford Bob's **"Merchants of Morality"** (FOREIGN POLICY, March/April 2002).

For political and historical perspectives on the underlying causes of civil wars, see Monty G. Marshall and Ted Robert Gurr's report **"Peace and Conflict 2003: A Global Survey of Armed Conflicts, Self-Determination Movements, and Democracy"** (College Park: Center for International Development and Conflict Management, 2003). See also Chester A. Crocker, Fen Osler Hampson, and Pamela Aall's, eds., *Turbulent Peace: The Challenges of Managing International Conflict* (Washington: United States Institute of Peace Press, 2001). For economic views on the sources of conflict, see Herschel Grossman's **"A General Equilibrium Model of Insurrections"** (*American Economic Review,* Vol. 81, No. 4, September 1991), where the author does not distinguish rebels and revolutionaries from "bandits or pirates." And Jack Hirshleifer postulates the Machiavellian theorem that no opportunity for profitable violence will go unexploited in *The Dark Side of the Force: Economic Foundations of Conflict Theory* (Cambridge: Cambridge University Press, 2001).

For links to relevant Web sites, access to the *FP* Archive, and a comprehensive index of related FOREIGN POLICY articles, go to www.foreignpolicy.com.

LOCAL WARS, GLOBAL CASUALTIES

Civil war is not just disastrous for the countries directly affected; it hurts the surrounding regions and often poses risks for even remote, seemingly unaffected nations. Within the country at war, combat-related deaths represent just a small if gruesome part of the costs: War-related economic ruin also intensifies poverty and disease. Throughout the region, economic growth declines and investment flows dry up. Disease spreads across borders through the flow of refuges. And higher military spending induced by real or potential civil war can fuel pointless regional arms races.

Finally, civil war creates territories beyond the control of recognized governments. These no-go areas can be damaging to the international community. Around 95 percent of the global production of hard drugs is located in civil war countries. Witness how sources of supply shift in response to the changing pattern of conflict: As the Shining Path guerrillas were defeated in Peru in the early 1990s, drug production shifted to territory held by the FARC, or the Revolutionary Armed Forces of Colombia. These lawless areas also provide safe havens and training for international terrorists.

Over the last several decades, national, regional, and global organizations seeking to end or prevent civil wars have often focused on the wrong challenges, or on the right challenges but at the wrong time. Certainly, no single, magic policy will fix the problem; a range of initiatives is urgently required across a broad front. But if governments and multilateral organizations can help curb rebel financing and armament, accelerate the economic development of the countries most at risk, and provide an effective military presence in post-conflict settings, the global incidence of civil war will decline dramatically. These are viable objectives, and they are likely much cheaper than the long-term consequences of continued conflict and neglect.

Paul Collier is director of the World Bank's development research group.

From Petro to Agro: Seeds of a New Economy

by Robert E. Armstrong

Overview

Winston Churchill is said to have stopped predicting future events because the future was just "one damned thing after another." Nonetheless, we need to keep an eye on the future and speculate as to what the next damned thing might be. One candidate is the changing raw material base for the economy.

Today, the hydrocarbon molecule is the basic unit of commerce. In a biobased economy, genes will replace petroleum. So, just as we currently demand assured access to sources of hydrocarbon molecules (oil), in the near future we will demand assured access to a broad-based, diverse supply of genes (plants and animals). This shift has security implications. Relations with oil-rich countries will be of less importance, and relations with gene-rich states—mostly the biodiverse regions along the equator—will assume greater significance. Conflicts may arise between gene-rich, technology-poor countries that control the basic raw materials of a biobased economy and gene-poor, technology-rich nations that control the production methods.

American instruments of power will be challenged to meet the demands of a biobased economy. We already see diplomatic challenges with the United Nations Framework Convention on Biological Diversity and controversy with Europe over genetically modified crops. Informational and economic challenges and opportunities will likewise appear. It may be challenging for U.S. land forces, especially the Army, to meet the demands of securing access to large supplies of new genetic material.

Agriculture will become increasingly important as a part of the Nation's industrial base, as it offers the most economical way to produce large quantities of biological materials. Homeland defense will have to consider heartland defense, as agricultural fields will assume the same significance as oil fields.

The Age of Geology

For much of the last century, and particularly since the end of World War II, petroleum has been the primary raw material for U.S. industrial and consumer needs. As a nation, our petroleum use is twice that of our consumption of either coal or natural gas and four times greater than use of nuclear or renewable energy sources.[1]

The bulk of our petroleum use goes to meet energy demands, with approximately 90 percent of a barrel of crude oil going to gasoline, diesel, and other fuels. Since 1949, however, the industrial consumption of petroleum for nonfuel use has increased nearly seven-fold.[2] The chemical industry, for example, relies on petroleum for more than 90 percent of its raw materials to manufacture its myriad of products, ranging from plastics, refrigerants, and fertilizers to detergents, explosives, and medicines. Virtually everything requires petroleum or petroleum derivatives for its manufacture.

We are beginning to see a shift from petroleum, however. As the 20th century was ending, Michael Bowlin, then-president of the American Petroleum Institute, who was also chairman and chief executive officer of Arco, told industry executives that the world was entering "the last days of the Age of Oil."[3] Estimates of the remaining life of the reserves vary widely, but many experts agree that worldwide oil production will peak between 2010 and 2020. Even if we agree with those who hold that the petroleum supply may be renewable, environmental pressures and economic incentives will remain that will move us to newer technologies.[4] Far from repeating the apocalyptic warnings of the 1960s and 1970s about the end of oil, Bowlin pointed to new technologies that will replace petroleum.

The Age of Biology

Prominent among the replacements for petroleum will be products developed from biological sources. Using biological materials, that is, plants and animals, as raw materials for industrial and consumer products is not a new idea. Before the rise of cheap oil, agriculture was the dominant source of our raw materials. Indeed, when the U.S. Department of Agriculture was established in 1862, its motto proclaimed, "Agriculture is the foundation of manufacture and commerce." Even today, agriculture supplies raw materials for industry; for example, about 8 percent of the U.S. corn crop goes to industrial uses rather than directly meeting food or feed requirements.[5] Moreover, the agricultural industry offers the most cost-effective way to manufacture large volumes of biological materials.

In its vision statement for the 21st century, the National Agricultural Biotechnology Council forecasts agriculture to be the source of not only our food, feed, and fiber, but also our energy, materials, and chemicals.[6] In a 1999 report on biobased industrial products, the National Research Council noted that U.S. farmers already generate annually about 280 million tons of waste biomass—leaves, stalks, and partially used plant portions. That is more than sufficient material to serve as feedstock for all of the domestic industrial chemicals that can be readily manufactured from agricultural sources.[7]

the United States clearly has the production capacity to process the raw materials for a biobased economy

Domestically, resources needed for food and feed production will not compete with resources required to grow our industrial raw materials. Our natural resource base of land and water is more than adequate to meet the demand. The United States has the largest amount of arable land per capita of any country in the world (1.73 acres for the United States versus 0.99 for other developed countries; the developing world average is only 0.49 acres).[8] Additionally, through the U.S. Department of Agriculture Conservation Reserve Program, 35 million acres are left fallow each year, some of which could be used to grow crops specifically for biomass.

While water does present some local and regional challenges, groundwater depletion rates in the United States have slowed overall during the last 20 years. As the need for affordable water increases, improvements in irrigation technology and the development of new water resources are likely to follow. One estimate suggests that improvements in irrigation technology alone can reduce the anticipated worldwide demand for additional water resources by one-half during the next 25 years.[9] Thus, while a concern, water availability is not likely to present a barrier to expanded agricultural production.

Technological innovations in agricultural production probably will continue to increase yields. Corn yields, for example, gained an average of 1.0 bushel per acre per year during the last century. In the last half of the century, the average increase was 1.8 bushels per acre per year. Depending on soil characteristics and water availability, even something as simple as the spacing between cornrows can be used to maximize yields. In the late 1990s, corn yields in the United States averaged 134 bushels per acre. Some researchers believe that within the next 20 years, technology and cultural practices can increase yield averages to nearly 260 bushels per acre.[10]

The United States clearly has the production capacity to produce and process the raw materials for a biobased economy. Still, in most current industrial practices, the cost of the conversion process—turning biomass into energy, materials, and chemicals—is not competitive with petroleum. Consider five common industrial products: adhesives, acetic acid, pigments, inks, and plastics. A comparison of their cost when derived from petroleum with their cost when derived from plant materials generally shows not only a greater cost for plant-derived materials but also great disparity in the spread, based on which product is being considered.[11]

A key problem with making cost comparisons is that the production costs are based on using existing facilities designed for petroleum feedstocks. When using biomass, some of the end products can be made through direct physical or chemical processing; others can be produced indirectly through fermentation (using microbial agents) or by enzymatic processing. What is needed are "biorefineries."[12] Like an oil refinery, a biorefinery would take carbon and hydrogen and produce desired products. The biorefinery's economic advantage will emerge from its dual capability. Along with the desired end products, foods, feeds, and biochemicals could be produced.

Prototypes of the biorefinery already exist in our industrial base in the form of corn wet mills, soybean processing facilities, and pulp and paper mills. While the prototypes of full-scale biorefineries are only in the planning stage, two facilities designed for specific biobased end products are presently coming online. Earlier this year, Cargill Dow opened a $300-million facility in Nebraska to manufacture a bioplastic made from sugars found in corn kernels.[13] As the technology improves, the company plans to extract the sugars from less costly agricultural waste, such as corn stalks, wheat straw, and rice hulls.

Similarly, DuPont announced last year that it had successfully manufactured a key ingredient for one of its new polymers, using corn sugars instead of petrochemicals.[14] The conversion was done in a pilot plant using a fermentation technology. Anticipating the eventual cost competitiveness of biological materials, the company has built a full-scale production plant for the polymer and designed it to use either petrochemical feedstocks or biobased feedstocks.

Is this biobased economy just a vision with a few immediate examples, or is there a long-term probability for its success? In its 1999 report on biobased industrial products, the National Research Council argued that a competitively priced, biobased products industry eventually would replace much of the petrochemical industry. As an intermediate goal, the report suggested that by 2020, a biobased economy could provide 25 percent of the 1994 levels of the Nation's organic carbon-based industrial feedstock chemicals and 10 percent of liquid fuels. The report suggested that, ultimately, 90 percent of the U.S. organic chemical consumption and 50 percent of our liquid fuel needs could be met by a domestic biobased economy.

Article 8. From Petro to Agro: Seeds of a New Economy

Production Costs (dollars per pound)		
Product	Petroleum-derived	Plant-derived
Adhesives	1.65	1.40
Acetic acid	0.33	0.35
Pigments	2.00	5.80
Inks	2.00	2.50
Plastics	0.50	2.00

In this new economy, plants and animals will be specifically bred and farmed to produce desired raw materials. For example, if an industrial process requires a chemical to have certain tolerances to heat, a protein may be available to provide that tolerance. The protein, which would be the product of a gene, could be derived from plants. If the protein occurs naturally in animals or in plant species that are not easily farmed, genetic engineering offers the ability to transfer the gene to a plant species more suited to agricultural production. (The product of moving genes from one species to another is called a *transgenic*.) Once introduced into an agriculturally desirable plant, the protein can then be produced more cost-effectively and made available on a commercial scale.

Fueling the Biobased Economy

As the biobased economy matures and issues of production and processing are improved, the demand for new products will grow. New products will require new raw materials. In a biobased economy, the basic raw material will be genes, and novel genes will be the source of novel products. Thus, as we shift from an economy based on geology to one based on biology, the basic unit of commerce will shift from the hydrocarbon molecule to the gene. The Quadrennial Defense Review 2001 cites access to key markets and strategic resources as part of our enduring national interests.[15] Just as we currently demand assured access to sources of hydrocarbons, in the near future we will demand assured access to a broad-based, diverse supply of genes.

As with any resource vital to our economy, the location of large supplies of genes will become important to our national security concerns. In our petroleum-based world, the resource is concentrated in various pockets distributed worldwide in nearly all climate regions. Obviously, genes are distributed worldwide, as there is life in every nook and cranny of this planet. However, the overwhelming majority of genes are concentrated in the equatorial regions.

> **as we shift from an economy based on geology to one based on biology, the basic unit of commerce will shift from the hydrocarbon molecule to the gene**

Biologists refer to a region's biodiversity when commenting on the range of life forms present. The more life forms present (that is, the more genes present), the greater the biodiversity. The general biological principle of *latitudinal diversity gradient* contends the closer to the equator, the greater the biodiversity. The amount of solar energy present, the lack of seasonal climate fluctuations, and the expanse of land explain the gradient's existence. By way of illustration, consider the results of a study that used comparable sized plots of land at different latitudes to compare the number of different bird species found at each latitude: Greenland, 56 species; New York state, 195 species; Colombia, 1,525 species. Plants show a similar degree of biodiversity. For example, in all of Canada and the United States, there are only 700 native tree species. In one census involving about 25 acres in Borneo, more than 1,000 different tree species were cataloged.[16]

International Implications

In a biobased world, our relations with Ecuador (to use a representative country that takes its very name from the equator) will be more important than those with Saudi Arabia. The United States must consider what controversies could arise over another nation's genetic treasure and how best to secure access and provide compensation to the regional owners. These are not new issues.

A classic example that illustrates the potential issues is the rosy periwinkle plant of Madagascar. In the early 1950s, a plant biologist working for the U.S. drug firm Eli Lilly extracted two cancer-fighting compounds from the flower. During the course of the patents on the two compounds, Lilly earned hundreds of millions of dollars from the sale of the drugs. Madagascar received no compensation whatsoever.[17]

By the early 1990s, two documents were ready for international agreement that sought to address cases like that of the periwinkle, among other things. The Trade-Related Aspects of Intellectual Property Rights (TRIPS) Agreement—part of the Final Act of the Uruguay Round of Multilateral Trade Negotiations—sought to strengthen international intellectual property protection in order to promote world trade. The United Nations Framework Convention on Biological Diversity, known commonly as the Biodiversity Treaty, sought to preserve agrarian societies and promote sustainable development.

Throughout the Uruguay Round, the United States strongly supported the TRIPS notion of protecting international intellectual property. Emblematic of the problems associated with that position were the 1993 riots in India directed against W.R. Grace, a U.S. chemical firm. Indian farmers were protesting that Grace had a patent on an insecticide derived from the neem tree, even though the farmers had a traditional method to extract the compound from the leaves. Although Grace's process gave the compound a shelf life and allowed it to be transported to areas where neem trees were not available, the farmers accused Grace of "gene piracy."

The Biodiversity Treaty was seen by many to be in conflict with the U.S. position on TRIPS. The Biodiversity Treaty sought to address the issue of the biodiverse-rich underdeveloped countries seeking compensation for the resources taken and used by the technology-rich developed countries. Provisions of the treaty require biodiverse-rich countries to provide access to genetic material in return for the developing countries providing a fair and equitable share of the benefits. U.S. pharmaceutical and biotechnology firms initially opposed the treaty. Eventually, however, they dropped their opposition, out of fear that it might ultimately preclude their exploration for genetic resources in underdeveloped countries. The treaty was signed by President William Clinton in 1993 but was never ratified.[18]

Although the treaty was not ratified, the business sector moved forward with an agreement that serves as a model for such arrangements. In 1991, Merck and Company signed an agreement with the Costa Rican Institutio Nacional de Biodiversidad (INBio) for a 2-year renewable contract, in which INBio supplied Merck with extracts from plants, insects, and micro-organisms for its drug-screening program. In exchange, Merck paid INBio $1,135,000 and royalties on any resulting commercial products.

Thus, even a decade ago, the business sector was quite aware of the potential for genes as raw materials. This is especially true in the pharmaceutical industry at the moment, as about one-fourth of all prescription drugs contain an active ingredient derived from plants.

In a biobased economy, with many players seeking access to the biodiversity treasures of developing countries, the possible international scenarios that might arise are limitless: conflicts between developed countries over who had access to what gene at what time; conflicts between developing and developed countries over access to genes and compensation; conflicts between developing countries over territory, and thus ownership, of particular stores of genes.

In this context, a serious dilemma could surface if a state set out to destroy large amounts of diverse genetic material. This is not a hypothetical situation. It is estimated that some 31 million hectares of rainforest are destroyed annually.[19] Article 3 of the Biodiversity Treaty states that countries have the "sovereign right to exploit their own resources pursuant to their own environmental policies."[20] If genes were the basic unit of commerce, would we tolerate another state's environmental policies that allowed for the continued destruction of the rainforest?

a serious dilemma could surface if a state set out to destroy large amounts of diverse genetic material

Another likely point of international friction will be the use of transgenics. Moving genes from one species to another provides for tremendous diversity and the opportunity to create new products. It also raises safety and ethical concerns about introducing such genetically modified organisms (GMOs) into the environment. (A distinction is drawn between GMOs that are nonliving end products that would have no effect on the environment—for example, the heat tolerance protein theorized above—and living modified organisms [LMOs], such as seeds, that may have some environmental consequences.)

Current Department of Agriculture figures cite a continuing increase in acreage planted with genetically modified crops in the United States. In 2002, nearly one-third of all corn acreage will use genetically modified (GM) seed, while GM cotton will account for just over 70 percent of acreage and GM soybeans will be planted in approximately three-fourths of soybean fields.[21] The use of GMOs only will increase as the biobased economy matures, and, likewise, the potential for disputes will increase.

These are not hypothetical issues for the distant future but are present day concerns. The European Union (EU), for example, has had a moratorium on approving the importation of GM crops for the past 4 years. While it appears that the EU may be moving toward approval by summer 2003, the moratorium has been a cause of controversy between the EU and the United States. As recently as February 2002, the Secretary of Agriculture noted that the United States had not ruled out the possibility of filing a formal complaint with the World Trade Organization (WTO).[22] At the same time, however, environmental activists in Europe are encouraging EU governments to ban GM crops of beet and oilseed rape. In Australia, the Insurance Council of Australia has stated its reluctance to insure farmers, biotechnology companies, or food companies in cases involving GMOs.[23]

Significant multilateral international efforts have been made to address specific concerns surrounding LMOs. In January 2000, the Biosafety Protocol to the Biodiversity Treaty was signed. Known as the Cartagena Protocol on Biosafety, it is the first protocol to the Biodiversity Treaty. Its intent is to provide countries the chance to obtain information about LMOs before they are imported. Moreover, it acknowledges each country's right to regulate bioengineered organisms and provides a framework to help the developing world to protect its biodiversity further. Although the United States is not a party to the Biodiversity Treaty and thus cannot be a party to the Protocol, it participated in the negotiations as a member of the so-called Miami Group, a coalition of leading agricultural exporters that included Argentina, Australia, Canada, Chile, and Uruguay.[24]

While the various treaties and scenarios described above depict potential conflicts, not all international implications of a biobased economy will be filled with peril. For example, consider the implications for job creation. As a raw material, petroleum has considerably more energy per unit volume than biological materials. Thus, it is economical to transport petroleum from its source to distant refineries for processing and then further to ship the refinery products for use as end products or industrial intermediates. With biological materials, however, the economics will not support shipping the raw materials much farther than 250 to 300 miles from their point of origin. Biorefineries will have to be built close to the source of their raw materials. A regionalized agriculture will likely develop, with certain areas growing specific crops to supply regional biorefineries. Additional processing and manufacturing of value-added biologically based products can economically take place farther from a biorefinery, but there will be limits to the distances involved. The significance is the likely creation of nonfarming jobs in rural areas.

Urbanization in the developing world is often noted as a major issue of strategic concern for the 21st century. Currently, there are approximately 40 cities in the world with populations of 5 million or more. By 2015, it is anticipated that nearly 25 more will join the ranks. Only 11 of these 65 will be in the developed world.[25] Moreover, the demographic structure of societies in developing countries is heavily weighted toward people 25 years of age and younger. Unemployment among large numbers of young urban males in developing countries is frequently cited as a root cause of the terrorism that we are fighting today. A biobased economy ultimately could help stem the flow of urbanization and provide rural employment opportunities.

Domestic Implications

Just as new international issues will surface as a result of our transition to a biobased economy, new domestic considerations will likewise arise. For example, most homeland defense planning currently focuses on the protection of urban populations and infrastructure, while the safeguarding of agricultural areas does not receive much consideration.[26] Agriculture simply does not enter into the thinking of most people. Throughout most of the last century (from about 1930 to 1999), agriculture as a percent of U.S. employment declined nearly 90 percent—from 23 percent to 2.6 percent. The number of farms declined from 6.3 million to 2.2 million. Agriculture was not even included among the eight critical national infrastructures in Presidential Decision Directive (PDD) 63, "Critical Infrastructure Protection." Interestingly, however, agriculture is included as a subgroup of the Weapons of Mass Destruction Preparedness Group resulting from PDD 62, "Combating Terrorism."

In fairness, it has not seemed particularly necessary to include agriculture as critical infrastructure, since croplands have not surfaced as likely terrorist targets. Terrorists usually aim to score immediate to near-term effects by striking high-profile targets. While a present-day attack on our field crops could have a large economic price tag, it certainly would not affect our ability to feed ourselves. Food is plentiful worldwide, and the marketplace easily could meet any immediate or near-term demands. (Even with the growing world population, per capita food production has actually increased during the last 30 years from 2,360 calories per day to 2,740 calories per day.[27])

agriculture was not even included among the eight critical national infrastructures in Presidential Decision Directive 63

In addition to field crops, farm animals, food in the processing or distribution chain, food at wholesale or retail establishments, and agricultural facilities are all potential targets.[28] Presently, an attack on any link in the chain would result in large economic losses, as well as likely loss of human and animal life. It is estimated that a natural outbreak of foot and mouth disease on just 10 farms would result in a $2 billion loss.[29] Losses from last year's outbreak of foot and mouth disease in the United Kingdom were estimated at $30 billion.[30] However, if we relied upon agriculture to provide the raw materials for our economy, the potential disruption could be orders of magnitude greater.

Consider this hypothetical scenario. What if, as the National Research Council report suggests, we did derive 50 percent of our liquid fuels from agriculture? At present, agriculture provides only 1.2 percent of our "gasoline" supply in the form of ethanol, much of which is blended with gasoline as an oxygenate to reduce seasonal pollution effects. As new biotechnologies improve the processing of biomass, ethanol will become an economically viable option, and it will become a larger source of our liquid fuel supply. At that point, destruction of a large portion of U.S. farmlands would be tantamount to an invasion of Kuwait.

The whole issue of agricultural bioterrorism is complex, but for the purpose of this argument, let us focus solely on croplands. How vulnerable are our croplands? In 1970, without planning or assistance from any organized terrorist group, a naturally occurring epiphytotic, an epidemic in the plant world, destroyed 15 percent of the U.S. corn crop with an estimated value of $1 billion.[31] Although we have diversified the genetic base of corn in an effort to avoid another such disaster, crops are still vulnerable to disease. Any number of organisms, including various molds, fungi, viruses, and bacteria, can cause epiphytotics. These organisms are easily grown in laboratories, at no threat to humans, and can be transported worldwide without detection.

At present, our crops present a relatively simple target set for anyone wishing to do them harm. The U.S. crop base is fairly uniform, with 8 of every 10 acres planted to just 3 crops: corn, wheat, or soybeans. There is genetic diversity within each crop, offering some disease resistance. Predicting the actual loss for any given attack would be based on several assumptions, as epiphytotics are dependent on multiple variables. Moisture and temperature are the most complex variables involved and are extremely difficult to predict in any long-term fashion.[32] Nonetheless, well-coordinated simultaneous attacks in many areas, using multiple pathogens, would no doubt result in significant losses. (Significant in this case could be analogous to the 1970 corn epiphytotic. This would have the net effect of reducing our annual supply of raw materials by 15 percent.)

The U.S. Department of Agriculture is responsible for the Animal and Plant Health Inspection Service (APHIS). Under APHIS, the Plant Protection and Quarantine (PPQ) Division is charged with protecting our crops from the national and international spread of diseases and pests. Since 1982, PPQ has conducted a nationwide survey on crop health and is responsible for dispatching rapid response teams to control any disease outbreak. In the face of a concerted attack, however, one could envision PPQ requiring the assistance of large numbers of people to help man the response teams. It is not inconceivable that military troops could be requested, in much the same manner as they are today to aid with firefighting.

From a plant protection perspective, the shift to a biobased economy will have some positive aspects. To provide new materials for industry, there will be a demand for new genes and their products. If novel genes are found in plants that can be easily grown in this country, then their direct cultivation would be the preferred method rather than creating a transgenic with corn, wheat, or soybeans. With direct cultivation, the overall U.S. crop base would be broadened and thus provide a more challenging target set for terrorists. Also, the construction of regional biorefineries would complicate targeting more than the current groupings of petroleum refineries.

While a biobased economy will no doubt bring the United States the same benefits of slowing urbanization and rural revitalization as anticipated for the developing world, the net effect most likely would be marginal. We will remain a predominantly urban society. As we reconsider the terrorist threat in the wake of September 11, however, it is important to note that agriculture will assume a greater significance as a potential target.

Challenges to U.S. Instruments of Power

Converting to a biobased economy will present new but not totally unfamiliar challenges on all fronts. This is not the first time we have developed and used new resources. Nor will it be the first time we have sought to obtain resources from other nations or wanted to trade finished products. None of these changes will happen quickly or without warning. Nonetheless, it is worth considering some possible effects on our diplomatic, informational, economic, and military instruments of power.

Diplomatic and Informational

Already, diplomatic challenges are being presaged by topics such as the Biodiversity Treaty and possible WTO action for our GMOs. Such issues may well become the norm, requiring a diplomatic corps well trained in scientific and technical skills. Water warrants some extended discussion, as it will be at the heart of diplomatic concerns in the 21st century, regardless of the world's resource base. A biobased economy, though, may well intensify the issue.

Globally, the renewable fresh water supply has fallen by nearly two-thirds in the last 50 years. During that same period, the human population has increased nearly 250 percent. Two-thirds of the world's water demands are for agricultural use, and while irrigated agriculture accounts for only 20 percent of farmland, 45 percent of the world's food supply is grown on irrigated land. By 2025, it is estimated that nearly 3 billion people—40 percent of the projected world's population—will find it difficult or impossible to satisfy basic water needs.

> water will be at the heart of diplomatic concerns in the 21st century

The potential international points of conflict over water are also significant. Two or more countries share 261 of the world's rivers. Some 51 countries, within 17 international river basins, are at risk of water disputes during the next decade. An analysis of 1,831 international water-related disputes over the last 50 years revealed that about one-fourth resulted in violence.[33]

Although water will be a problem, it will not be an insurmountable one. In a 1999 National Academy of Science report on the future of water in the Middle East, it was noted that additional supplies could be obtained by using a variety of techniques. Some involve improved management of watersheds and collection of water that now is lost as runoff. Other techniques use current technologies and include wastewater reclamation and desalination. Some of these can be made even more productive and economical with further improvement. Conservation still remains a significant factor in extending water supplies. Between 1985 and 1993, for example, Israel reduced its water consumption by more than 200 million cubic meters per year, almost entirely through improvements in irrigation and water delivery restrictions.[34]

Former U.S. Senator Paul Simon is a strong advocate of desalination. In his 1998 book on the world's coming water crisis, he noted the progress being made in desalination technologies and use. About 11,000 plants are in operation in more than 125 countries. Desalination is most widely used in the Middle East, which accounts for about 60 percent of the world's plants. In fact, Saudi Arabia built the first modern desalination plant in the late 1930s. To be certain, the economics of desalination are still not competitive, especially for agriculture, but continued development will ultimately drive down the price. That will be especially true as the price of water from other sources rises.[35]

The informational element of the biobased economy is of particular interest and is worthy of a separate study. It is probably unprecedented that both government and business sources are being required by the general population to provide such large amounts of detailed technical information on procedures and products. This issue will only become more complicated, as nontechnical societies will demand data, and bioethics considerations will have to consider differing cultural views.

Economic

The economic forces of globalization at work today will not be affected by the biobased economy (with the possible exception of urbanization, as previously discussed). Thomas Friedman points out that the driving force of globalization is free market capitalism.[36] While a discussion of agricultural trade may well question how much it follows the rules of a truly free market, it is instructive to note our position in that arena. According to the Department of Agriculture Economic Research Service, the United States accounted for 19.2 percent of the world's agricultural exports in 2000. In that same year, we accounted for 13.7 percent of worldwide agricultural imports.[37]

Friedman also notes that globalization has its own set of defining technologies, which includes computerization, miniaturization, digitization, satellite communications, fiber optics, and the Internet. Those are the same technologies American farmers use in a technique called *precision agriculture*, which enables them to integrate all available data and to make the most efficient and economical decisions concerning a crop. For example, using data collected from field sensors, a farmer may detect a developing pest problem. Rather than treating an entire field, as would have been the solution in the past, very targeted treatments can be applied, saving time and money.

In an economy dominated by products derived from agriculture, it is unlikely that we would lose our position of trade or production capability. In fact, our position only will be strengthened, as the defining technologies used in agriculture are the very ones also used in biotechnology. Thus, we would likely achieve an economic advantage through our combined biotechnology skills—which will allow us to identify and use novel genes more quickly—and our agricultural production capacity.

The National Corn Growers Association, an industry trade group, has coined the phrase that the United States is the "Saudi Arabia of Corn." In a globalized, biobased economy, we will be the new Persian Gulf.

Military

Of all the instruments of national power, the military is the one most likely to be affected by a shift in our resource base. The instruments of diplomacy, information, and economics do not require long lead times to research, develop, and acquire their tools of the trade. Nor are the consequences potentially as serious if an initial misstep is made in exercising one of those instruments of power. The international consequences of launching military operations, however, can be long lasting and potentially fatal to those directly involved.

It can be argued that there is less likelihood of exercising the military instrument of power in a biobased economy than in our current petroleum-based economy. That may be true, especially in terms of needing to ensure a daily supply of new raw material—genes rather than oil. Nonetheless, demand for new raw material will remain considerable. Novel genes will be the source of novel products in the biobased economy. While the other instruments of power may play a greater role in securing access to novel genes, the military must still be prepared to operate in areas of enduring U.S. interest.

> **while the other instruments of power may play a greater role in securing access to novel genes, the military must still be prepared to operate in areas of enduring U.S. interest**

To be certain, the U.S. military is not ignoring biology and is particularly aware of the potential benefits of new biological technologies. In a 2001 National Research Council report on future biotechnology opportunities for the Army, the five areas of sensors, electronics and computing, materials, logistics, and therapeutics were identified as significant.[38] Additional work done by the Office of Net Assessment (Office of the Secretary of Defense) has further identified a number of specific areas that could enhance soldier performance and provide greater protection on the battlefield. One particularly intriguing notion is the idea to use biological evolutionary processes to create more effective battlefield systems.[39]

Advances in any or all of these areas will benefit the Nation's 21st-century military, regardless of how it is organized or which wars it is prepared to fight. However, the likely missions and the force structure required for those missions need to be considered.

Special Challenges for the Land Force

The Army's current transformation process and its efforts to find the proper mix of heavy and light forces should be considered in the context of the country's transformation to a biobased economy. Gas and oil undeniably will remain important resources through the first part of this century. Given the geography of a scenario envisioning armor and infantry battling over petroleum and gas resources—a possible replay of Operation *Desert Storm* or an even larger future conflict over Eurasia's untapped gas and oil—multi-ton vehicles, whether part of a legacy force or new acquisitions, are no doubt indispensable. Thirty years from now, however, biology may well have simultaneously reduced the need for fossil fuels and increased the need for access to highly diverse genetic resources.

The Army's transformation to its new Objective Force also should be completed by then. Of its many attributes, the new force is to be agile and deployable. Legacy vehicles such as the multi-ton Abrams tank and Bradley infantry fighting vehicle will still be supplementing the force until about 2030. The Interim Force will be fielded as a bridge between our current force and the final Objective Force that will also have multi-ton armored vehicles. An important question to ask is whether such an army will have the necessary equipment to conduct a forcible entry into an equatorial region to secure the genetic resources contained in a given 5,000-square-mile patch of rainforest. The significance of the question lies in the long lead time needed for research, development, and acquisition of weapons systems.

The Chief of Staff of the Army, General Eric Shinseki, recently made the following comment:

> *We [the Army] must be able to project power anywhere in the world, not just in the easily accessible areas with multiple air and seaports of debarkation, but in the most remote, desolate, landlocked and infrastructure-poor areas as well...The one map in the Chief's office is a map of the Caspian Basin. The reason the map is there ... [is that] ... it's the one part of the world we didn't know how to get to, and nobody else could, either. For the last two years, the Army War College has conducted its war game in the Caspian Basin.*[40]

General Shinseki certainly has the right idea. He simply needs to add a map of the Amazon Basin to his wall.

Despite all the possible considerations surrounding the shift from geology to biology, it is important to recognize that "war's ultimate stakes," as retired U.S. Army War College Commandant Major General Robert Scales calls them, will not change for the warfighter, whether organized as heavy, light, or something in between. Land, people, and resources will remain the primary focus of warfare.[41] Moreover, it is important to note that biological technologies will not provide the next and final wave for the revolution in military affairs and deliver some sort of ultimate weapon to ensure our future victories. Again, to quote General Scales: "2,500 years of history confirm that ambiguity, miscalculation, incompetence, and above all chance will continue to dominate the conduct of war. In the end, the incalculables of determination, morale, fighting skill, and leadership, far more than technology, will determine who wins and loses."[42]

Despite the war on terrorism, we are at one of those periods in history in which we are not burdened by pressures of such imminent danger that our very existence is threatened. We have time to ponder the distant future. We have an opportunity to shape our relationships with those countries that will be strategically important to us. We have an opportunity to invest in those technologies that will be important to our economic advantage in a biobased economy.

We also have the chance to prepare now for the type of warfare that we may encounter in a world run on biological resources. Sir Michael Howard, the noted historian and military theorist, commented, "You can rest assured, whatever doctrine you are working on in peace time, you probably have it wrong. What counts is making sure that you're not so wrong that you can't get it right when the time comes."[43] Before progressing much further in our current thinking, pausing to take stock of the next damned thing may prove to be a damned smart thing.

Notes

1. Petroleum Industry Analysis Brief, accessed at <http://www.eia.doe.gov/emeu/mecs/iab/petroleum/index.html>.
2. Energy Information Administration/Annual Energy Review 2000, "Table 5.12b Petroleum Consumption by the Industrial Sector, 1949–2000," accessed at <http://www.eia.doe.gov/emeu/aer/pdf/pages/sec5_29.pdf>.
3. DIGEST, compiled from reports by the Associated Press, Bloomberg News, Dow Jones Service, Reuters, and *Washington Post* staff writers, accessed at <http://search.washingtonpost.com/wp-srv/wplate/1999-02/10/1741-021099-idx.html>.
4. Commentary, "Potential Oil Supply Refill?" The *Washington Times*, Wednesday, May 29, 2002, A12.
5. Sam Willet, National Corn Growers Association, personal communication.
6. National Agricultural Biotechnology Council, *Vision for Agricultural Research and Development in the 21st Century*, December 14, 1998.
7. National Research Council, *Biobased Industrial Products: Priorities for Research and Commercialization* (Washington, DC: National Academy Press, 1999).
8. Margot Anderson and Richard Magleby, *Agricultural Resources and Environmental Indicators, 1996–1997*, Agricultural Handbook no. 712, July 1997, U.S. Department of Agriculture, Economic Research Service, chapter 7.1, 7.
9. Ibid.,7.
10. W.E. Larson and V.B. Cardwell, "Corn Production: A Guide to Profitable and Environmentally Sound Management. Future Expectations," accessed at <http://citv.unl.edu/cornpro/html/future/future.html>.
11. Department of Energy, *Plant/Crop-Based Renewable Resources 2020: A Vision to Enhance U.S. Economic Security through Renewable Plant/Crop-Based Resource Use*, January 1998.
12. Section 9003, Biorefinery Development Grants of the Farm Security and Rural Investment Act of 2002, authorizes development grants to build biorefineries to "develop transportation and other fuels, chemicals, and energy from renewable sources."
13. Terence Chea, "From Fields to Factories: Plant-Based Materials Replace Oil-Based Plastics, Polyesters," *The Washington Post*, May 3, 2002, E1, accessed at <http://www.washingtonpost.com/wp-dyn/articles/A24747-2002May2.html>.
14. "DuPont Makes Key Polymer Ingredient from Corn Instead of Petroleum," DuPont news release, May 1, 2001, accessed at <http://www.dupont.com/corp/news/releases/2001/nr05_01_01.html>.
15. Department of Defense, *Quadrennial Defense Review Report* (Washington, DC: Department of Defense, September 30, 2001), accessed at <http://www.defenselink.mil/pubs/qdr2001.pdf>.
16. Edward O. Wilson, *The Diversity of Life* (Cambridge, MA: Belknap Press, 1992), 195–197.
17. Richard Stone, "The Biodiversity Treaty: Pandora's Box or Fair Deal?" Science, June 19, 1992.
18. Charles R. McManis, "The Interface between International Intellectual Property and Environmental Protection: Biodiversity and Biotechnology," *Washington University Law Quarterly* 76, no. 1 (Spring 1998), 255–280.
19. "Rates of Rainforest Loss," Rainforest Fact Sheets, Rainforest Action Network, May 8, 2002, accessed at <http://www.ran.org/info_center/factsheets/04b.html>.
20. McManis, 258.
21. "U.S. Department of Agriculture Figures Show Rise in GM Crop Plantings," *Chemical Week* 164, no. 15, 36.
22. "EU Says it Will Approve GMO Imports as Early as October," *Food and Drink Weekly* 8, no. 10, 1.
23. "Pressure on EU to Ban GM Crops," A Northern Light Special Collection Document, Document ID FE20020404310000906, accessed at <http://www.allafrica.com/>.
24. U.S. Department of State, Office of the Spokesman, "Fact Sheet: The Cartagena Protocol on Biosafety" (Washington, DC: U.S. Department of State, February 16, 2000), accessed at <http://usinfo.state.gov/topical/global/biotech/00021601.htm>.
25. "Millennium in Maps, Population" (map supplement), *National Geographic* no. 4 (October 1998).
26. Henry S. Parker, *Agricultural Bioterrorism: A Federal Strategy to Meet the Threat*, McNair Paper 65 (Washington, DC: National Defense University Press, 2002), 12.
27. "Can the Planet Produce Enough Food to Feed the Billions Who Will Be Born in the Future?" *National Geographic* no. 4 (October 1998), 59.
28. Parker, 12.
29. Floyd P. Horn and Roger G. Breeze, "Agriculture and Food Safety," in Food and Agricultural Security: Guarding against Natural Threats and Terrorist Attacks Affecting Health, National *Food Supplies, and Agricultural Economics*, ed. Thomas W. Frazier and Drew C. Richardson, Annals of the New York Academy of Natural Sciences 894 (New York: New York Academy of Sciences, 1999), 9–17.
30. Lawrence Alderson, "Foot-and-Mouth Disease in the United Kingdom 2001: Its Cause, Course, Control and Consequences," accessed at <http://www.warmwell.com/aldersonsept3.html>.
31. Paul Rogers, Simon Whitby, and Malcolm Dando, "Biological Warfare Against Crops," *Scientific American* 280, no. 6 (June 1999), 72.
32. "Concepts of Epiphyt," accessed at <http://www.pa.ipw.agrl.ethz.ch/~w3pa/models/epiphyt/concepts.html>.
33. Robert Toguchi, seminar remarks at *The Role of American Military Power*, held by the Association of the U.S. Army, April 2002.
34. U.S. National Academy of Sciences, et al., *Water for the Future: The West Bank and Gaza Strip, Israel and Jordan* (Washington, DC: National Academy Press, 1999).
35. Paul Simon, *Tapped Out: The Coming World Water Crisis and What We Can Do About It* (New York: Welcome Rain Publishers, 1998).
36. Thomas L. Friedman, *The Lexus and the Olive Tree* (New York: Farrar, Straus and Giroux, 2000).
37. U.S. Department of Agriculture, Economic Research Service, Briefing Room, "U.S. Agricultural Trade: Global Agricultural

Trade," accessed at <http://www.ers.usda.gov/briefing/agtrade/commoditytrade.htm>.

38. Board on Army Science and Technology, National Research Council, *Opportunities in Biotechnology for Future Army Applications* (Washington, DC: National Academy Press, 2001).

39. Jerry Warner, et al., *Exploring Biotechnology, Opportunities for the Department of Defense*, Critical Review and Technology Assessment Report, January 31, 2002.

40. General Eric K. Shinseki, Chief of Staff of the Army, seminar remarks at *The Role of American Military Power*, held by the Association of the U.S. Army, November 2001.

41. Robert Scales, *Future Warfare* (Carlisle Barracks, PA: U.S. Army War College, 1999), 26.

42. Ibid., 25.

43. Michael Howard, "Military Science in an Age of Peace," *Journal of the Royal United Services Institute for Defence Studies* no. 119 (March 1974), 3–9.

Robert E. Armstrong is a senior research fellow in the Center for Technology and National Security Policy at the National Defense University. Dr. Armstrong may be contacted via e-mail at armstrongre@ndu.edu or by phone at (202) 685–2529.

From *Defense Horizons*, October 2002. National Defense University Press

UNIT 3
Weapons of Mass Destruction

Unit Selections

9. **Ex-Inspector Says C.I.A. Missed Disarray in Iraqi Arms Program**, James Risen
10. **The Nuclear Crisis on the Korean Peninsula: Avoiding the Road to Perdition**, Selig S. Harrison
11. **N. Korea Shops Stealthily for Nuclear Arms Gear**, Joby Warrick
12. **Nuclear Nightmares**, Bill Keller
13. **Towards an Internet Civil Defence Against Bioterrorism**, Ronald E. LaPorte et al.

Key Points to Consider

- What type of weapons or methods do you believe future terrorists are most likely to use in attempting to launch attacks in the United States?
- Today there is an intense debate about whether the failure to find weapons of mass destruction in Iraq to date is a decision failure of Bush and his closest advisors, an intelligence failure, or a situation where the jury should remain undecided until U.S. weapon inspectors have completed their investigation. What is your position on this issue?
- Explain why current efforts to strengthen existing or new arms control agreements are not likely to help slow proliferation of weapons of mass destruction.
- What measures should nation-states or international organizations take to prevent future deals, such as past nuclear barter deals between North Korea and Pakistan and between Pakistan and Libya, from occurring?
- Given the recent evidence of a black market involved in the sale of nuclear blueprints, parts, and energy sources to the highest bidder, what additional measures would you propose to attempt to prevent terrorists or other nation-states from obtaining or developing weapons of mass destruction?
- Are there any additional homeland defense measures that should be taken to prepare for a future nuclear or biological attack in the United States?
- A national or international program? Explain which one of Laporte et al's proposals would be more effective.

 **Links: www.dushkin.com/online/**
These sites are annotated in the World Wide Web pages.

National Defense University Website
http://www.ndu.edu

U.S.-Russia Developments
http://www.acronym.org.uk/start

The Bulletin of the Atomic Scientists
http://www.bullatomsci.org

Federation of American Scientists
http://www.fas.org

ISN International Relations and Security Network
http://www.isn.ethz.ch

The RMA Debate: Terrorism and Counter-terrorism
http://www.comw.org/rma/fulltext/terrorism.html

Terrorism Research Center
http://www.terrorism.com

The chilling prospect that the terrorists who attacked America on September 11, 2001 might have been carrying a nuclear, biological, chemical, or dirty bomb is a serious concern for U.S. authorities. The deaths of five and the possible exposure of thousands of Americans to anthrax spores from letters delivered through the U.S. Postal Service as well as the discovery of notes and training manuals at former al Qaeda training camps that contained information about chemical, biological, and nuclear (CBN) weapons dramatically underscored how far proliferation of weapons of mass destruction has progressed. Many government officials and policy experts were surprised again in 2003 at the extent of proliferation after Libya acknowledged the existence of a covert nuclear program and the fact that they had purchased Chinese blueprints for building a nuclear bomb from the former head of Pakistan's nuclear program. How to protect citizens from weapons of mass destruction—chemical (WMD), biological, and nuclear (CBN)—and prevent additional proliferation is a central security concern for policy makers worldwide.

Scholars and practitioners, however, also continue to disagree about whether nation-states, terrorists, or lone deviants are most likely to use weapons of mass destruction against civilians in the future. Consequently, there is even less agreement now than in past years about how best to deter or counter weapons of mass destruction and whether the spread of CBN weapons encourages or acts as a disincentive to interstate war. The failure of U.S. weapons inspectors to find an active WMD program in Iraq after the U.S. military intervention adds additional layers of controversy to policy debates about how best to monitor, deter, or prevent countries from resurrecting old or pursuing new covert weapons of mass destruction programs. In "Ex-Inspector Says C.I.A. Missed Disarray in Iraqi Arms Program," James Risen describes why the former senior U.S. official in charge of searching for weapons of mass destruction in Iraq after the war, David Kay, concluded that US intelligence missed signs of the chaos in the Iraqi WMD program.

During the Cold War, four major nuclear arms control agreements were developed: the Moscow-Washington negotiations that led to the Strategic Arms Limitation Treaty (SALT) and the Anti-Ballistic Missile (ABM) Treaty of 1972; the Nonproliferation Treaty (NPT); and the grandfather of them all, the Atmospheric Test Ban Treaty negotiated during the Eisenhower administration. Collectively, these agreements were core components of the "nuclear arms control regime". This regime suffered a major setback in December 2001 when U.S. President Bush announced that the United States planed to unilaterally withdraw from the 1972 Anti-Ballistic Missile (ABM) Treaty in order to pursue the development of an anti-ballistic missile system. Bush justified the U.S. policy change on the grounds that a new ABM system was required to deter future missile attacks by rouge states and possibly even by terrorist groups. The decision triggered intense criticism at home and abroad. Many worried that the world was on the eve of a new arms race. Additional concerns were expressed after the United States announced that the U.S. would not support multinational efforts to bolster the enforcement mechanism of the 1972 Biological Weapons Convention. Several countries stopped work on national legislation to implement a strengthened international enforcement protocol after the United States government withdrew support.

As faith in arms control weakened, several nation-states reconsidered their national strategies as well. Today at least 25 countries either have, or are in the process of developing weapons of mass destruction. Two dozen are researching, developing, or stockpiling chemical and/or biological weapons. Such activities make it difficult to see how the genie of nuclear weapons technology will be kept in the bottle. How real the threat of a regional war that engulfs the world was dramatically illustrated at the end of 2001 when two declared nuclear powers, India and Pakistan, massed troops on their shared borders, recalled diplomats, and issued thinly veiled threats about what might happen in the event of another armed conflict. India and Pakistan subsequently agreed to a series of measures designed to ensure that future conflicts over Kashmir would not escalate into nuclear war. However, repeated assassination attempts on the life of Gen. President Musharraf of Pakistan, in 2003, underscored how fragile the current government may be and raises the possibility that radical Islamic fundamentalists may be running countries—in the future—that have weapons of mass destruction.

In order to pursue the War on Terrorism and military intervention in Iraq, senior U.S. officials made a series of comprises and trade-offs that impacted WMD counter-proliferation efforts in other parts of the world. Since initiating military campaigns in Afghanistan and Iraq, the United States government has downplayed the threat posed by several other conflicts, including North Korea's unilateral termination of the 1994 framework agreement. According to the agreement, North Korea would not extract enriched uranium from a nuclear reactor at Yongbyon in order to build nuclear bombs. In exchange for this agreement, North Korea would get financial aid to purchase energy from alternative energy sources in 2002. Next, North Korean leader, Kim Jong Il, requested that UN Weapons inspectors leave the country and unilaterally withdrew North Korea from the Nuclear Non-proliferation Treaty. To contain the Korean nuclear crisis, the Bush administration backed away from a longstanding declaration that it would not tolerate a North Korean nuclear arsenal, and announced instead, that the United States was willing to engage North Korea in informal talks. A few days later the U.S. government said that bombing North Korea was no longer an immediate option since the country had acquired nuclear weapons.

In "The Nuclear Crisis on the Korean Peninsula: Avoiding the Road to Perdition," the Task Force on U.S. Korean Policy outlines an approach that is similar to current U.S. policy towards North Korea. Instead of confrontation, the Task Force on U.S. Korea Policy recommends a policy of constructive engagement with friends and allies in the region to help exert maximum pressure on North Korea and pursue a long-term policy of liberalizing the North Korean system. With substantial diplomatic help from China and other nation-states, the U.S. shift to a multilateral approach quickly succeeded in jumpstarting new six party negotiation talks during 2003. The talks, however, ended in February 2004 without a resolution to the crisis. North Korea had put nuclear disarmament on the bargaining table but then accused the United States of blocking progress towards a new agreement.

During the fall of 2003 an intercepted French cargo ship was found to be carrying aluminum pipes that were ultimately destined for North Korea's secret nuclear bomb program. Joby Warrick in "N. Korea Shops Stealthily for Nuclear Arms Gear," details how the captured ship provides a "glimpse into the shadowy world of weapons proliferation, in which missile parts and bombs materials circle the globe undetected, secreted away in cargo ships and suitcases."

Even more concerns were expressed after another cargo ship was intercepted with sophisticated centrifuge parts earmarked for delivery to Libya. The U.S. and several European nation-states pressured Libya to disclose a covert nuclear program and invited international inspectors into the country. By the end of the year, UN International Atomic Energy Weapons inspectors had visited four sites related to the secret nuclear program. The U.S. was allowed to quickly remove large amounts of documentation and equipment related to the covert WMD programs out of the country. As the investigation proceeded in 2004, outsider investigators learned that Libya had been able to obtain blueprints for older Chinese nuclear bombs in addition to sophisticated parts needed to build a bomb from the former head of Pakistan's nuclear bomb program. Today, no one knows how many other countries received the same blueprints and parts through a black market ring that included middlemen, a private company in Malaysia, and several European business suppliers.

The recent disclosures indicate that extensive nuclear proliferation of knowledge, critical components, and sources of energy for nuclear bombs are greater than most experts had realized. Current conditions reinforce the view of many nuclear proliferation experts that a nuclear attack will occur in the United States. Bill Keller, in "Nuclear Nightmares," explains why the most hotly disputed questions relate to when and how an attack will occur. Given the difficulties involved in obtaining adequate supplies of missile materials needed for a nuclear bomb, many experts now predict that terrorists are most likely to use radiation and other nuclear materials to cause disruptions, terror, and deaths.

The recent recognition that Americans may be the victim of a WMD on domestic soil in the near-term future led the Bush administration to form a new agency, the new Department of Homeland Security (DHS), in 2003. Billions of dollars have already been allocated to tackle the multi-faceted problems involved in defending against the threat of biological weapons. The initial budget of the new Department of Homeland Security was $55 million. The budget by fiscal year 2004 had increased 135%, pending Congressional approval. Civilian agencies, such as the National Institute of Health, are also playing a central role by building a series of new biosecurity labs (BSL) to study deadly pathogens and train future biowarfare specialists. By 2003, the Bush administration had authorized $274 million to combating bioterrorism, track infectious diseases, and to develop new research on vaccines. Currently the recently formed Department of Homeland Security (DHS) is monitoring 30 cities for biological pathogens. By 2005, 60 more cities will be monitored.

While no one really knows when or if another biological attack is in the offing, most individuals agree that it is prudent to prepare for this awful eventuality. The Bush administration decision to vaccinate essential government workers for smallpox is one recent program effort to be prepared for future contingencies. Ronald Laporte and his associates warn that there is little evidence that the large resources currently being put into bioterrorism preparedness will work. Instead, citizens in countries throughout the world must face the disturbing fact that it is very difficult to predict and guard against a bioterrorism attack. There are too many targets, too many means to penetrate the targets, the occurrences are too rare, and the bioterrorists are too crafty. Instead of building an inflexible Maginot line of defense as the U.S. is now attempting to do, Laporte and his associates in, "Towards an Internet Civil Defence against Bioterrorism," advocate relying upon an ever alert and flexible electronic-matrix of civil defense.

Many analysts throughout the 1990s had wondered why terrorists hadn't used biological weapons. Earlier incidents involving terrorist attempts to use biological weapons suggested that the requirement of having sophisticated high tech delivery methods was an obstacle to terrorists interested in causing mass casualties. However, the anthrax letter attacks that led to the death of 5, and placed more than 3,000 Americans at risk for possible anthrax infection may have signaled that this barrier will no longer prevent bioterrorist attacks. A recent series of terrorist incidents involving small amounts of ricin found in congressional and other government buildings warned the U.S. government not to implement new regulations on truckers. These incidents, like thousands of earlier ones, also underscore the fact that biological and chemical agents will also be used to disrupt daily life and instill terror by a variety of terrorist groups with varying political and personal agendas.

Ex-Inspector Says C.I.A. Missed Disarray in Iraqi Arms Program

By JAMES RISEN

WASHINGTON, Jan. 25 — American intelligence agencies failed to detect that Iraq's unconventional weapons programs were in a state of disarray in recent years under the increasingly erratic leadership of Saddam Hussein, the C.I.A.'s former chief weapons inspector said in an interview late Saturday.

The inspector, David A. Kay, who led the government's efforts to find evidence of Iraq's illicit weapons programs until he resigned on Friday, said the C.I.A. and other intelligence agencies did not realize that Iraqi scientists had presented ambitious but fanciful weapons programs to Mr. Hussein and had then used the money for other purposes.

Dr. Kay also reported that Iraq attempted to revive its efforts to develop nuclear weapons in 2000 and 2001, but never got as far toward making a bomb as Iran and Libya did.

He said Baghdad was actively working to produce a biological weapon using the poison ricin until the American invasion last March. But in general, Dr. Kay said, the C.I.A. and other agencies failed to recognize that Iraq had all but abandoned its efforts to produce large quantities of chemical or biological weapons after the first Persian Gulf war, in 1991.

From interviews with Iraqi scientists and other sources, he said, his team learned that sometime around 1997 and 1998, Iraq plunged into what he called a "vortex of corruption," when government activities began to spin out of control because an increasingly isolated and fantasy-riven Saddam Hussein had insisted on personally authorizing major projects without input from others.

After the onset of this "dark ages," Dr. Kay said, Iraqi scientists realized they could go directly to Mr. Hussein and present fanciful plans for weapons programs, and receive approval and large amounts of money. Whatever was left of an effective weapons capability, he said, was largely subsumed into corrupt money-raising schemes by scientists skilled in the arts of lying and surviving in a fevered police state.

"The whole thing shifted from directed programs to a corrupted process," Dr. Kay said. "The regime was no longer in control; it was like a death spiral. Saddam was self-directing projects that were not vetted by anyone else. The scientists were able to fake programs."

In interviews after he was captured, Tariq Aziz, the former deputy prime minister, told Dr. Kay that Mr. Hussein had become increasingly divorced from reality during the last two years of his rule. Mr. Hussein would send Mr. Aziz manuscripts of novels he was writing, even as the American-led coalition was gearing up for war, Dr. Kay said.

Dr. Kay said the fundamental errors in prewar intelligence assessments were so grave that he would recommend that the Central Intelligence Agency and other organizations overhaul their intelligence collection and analytical efforts.

Dr. Kay said analysts had come to him, "almost in tears, saying they felt so badly that we weren't finding what they had thought we were going to find—I have had analysts apologizing for reaching the conclusions that they did."

In response to Dr. Kay's comments, an intelligence official said Sunday that while some prewar assessments may have been wrong, "it is premature to say that the intelligence community's judgments were completely wrong or largely wrong—there are still a lot of answers we need." The official added, however, that the C.I.A. had already begun an internal review to determine whether its analytical processes were sound.

Dr. Kay said that based on his team's interviews with Iraqi scientists, reviews of Iraqi documents and examinations of facilities and other materials, the administration was also almost certainly wrong in its prewar belief that Iraq had any significant stockpiles of illicit weapons.

"I'm personally convinced that there were not large stockpiles of newly produced weapons of mass destruction," Dr. Kay said. "We don't find the people, the documents or the physical plants that you would expect to find if the production was going on.

"I think they gradually reduced stockpiles throughout the 1990's. Somewhere in the mid-1990's, the large chemical overhang of existing stockpiles was eliminated."

While it is possible Iraq kept developing "test amounts" of chemical weapons and was working on improved methods of production, he said, the evidence is strong that "they did not produce large amounts of chemical weapons throughout the 1990's."

Regarding biological weapons, he said there was evidence that the Iraqis continued research and development "right up until the end" to improve their ability to produce ricin. "They were mostly researching better methods for weaponization," Dr. Kay said. "They were maintaining an infrastructure, but they didn't have large-scale production under way."

He added that Iraq did make an effort to restart its nuclear weapons program in 2000 and 2001, but that the evidence suggested that the program was rudimentary at best and would have taken years to rebuild, after being largely abandoned in the 1990's. "There was a restart of the nuclear program," he said. "But the surprising thing is that if you compare it to what we now know about Iran and Libya, the Iraqi program was never as advanced," Dr. Kay said.

Dr. Kay said Iraq had also maintained an active ballistic missile program that was receiving significant foreign assistance until the start of the American invasion. He said it appeared that money was put back into the nuclear weapons program to restart the effort in part because the Iraqis realized they needed some kind of payload for their new rockets.

While he urged that the hunt should continue in Iraq, he said he believed "85 percent of the significant things" have already been uncovered, and cautioned that severe looting in Iraq after Mr. Hussein was toppled in April had led to the loss of many crucial documents and other materials. That means it will be virtually impossible to ever get a complete picture of what Iraq was up to before the war, he added.

"There is going to be an irreducible level of ambiguity because of all the looting," Dr. Kay said.

Dr. Kay said he believed that Iraq was a danger to the world, but not the same threat that the Bush administration publicly detailed.

"We know that terrorists were passing through Iraq," he said. "And now we know that there was little control over Iraq's weapons capabilities. I think it shows that Iraq was a very dangerous place. The country had the technology, the ability to produce, and there were terrorist groups passing through the country—and no central control."

C.I.A. Missed Signs of Chaos

But Dr. Kay said the C.I.A. missed the significance of the chaos in the leadership and had no idea how badly that chaos had corrupted Iraq's weapons capabilities or the threat it raised of loose scientific knowledge being handed over to terrorists. "The system became so corrupt, and we missed that," he said.

He said it now appeared that Iraq had abandoned the production of illicit weapons and largely eliminated its stockpiles in the 1990's in large part because of Baghdad's concerns about the United Nations weapons inspection process. He said Iraqi scientists and documents show that Baghdad was far more concerned about United Nations inspections than Washington had ever realized.

"The Iraqis say that they believed that Unscom was more effective, and they didn't want to get caught," Dr. Kay said, using an acronym for the inspection program, the United Nations Special Commission.

The Iraqis also feared the disclosures that would come from the 1995 defection of Hussein Kamel, Mr. Hussein's son-in-law, who had helped run the weapons programs. Dr. Kay said one Iraqi document that had been found showed the extent to which the Iraqis believed that Mr. Kamel's defection would hamper any efforts to continue weapons programs.

In addition, Dr. Kay said, it is now clear that an American bombing campaign against Iraq in 1998 destroyed much of the remaining infrastructure in chemical weapons programs.

Dr. Kay said his team had uncovered no evidence that Niger had tried to sell uranium to Iraq for its nuclear weapons program. In his State of the Union address in 2003, President Bush reported that British intelligence had determined that Iraq was trying to import uranium from an African nation, and Niger's name was later put forward.

"We found nothing on Niger," Dr. Kay said. He added that there was evidence that someone did approach the Iraqis claiming to be able to sell uranium and diamonds from another African country, but apparently nothing came of the approach. The original reports on Niger have been found to be based on forged documents, and the Bush administration has since backed away from its initial assertions.

Dr. Kay added that there was now a consensus within the United States intelligence community that mobile trailers found in Iraq and initially thought to be laboratories for biological weapons were actually designed to produce hydrogen for weather balloons, or perhaps to produce rocket fuel. While using the trailers for such purposes seems bizarre, Dr. Kay said, "Iraq was doing a lot of nonsensical things" under Mr. Hussein.

The intelligence reports that Iraq was poised to use chemical weapons against invading troops were false, apparently based on faulty reports and Iraqi disinformation, Dr. Kay said.

When American troops found that Iraqi troops had stored defensive chemical-weapons suits and antidotes, Washington assumed the Iraqi military was poised to use chemicals against American forces. But interviews with Iraqi military officers and others have shown that the Iraqis kept the gear because they feared Israel would join an American-led invasion and use chemical weapons against them.

Role of Republican Guards

Dr. Kay said interviews with senior officers of the Special Republican Guards, Mr. Hussein's most elite units, had suggested that prewar intelligence reports were wrong in warning that these units had chemical weapons and would use them against American forces as they closed in on Baghdad.

The former Iraqi officers reported that no Special Republican Guard units had chemical or biological weapons, he said. But all of the officers believed that some other Special Republican Guard unit had chemical weapons.

"They all said they didn't have it, but they thought other units had it," Dr. Kay said. He said it appeared they were the victims of a disinformation campaign orchestrated by Mr. Hussein.

Dr. Kay said there was also no conclusive evidence that Iraq had moved any unconventional weapons to Syria, as some Bush administration officials have suggested. He said there had been persistent reports from Iraqis saying they or someone they knew had seen cargo being moved across the border, but there is no proof that such movements involved weapons materials.

Dr. Kay said the basic problem with the way the C.I.A. tried to gauge Iraq's weapons programs is now painfully clear: for five years, the agency lacked its own spies in Iraq who could provide credible information.

During the 1990's, Dr. Kay said, the agency became spoiled by on-the-ground intelligence that it obtained from United Nations weapons inspectors. But the quality of the information plunged after the teams were withdrawn in 1998.

"Unscom was like crack cocaine for the C.I.A.," Dr. Kay said. "They could see something from a satellite or other technical intelligence, and then direct the inspectors to go look at it."

The agency became far too dependent on spy satellites, intercepted communications and intelligence developed by foreign spies and by defectors and exiles, Dr. Kay said. While he said the agency analysts who were monitoring Iraq's weapons programs did the best they could with what they had, he argued that the agency failed to make it clear to American policy makers that their assessments were increasingly based on very limited information.

"I think that the system should have a way for an analyst to say, 'I don't have enough information to make a judgment,' " Dr. Kay said. "There is really not a way to do that under the current system."

He added that while the analysts included caveats on their reports, those passages "tended to drop off as the reports would go up the food chain" inside the government.

As a result, virtually everyone in the United States intelligence community during both the Clinton and the current Bush administrations thought Iraq still had the illicit weapons, he said. And the government became a victim of its own certainty.

"Alarm bells should have gone off when everyone believes the same thing," Dr. Kay said. "No one stood up and said, 'Let's examine the footings for these conclusions.' I think you ought to have a place for contrarian views in the system."

Finds No Pressure From Bush

Dr. Kay said he was convinced that the analysts were not pressed by the Bush administration to make certain their prewar intelligence reports conformed to a White House agenda on Iraq.

Last year, some C.I.A. analysts said they had felt pressed to find links between Iraq and Al Qaeda to suit the administration. While Dr. Kay said he has no knowledge about that issue, he did not believe that pressure was placed on analysts regarding the weapons programs.

"All the analysts I have talked to said they never felt pressured on W.M.D.," he said. "Everyone believed that they had W.M.D."

Dr. Kay also said he never felt pressed by the Bush administration to shape his own reports on the status of Iraq's weapons. He said that in a White House meeting with Mr. Bush last August, the president urged him to uncover what really happened.

"The only comment I ever had from the president was to find the truth," Dr. Kay said. "I never got any pressure to find a certain outcome."

Dr. Kay, a former United Nations inspector who was brought in last summer to run the Iraq Survey Group by George J. Tenet, the director of central intelligence, said he resigned his post largely because he disagreed with the decision in November by the administration and the Pentagon to shift intelligence resources from the hunt for banned weapons to counterinsurgency efforts inside Iraq. Dr. Kay is being succeeded by Charles A. Duelfer, another former United Nations inspector, who has also expressed skepticism about whether the United States will find any chemical or biological weapons.

Dr. Kay said the decision to shift resources away from the weapons hunt came at a time of "near panic" among American officials in Baghdad because of rising casualties caused by bombings and ambushes of American troops.

He added that the decision ran counter to written assurances he had been given when he took the job, and that the shift in resources had severely hampered the weapons hunt.

He said that there is only a limited amount of time left to conduct a thorough search before a new Iraqi government takes over in the summer, and that there are already signs of resistance to the work by Iraqi government officials.

From the *New York Times*, January 26, 2004, pp. 1–7. © 2004 by The New York Times Company. Reprinted by permission.

The Nuclear Crisis on the Korean Peninsula: Avoiding the Road to Perdition

THE TASK FORCE ON U.S. KOREA POLICY

"Confrontational United States policies toward North Korea, adopted unilaterally, would not only exacerbate the nuclear crisis but also undermine United States relations with Northeast Asia as a whole.... The United States would end up with the worst of both worlds: a nuclear-capable North Korea and severely strained relations with key powers important to United States interests globally as well as regionally. Conversely, by pursuing constructive engagement in concert with its friends and allies in the region, the United States would maximize the pressure on North Korea for an acceptable nuclear settlement and promote the long-term United States objective of liberalizing the North Korean system."

On October 4, 2002, North Korea acknowledged that it had initiated a clandestine program to produce enriched uranium despite a pledge not to do so in Article 3, Section 2 of the 1994 Agreed Framework. This revelation has set in motion an escalating confrontation with the United States in which North Korea has withdrawn from the nuclear Non-Proliferation Treaty, expelled International Atomic Energy Agency (IAEA) inspectors, and moved to restart the plutonium-production program frozen under the 1994 accord.

The Task Force on U.S. Korea Policy unanimously agrees that a resumption of plutonium production by North Korea and its success in developing a weapons-grade uranium-enrichment capability would be likely to have a disastrous impact on the stability and security of Northeast Asia and on the global nonproliferation regime. Such an outcome could touch off a regional arms race, driving South Korea, Japan, and Taiwan to reconsider the development of nuclear weapons. The task force recommends urgent diplomatic initiatives by the United States to test whether North Korea is in fact prepared for a verifiable end to all aspects of its nuclear weapons development, including both bilateral United States–North Korea negotiations and a broader multilateral process. At the same time, the task force warns that the American effort to prevent a nuclear-armed North Korea is likely to succeed only if the United States acts in concert with South Korea, Japan, China, Russia, and the EU, and only if the resolution of the nuclear issue is addressed together with the pursuit of four other directly related issues: normalizing United States economic and political relations with North Korea, guaranteeing the security of a nonnuclear North Korea, promoting the reconciliation of North and South Korea, and drawing North Korea into economic engagement with its neighbors.

The task force emphasizes the need for a flexible American response to the rapid change now taking place in South Korean attitudes toward relations with the North and the impact of United States policies on North–South reconciliation. In seeking to resolve the nuclear issue, the United States should give great weight to the views of South Korea and Japan regarding the terms of a settlement and the best way to achieve one. South Korea and Japan would bear the brunt of any military conflict with the North resulting from mishandling of the nuclear issue, and would be most directly affected if the North should progress from its present nascent nuclear capability to the actual operational deployment of nuclear weapons.

The task force points out that South Korea, Japan, China, and Russia have all urged the United States to link the resolution of the nuclear issue with the sustained pursuit of constructive engagement with North Korea. All four have proved reluctant to squeeze North Korea economically. Constructive engagement, they believe, will encourage reform of the autarkic, overcentralized North Korean economic system, reducing the chances of an economic collapse that would lead to a destabilizing refugee exodus into neighboring countries. All four have indicated their support for a United States security guarantee to North Korea and the normalization of United States–North Korean relations as essential components of a settlement in which North Korea verifiably dismantles its nuclear weapons program. Russian President Vladimir Putin and Chinese President Jiang Zemin, in their December 2, 2002 joint statement in Beijing, called for "equal dialogue" between the United States and North Korea, explicitly linking the nuclear issue and the normalization of relations.

The task force warns that confrontational United States policies toward North Korea, adopted unilaterally, would not only exacerbate the nuclear crisis but also undermine United States relations with Northeast Asia as a whole, especially with South Korea, jeopardizing the future of the United States–South Ko-

Article 10. The Nuclear Crisis on the Korean Peninsula: Avoiding the Road to Perdition

rean alliance. The United States would end up with the worst of both worlds: a nuclear-capable North Korea and severely strained relations with key powers important to United States interests globally as well as regionally. Conversely, by pursuing constructive engagement in concert with its friends and allies in the region, the United States would maximize the pressure on North Korea for an acceptable nuclear settlement and promote the long-term United States objective of liberalizing the North Korean system. Regional economic interaction would gradually make North Korea's closed society more penetrable. In a more porous North Korea, the economic reforms now beginning there would be accelerated, leading in time to a diffusion of economic power that would loosen totalitarian political controls and moderate human rights abuses.

Members of the task force are divided in their assessment of North Korean intentions. A majority feels that North Korea is using its nuclear weapons program as a bargaining chip and would be prepared to give it up in return for economic benefits and security assurances, but would go ahead with nuclear weapons development in the absence of sufficient inducements. A minority argues that North Korea is determined to become a nuclear power or, at the very least, to keep other powers guessing and is unlikely to accept inspection safeguards adequate to verify a complete cessation of its nuclear weapons development. Most members agree that North Korean intentions cannot be fully tested through piecemeal negotiations limited to the nuclear issue alone. To put North Korean intentions to a definitive test, the task force concludes, it would be necessary for the United States to join with South Korea, Japan, China, and Russia in negotiating a broad regional accommodation with North Korea that would guarantee its sovereignty and military security and would promote its economic development in return for an end to its nuclear program. The inclusion of Russia is essential and would mark a departure from past United States regional strategy, which has focused on trilateral cooperation with South Korea and Japan, as military allies, and has included China, but not Russia, in a series of Geneva security dialogues from 1997 to 2000.

Such a regional accommodation would be a logical outgrowth of the policies announced by South Korea, Japan, China, and Russia for dealing with the present crisis. But it would require a new readiness on the part of the United States to coexist with North Korea, notwithstanding its totalitarian system, seeking gradual change there and putting aside the hopes for the collapse of the Pyongyang regime that have been expressed by many United States officials in both the Clinton and Bush administrations.

The task force believes that negotiations are urgent. North Korea has said that it is ready to negotiate a verifiable end to all nuclear weapons development if the United States will make a formal commitment in writing to "respect its sovereignty," diplomatic language for not seeking to overthrow its government; not to attack it; and "not to hinder" its economic development. To accept this offer, the United States has responded, would be to submit to "blackmail," and North Korea must first dismantle its nuclear weapons program under adequate safeguards as a precondition for negotiations embracing other issues.

The dictionary definition of blackmail is "extortion by intimidation." Negotiating with North Korea to achieve United States goals would not be submitting to blackmail because both North Korea and the United States have adopted a threatening posture toward the other. In North Korean eyes, it is plausible that the United States, with nearly 10,000 nuclear weapons and overwhelming superiority in airpower, might stage a preemptive strike. This anxiety has been exacerbated by the rationale for preemptive action against potential security threats presented in the September 20 United States National Security doctrine. As former Defense Secretary William Perry has observed, the reason that North Korea wants nuclear weapons "is security, is deterrence. Whom would they be deterring? They would be deterring the United States. We do not think of ourselves as a threat to North Korea, but I truly believe they consider us a threat to them."[1]

To achieve its objectives in Korea, the United States should be sensitive to North Korea's feelings of insecurity and adopt policies that address North Korean concerns. Such policies would not be submission to blackmail but rather the exercise of prudent realism in the pursuit of United States interests. Nor would it place the United States in the position of a supplicant. Members envisage closely synchronized diplomatic steps by both sides that would not require either to make an unreciprocated first move to break the stalemate. Detailed scenarios spelling out such steps and follow-up action culminating in a seven-power regional conference (the United States, China, Russia, Japan, South Korea, North Korea, and the EU) are presented in the recommendations that follow.

Some Bush administration officials argue that there is no point in testing North Korean intentions because Pyongyang demonstrably cannot be trusted to honor any bilateral or multilateral commitments it might make to dismantle its nuclear program. The fact that it has initiated a uranium-enrichment program inconsistent with the Agreed Framework is cited in support of this argument. As noted earlier, in Article 3, Section 2 of the Agreed Framework, North Korea did pledge that it would "consistently take steps to implement" the 1991 North–South Joint Declaration on the Denuclearization of the Korean Peninsula, which explicitly barred the development of uranium-enrichment facilities. While condemning the North for its failure to live up to this obligation, the task force pointed out that North Korea did honor the operative provisions of the 1994 accord providing for the suspension of its plutonium-production facilities. Moreover, the Clinton administration, faced with domestic United States political opposition and confident that the Pyongyang regime would collapse anyway, also failed to honor key provisions of the accord: Article 1, Section 1, which envisaged the installation of 2,000 megawatts of nuclear-powered electricity-generating capacity "by the target date of 2003," and Article 2, which provided for the "full normalization of political and economic relations."

In North Korean eyes, the United States received up front what it wanted from the 1994 accord—the suspension of a North Korean plutonium program that could otherwise have produced up to 30 nuclear weapons a year—while North Korea collected only unfulfilled promises, with the exception of the

500,000 tons of heavy oil annually pledged in Article 1, Section 2. Thus, without condoning North Korean duplicity in starting an enrichment program, this breach of the accord does not, in itself, establish that North Korea cannot be trusted to carry out a new accord, particularly if it involves not only the United States but also powerful neighbors in Northeast Asia on whom it depends for economic support.

KEEPING THE THREAT IN PERSPECTIVE

In approaching negotiations, it is important to keep the North Korean nuclear threat in perspective, distinguishing between shortterm and longterm dangers and setting priorities accordingly. Advocates of preemptive military action often exaggerate existing and potential North Korean capabilities to bolster their case.

There is indeed a short-term danger that North Korea could produce sufficient plutonium for four to six nuclear weapons, within six or eight months, from the 8,000 spent fuel rods at Yongbyon that have been in storage under the 1994 accord. The expulsion of IAEA inspectors in December 2002 has left the status of these fuel rods uncertain. Getting them out of the country, as envisaged in the 1994 accord, and getting inspectors and monitoring equipment back in, should be the top United States priority. The urgency of forestalling the reprocessing of the fuel rods is underlined by the possibility of transfers of fissile material to third parties.

It is commonly assumed that North Korea already had one or two plutonium-based nuclear weapons when the Agreed Framework was concluded. Yet the reality is that the United States does not know how much plutonium had been produced before 1994, and, in any case, whether it has been weaponized. General James Clapper, director of the Defense Intelligence Agency during the 1994 nuclear crisis and now director of the National Imagery and Mapping Agency, has confirmed this. "Personally as opposed to institutionally, I was skeptical that they ever had a bomb," Clapper has said. "We didn't have smoking gun evidence either way. But you build a case for a range of possibilities. In a case like North Korea, you have to apply the most conservative approach, the worst-case scenario."[2] The CIA has not made a formal assessment of North Korea's plutonium capabilities since November 1993, when a National Intelligence Estimate reportedly asserted that "it is more likely than not" that the North had "one, possibly two" nuclear "devices," as distinct from weapons.

As for a uranium-based nuclear weapons capability, according to a declassified CIA estimate in December 2002, the uranium-enrichment plant under construction "could produce enough weapons-grade uranium for two or more nuclear weapons per year when fully operational, which could be as soon as mid-decade." The threat of an operational enrichment capability is not imminent, leaving time to head it off through negotiations.

North Korea is tightly insulated from outside influences. All television and radio sets must be registered and have fixed channels. Only the top echelon of the Workers Party has more than an inkling of what the rest of the world is like.

In summary, priority should be given to ending the short-term threat that would be posed if North Korea were to reprocess the spent fuel rods and restart the Yongbyon reactor. If this threat is removed by "refreezing" the reactor, as North Korea has offered to do, and by resuming monitoring of the spent fuel rods until they are shipped out of the country, there will be ample time for a graduated process of tit-for-tat concessions leading to the full dismantlement of North Korean nuclear capabilities—a crucial milestone in sustaining and strengthening the global nuclear nonproliferation regime.

Over time, and with sufficient testing, North Korea might be able to develop and deploy missiles capable of delivering nuclear weapons to the continental United States. But projections of an imminent capability ignore the technical constraints on the North Korean missile program. Moreover, they obscure the fact that North Korea agreed in 1998 to observe a moratorium on missile testing while negotiations proceeded on the normalization of relations with the United States. North Korea offered to discontinue all testing, production, and deployment of missiles with a range over 500 kilometers (300 miles) as part of the broad normalization agreement under discussion during the last days of the Clinton administration.

The longest-range missile currently deployed by North Korea is the Nodong, which has an estimated range of 1,300 kilometers (800 miles) and can carry a payload of 1,500 pounds. Such a range would allow North Korea to target all of Japan. The only known North Korean flight test of the Nodong was in May 1993 (Pakistan, however, may have provided North Korea with information from the tests of its Ghauri missile, which is believed to consist mostly or entirely of North Korean components).

North Korea's only test of a longer-range missile occurred in August 1998, when the three-stage Taepodong 1 (TD-1) missile was launched in an attempt to place a small satellite in orbit. This effort was not successful due to a failure of the missile's third stage. The test did demonstrate for the first time the North's technical capability to launch missiles with multiple stages, as well as its access to solid fuel technology, which was used in the third stage that failed. The TD-1 cannot be considered operational, however, without successful flight tests. Moreover, North Korea has not flight-tested a reentry heat shield that would be required for a long-range missile. Such a heat shield is required for a missile intended to deliver a warhead to targets on the ground, but is not needed to launch a satellite into orbit.

Even if the TD-1 were successfully tested, it could at best deliver a small payload as far as Alaska or Hawaii. Theoretically, such a light payload would be enough for a limited chemical or biological attack, but not for delivering a nuclear warhead. Whether chemical and biological weapons can be delivered effectively by long-range missiles is debatable.

The Taepodong 2 (TD-2), a longer-range missile that North Korea is believed to be developing, has never been flight-tested, and the status of its development is uncertain. As David Wright of the Union of Concerned Scientists pointed out in a working paper for the task force, the TD-2 would differ significantly from any missile that North Korea has built or tested. It would be much bigger than the TD-1 and would generate greater thrust, so that the mechanical stresses on the body would be much more severe than on previous North Korean missiles. Moreover, for such a big missile, North Korea is expected to use a cluster of four engines in the large first-stage booster, which would increase the complexity of the missile. The TD-2 would have to include a third stage, successfully tested, to achieve the ranges usually attributed to it, but North Korea has not successfully tested such a stage. As was noted, North Korea has not tested a reentry heat shield at long distances. All these considerations call into question how quickly such a missile could be successfully tested and made operational.

Official estimates of the possible range of the TD-2 are controversial. In its December 2001 National Intelligence Estimate, the CIA projected that a two-stage TD-2 "could deliver a several-hundred-kilogram payload up to 10,000 kilometers [6,200 miles]—sufficient to strike Alaska, Hawaii, and parts of the continental United States," and that a third stage could increase the range to cover all of North America.

These range estimates assume that the technology used in the TD-2 would be significantly better than that used in the TD-1. In particular, they appear to assume that the body is made of significantly lighter materials and that the engines provide higher thrust. In the absence of lighter materials and higher thrust, a two-stage TD-2 might be able to reach parts of Alaska with a nuclear payload but appears unlikely to be able to reach the continental United States or even the main Hawaiian islands.

Extending the moratorium on missile flight testing should be the most urgent United States objective in missile negotiations.

Similarly, without these upgrades the range of the three-stage TD-2 would be sufficient to reach Alaska and Hawaii, but only the extreme northwest corner of the continental United States. Such a capability would be troubling, but is considerably less than the official estimates. Without flight-testing, which is verifiable using satellites, North Korea cannot develop an operational long-range missile capability. Hence, it is urgent to keep the current missiletesting moratorium in force and to resume missile negotiations with North Korea.

IF NEGOTIATIONS FAIL

The majority of the task force believes that sustained negotiations, pursued seriously, can lead to the denuclearization of North Korea. If negotiations should fail, the task force opposes preemptive United States military action against North Korea or its nuclear facilities. While military action might be able to destroy known facilities, it would unable to destroy nuclear material and facilities at unknown locations. Moreover, attacks on North Korea would likely lead to retaliatory attacks on United States bases in Japan and South Korea and to a flood of refugees into neighboring countries.

United States Secretary of State Colin Powell suggested on December 29 that the United States could live with a nuclear-armed North Korea, asking, "What are they going to do with another two or three nuclear weapons? If they have a few more, they have a few more." The task force deplores this statement, minimizing as it did the damaging impact that a nuclear-armed North Korea would have on Northeast Asian stability and on the global nonproliferation regime. To make such a statement before testing North Korean intentions seriously at the bargaining table casts doubt on the readiness of the administration for serious negotiations and its broader commitment to the goal of nonproliferation. Nevertheless, the task force agrees with his implicit assessment that United States strategic and tactical nuclear capabilities in the Pacific would deter the use of nuclear weapons by North Korea against the United States and its allies. Preemptive military action to destroy North Korean nuclear and missile facilities would not be warranted. Should North Korea develop nuclear weapons, it would do so with a variety of possible motivations. One could be to deter a United States preemptive strike. Others could be to increase its bargaining leverage with South Korea and other countries, to reduce its conventional forces for economic reasons, and to sell fissile material for cash. The least likely motivation would be to initiate offensive action that could invite its destruction. Therefore, the United States should respond to a nuclear-armed North Korea with renewed efforts to alleviate North Korean security concerns and to keep the door open for improved relations.

For example, while preemptive military action has always been an implicit option in United States strategic doctrine, the United States should avoid provocative public statements asserting the United States right to take such action, which would only strengthen the forces within North Korea supporting the development of nuclear weapons. At the same time, the United States should promote regional conventional and nuclear arms control initiatives designed to constrain North Korea's nuclear buildup and to prevent a regional nuclear arms race.

A minority of the task force believes one contingency in which preemptive action might have to be considered: if clear evidence establishes that North Korea is transferring fissile material to third parties, and if this evidence is made public when it does not compromise military operations.

The task force warns that a confrontational United States posture toward a nuclear-armed North Korea could lead to a rupture in the United States–South Korean alliance. The steady improvement now taking place in South Korean relations with North Korea, centering on economic interchange, is driven by powerful undercurrents of Korean nationalism, reinforced by a strong consensus in South Korea that a collapse of the North Korean state would impose unacceptable economic burdens on the South. Faced with a nuclear-armed North Korea, would South Korea slow down this economic interchange, risking the resurgence of the military tensions that marked North–South relations until the June 2000 North–South summit?

The task force believes that this cannot be taken for granted and that the South could respond instead with arms control initiatives designed to restrain the North's nuclear buildup. Some South Koreans do not consider a North Korean nuclear capability necessarily threatening to the South, since they believe its purpose would be to deter United States military action against North Korea, which the South does not want and, secondarily, to balance what is seen as the latent threat of a nuclear-armed Japan. At the same time, the task force points with concern to the significant minority sentiment in the South in favor of acquiring a plutonium-reprocessing capability that would give the South its own nuclear option. The danger that this sentiment will grow is one of the governing reasons why the task force urges a determined negotiating effort without delay to head off a North Korean nuclear weapons capability.

SOUTH KOREA

During the past two years, South Korea has increasingly perceived the confrontational United States posture toward North Korea as an obstacle to the achievement of a rapprochement with the North that would lead to reconciliation and eventual confederation or reunification. Before deciding on future policies, the United States should initiate urgent high-level consultations with South Korea. Looking ahead, whether or not North Korea develops nuclear weapons, the United States should seek to harmonize its policies with those of the South and to adapt the United States–South Korean alliance to the evolving environment in the peninsula as a whole.

Maintaining positive United States relations with South Korea, as a showcase of democratic values and economic dynamism in Asia, should be the linchpin of a larger United States effort to promote a peaceful, nuclear-free peninsula moving toward North–South amity. American policy should not be driven by exaggerated fears of an economically stagnant North Korea with a GNP one-twentieth that of the South.

Tensions over how to deal with North Korea between the United States and many South Koreans, especially those in the younger generation, have been magnified by the dramatic expansion of social and political consciousness in the South that has been under way since the overthrow of military rule in 1987. By supporting the Park Chung Hee and Chun Doo Hwan dictatorships, the United States delayed the advent of democratization. But the pervasive infusion of democratic values resulting from a half century of interaction with the United States following the Korean War led to the upheaval of 1987. Since then, as the economy has developed, generating significant economic inequities, more and more middle-class and low-income South Koreans have participated in social and political movements. Nongovernmental civic organizations have blossomed. Some have focused on protecting the interests of consumers and small businesses and on curbing the concentrated economic power of the chaebol, or conglomerates. Others have arisen as a response to air and water pollution and other pervasive environmental problems aggravated by rapid growth. Legal reforms that have transferred power from the central government to local municipalities have encouraged the proliferation of civil society groups concerned with cleaning up local government and fostering community social welfare programs. Local governments in the areas where United States bases are located have been energized as forums for local residents' concerns about the United States presence. (In recent years, the Internet and e-mail have made it easier to communicate and organize. South Korea, with its near-universal literacy, ranks third in the world in per capita Internet use and now has a level and intensity of mass social and political awareness found in few countries.)

Foreign and defense policy debates in South Korea have until recently been confined to a relatively small intellectual and political elite. But the broadened base of social and political activism since 1987 has changed that. Popular interest in foreign policy has burgeoned during the past two years, fueled by fears that the United States might drag the South into an unwanted war. President Roh Moo Hyun capitalized on these fears in his election campaign. Equally important, the perception has grown in South Korea that the United States rides roughshod over South Korean sovereignty, disregarding South Korea's wishes in shaping its policies toward the North and clinging to a dominant role in the management of the United States–South Korean alliance that is no longer militarily necessary.

Recent opinion surveys show majority support for a continued United States force presence in Korea in the near to medium term. But the United States faces festering opposition to its military presence, fed by a variety of factors, some old, some new. For many years, popular resentment has focused on prostitution near military bases, on the degree of extraterritorial immunity enjoyed by United States servicemen who commit crimes in South Korea, and on the conspicuous occupation of prime urban real estate by United States military installations. This resentment has been building up for years but has not been visible to the American public because former authoritarian regimes kept a tight lid on it. Now it is steadily growing, stimulated by rising nationalism as North–South relations improve. The American presence is widely perceived as an obstacle to progress in reducing military tensions. The primacy of the United States in the United States–South Korean Combined Forces Command is increasingly a target for critics on the right and left who call for a new command structure and the return of wartime operational control over South Korean forces to Seoul.

Advocates of preemptive military action often exaggerate existing and potential North Korean capabilities to bolster their case.

The United States should be prepared to consider changes in the size, location, and character of the United States presence that would make it less abrasive to South Korea, more compatible with its sovereignty, and more sensitive to the changing climate of North–South relations. As immediate priorities, the task force recommended changes in the Status of Forces Agreement that would give South Korea increased legal rights over United States service personnel and urged that some United States installations, such as Yongsan in Seoul, should be moved away from population centers, where they are a major irritant. The task force supports the efforts of both governments to develop the Land Partnership Program, recently put into effect to consolidate bases and return sizable areas of land used by United States forces to South Koreans.

In the context of declining North–South and North–United States tensions, the United States should consider reducing the size of the United States presence and pulling back its forward deployed forces from the demilitarized zone (DMZ) so that they would no longer play a "tripwire" role in which they would automatically become involved if war breaks out. South Korea's well-trained, well-equipped, and highly capable forces would then bear the brunt of an attack, with United States forces in a supportive role. Such pullbacks could either be made unilaterally or, if arms control negotiations with the North should become possible, in return for the negotiated pullback of forward-deployed North Korean forces.

Other possible changes that should be considered include bilateral United States and South Korean peace agreements with the North that would formally end the Korean War and replacement of the obsolete machinery created to monitor the 1953 armistice, which assumes an adversarial North–South and North–United States relationship. The task force envisages new machinery suited to a climate of improved relations, such as the trilateral Mutual Security Commission (North Korean, South Korean, and United States generals) proposed by North Korea. Over time, if tensions in the peninsula decline, the American presence could be progressively phased out, but the United States–South Korean mutual security agreement would remain in effect and United States equipment could be kept available for emergency use in South Korean bases after American forces leave. Significantly, North Korea has made clear that it might not object to a United States ground force presence in Korea in the context of normalized relations.

Unless and until North Korea's nuclear weapons program is dismantled under adequate inspection safeguards, the task force envisages the indefinite continuation of the United States nuclear umbrella over South Korea. By the same token, the United States should be prepared to join in a four-power agreement with Russia, China, and Japan not to use or deploy nuclear weapons in Korea as part of a broader, verifiable settlement that would bar North and South Korea alike from developing nuclear weapons.

In the absence of a flexible United States posture attuned to South Korean priorities, the United States–South Korean alliance is likely to unravel in the years ahead, and there is a long-term danger of xenophobic animosity toward the United States on both sides of the DMZ that would poison United States relations with the 70 million people of a unified Korea. The task force urges, in particular, that the United States identify itself unambiguously with the goal of North–South reconciliation and stop blocking economic interchange in key areas such as energy, mindful that many Koreans regard the United States as the principal culprit responsible for the division of the peninsula in 1945 and believe that the United States must help put the pieces together again. Already, the seeds of animosity have led to demonstrations against the United States presence in the South, reflecting a widening gulf between the older generation, which remembers the sacrifices of the United States in the Korean War, and the younger generation, which does not.

NORTH KOREA

The dramatic reversal in the relative economic and military strength of North and South over the past five decades makes it necessary for the United States to revise its cold war assumption that the North is committed to the forcible reunification of the peninsula. Even before China and Russia ended their petroleum and food subsidies in the early 1990s, touching off an economic crisis that still continues, the North Korean economy was in deep trouble. The South was growing at a much faster rate. Moreover, it had translated this growth into increased military power by purchasing sophisticated United States weaponry and by developing an extensive network of defense industries in partnership with American firms. Since 1990 Russia and China have forged much closer economic ties with the South than with the North and have stopped selling new military hardware to the North. Many members of the task force believe that with each passing year, as its aging tanks and planes deteriorate and its economic malaise persists, North Korea has grown more insecure and more fearful of a United States preemptive strike or United States–led pressures to bring about its collapse. In place of its 1950 dream of forcible reunification, the North is now obsessed with its very survival, acutely aware that it is too weak economically to sustain a protracted war.

Most members of the task force accept the judgment of many military analysts that the reason for North Korea's forward deployment of so much of its military power at the DMZ is to deter a United States attack. If the North should develop a nuclear weapons capability, the task force feels, its motivations could be varied. One could be deterrence of a United States attack. Others could include a belief that this would strengthen its standing and bargaining leverage with South Korea and other countries; a desire to sell fissile material for cash; increasing its military freedom of action at the DMZ; and reducing its defense spending burden by downgrading conventional forces in its security equation and shifting manpower from military forces to the civilian economy. North Korean leader Kim Jong Il told South Korean President Kim Dae Jung during their June 2000 summit that he would like to demobilize hundreds of thousands of soldiers to provide the labor needed for the South Korean factories to be built in the proposed North–South industrial complex at Kaesong.

Despite North Korea's economic problems and the political uncertainties inherent in a closed, repressive totalitarian system, the majority of task force members feel that it is unrealistic to expect a collapse of the North Korean state, and that the United States should deal with North Korea as it is. Since the death of Kim Il Sung in 1994, the North Korean system has survived with his basic unity intact. The quasi-religious nationalist mystique associated with his memory continues to evoke broad popular acquiescence in the totalitarian discipline imposed by the ruling armed forces and the Workers Party. This atmosphere of mystery grew initially out of his role as a guerrilla leader fighting Japanese colonial rule, but its durability lies primarily in vivid historical memories of shared sacrifices under his leadership during the Korean War. Totalitarian discipline is reinforced in the North by deeply rooted Korean traditions of political centralization and obedience to authority.

The task force condemns the abuses of human rights inflicted by the North Korean system but cautions against the assumption that these abuses will lead to its collapse. In contrast to Germany, the two Koreas fought a fratricidal war. West German Chancellor Willy Brandt did not have to overcome the bitter legacy of such a conflict when he initiated Ostpolitik. It was the network of contacts and economic linkages between East and West Germany made possible by Ostpolitik, that set the stage for the upheaval triggered in the East by Mikhail Gorbachev's relaxation of the Soviet grip, just as the North–South contacts initiated by Kim Dae Jung will gradually change the North Korean system in the decades ahead.

For all its repression, East Germany never achieved the Orwellian thoroughness of North Korea, where children begin to spend six days a week apart from their parents at the age of three. Unlike Eastern Europe, in which television, shortwave radios, and cassettes leapfrogged national frontiers even during the cold war, North Korea is tightly insulated from outside influences. All television and radio sets must be registered and have fixed channels. Only the top echelon of the Workers Party has more than an inkling of what the rest of the world is like.

Although the North Korean state may not implode or explode in the foreseeable future, as some predict, it could well erode over a period of years. Attempting to promote a collapse by further squeezing North Korea economically would accelerate refugee flows to China, which would aggravate the humanitarian crisis in Chinese border areas without destabilizing the North Korean regime. Equally important, it would undermine the nascent economic reform process in North Korea, strengthening hard-liners who oppose reform.

During the worst famine years of 1995 and 1996, the government's food procurement and distribution machinery broke down, and private farm markets mushroomed in the North Korean countryside. Instead of closing them down by force, Kim Jong Il chose to look the other way, which eased the food shortage in urban areas and stimulated a broader movement toward an unofficial market economy. Since then foreign food aid administrators have reported direct evidence of more than 300 private markets. The new markets have coaxed food into circulation that farmers would otherwise have held back from government procurement officers. Agricultural surpluses produced by cooperative farms also find their way into the new markets, along with a wide variety of illicit items such as food diverted from overseas aid stocks, consumer goods obtained in the cross-border black-market trade with China, assets and products of state enterprise stolen by corrupt officials, and goods produced by small family private enterprises, which were legalized two years ago.

In July 2002, as Bradley Babson, a senior consultant to the World Bank, has elaborated in his working paper for the task force, North Korea initiated significant reforms in prices, wages, and other aspects of its economic management that reflect the cautious movement toward a market economy now under way. Some of these reforms have backfired and some have not been fully implemented, fueling inflationary pressures. But they signal a clear recognition of the need for change and a desire to move toward a market economy.

The government now pays farmers half as much for rice and other basic food commodities as it did previously, while selling these commodities to consumers in ration shops at prices five times higher. The increase to consumers brings food prices to levels approaching those that prevail in unofficial private markets. Wage increases, differentiated by occupational categories, range from ten to twenty times higher than previous rates. These wage differentials for different occupations reflect preferences of the government that were formerly embodied in the rationing system, and not the relative values that would be placed on labor in a true free market. The government's announced intention is to retain the public distribution system to ensure that all citizens receive a minimum ration of food staples, with the scale of those rations still subject to food availability. Any surplus above this minimum would be available according to the ability to pay. Households will be charged for rent, transport, and utilities at much higher levels than previously. As in the past, social services such as education, health care, and child care are to be provided free, but the health care system, in particular, is starved for funds and is steadily deteriorating.

In addition to moving toward monetized economic transactions, North Korea is encouraging decentralized decision-making. State enterprises are now expected to be self-sustaining without state subsidies and will be free to set their own production plans and engage in commercial transactions with other state entities.

Another example of the significant reform initiatives to date is the fact that the regime is pursuing increased foreign investment in carefully contained enclaves such as the projected special enterprise zones at Sinuiju, Kaesong, and Wonsan. These are important first steps toward opening up to foreign investment. The flow of new technologies and management approaches to these enclaves eventually will have spillover effects on the rest of the North Korean economy. The overall pace of economic reform is not yet fast enough, and the scope broad enough, however, to resolve North Korea's grave economic difficulties.

One of the most important areas where an acceleration of incentive-based reforms is needed is agriculture. North Korea is a mountainous country with only 18 percent of its land arable. It has thus faced food insecurity since its inception and has always needed to import food despite ambitious irrigation, reclamation, and mechanization programs. These programs brought increases in production, but collectivized farming stifled individual incentives and impeded agricultural growth. The cutoff of Russian and Chinese oil in 1990 had immobilized fertilizer factories, tractors, and irrigation pumps even before the floods of 1995 and 1996 led to the humanitarian catastrophe of famine in many parts of the country.

The United States and private voluntary American agencies deserve high praise for their substantial contributions to alleviating the continuing food crisis in North Korea. The task force regrets that United States contributions to the UN World Food Program have been suspended until North Korea complies with United States demands for stricter monitoring of food aid distribution. While these demands should be pressed, they should not be a condition for continuing food aid. It should be remembered that North Korea has opened up 270 of its 310 counties to monitoring by aid agencies—a remarkable change from the rigidly closed North Korean society of earlier years. The task force feels that the children and older people who receive the bulk of American food aid should not be the victims of political tension between North Korea and the United States.

The long-term American objective should be to help North Korea move toward sustainable food security. The task force emphasizes that it is logical in economic terms for a country with so little arable land to import some of its food. An overall regeneration of the North Korean economy would lead to increased exports that would make commercial food imports affordable once again. For the present, United States food aid should continue through annual contributions to the World Food Program, supplemented by multiyear bilateral commitments as part of United States agreements with North Korea to end its nuclear and missile programs. Japan and South Korea should be encouraged to increase, not reduce, food aid.

Negotiating with North Korea to achieve United States goals would not be submitting to blackmail.

The task force notes that the acute food shortage in mountainous border areas, especially North Hamgyong province, has been primarily responsible for the exodus of North Korean migrants to China in recent years. As a credible study in the British medical journal *Lancet* has reported, there has been relatively little migration from the rest of North Korea, including three other border provinces adjacent to North Hamgyong that also have easy access to China.

The task force strongly criticizes the North Korean government for its persecution of some returning migrants. At the same time, it deplores efforts to exploit the migration issue for political purposes, such as organizing North Korean migrants to storm into foreign embassies in China. Those responsible for such tactics have said openly that their long-term objective is to generate escalating migration from North Korea that would lead to its collapse. More likely, the effect would be to impede cross-border traffic that could nurture change in North Korea. The task force warns that a collapse would send millions of new migrants into neighboring countries, greatly magnifying an already grave humanitarian problem. The recommendations that follow suggest new policies on the part of China, North Korea, and the international community that would mitigate the plight of the migrants already in China and reduce new migration.

Ultimately, as noted earlier, humanitarian needs and human rights abuses in North Korea can be effectively addressed only in the context of a liberalization of its economic system that gradually erodes totalitarian political controls. Liberalization will be a tortuous process in the absence of external aid, trade, and investment linkages that stimulate economic growth. While focusing on security issues in its recommendations, the task force warns that these issues cannot be resolved through policies addressed to North Korean security concerns alone. Security guarantees would have to be provided in conjunction with bilateral United States economic aid linked to United States–supported aid from multilateral financial institutions and to aid from North Korea's neighbors. North Korea is not likely to dismantle its nuclear and missile programs under adequate verification unless the United States, South Korea, Japan, China, and Russia reciprocate with cooperative economic measures beneficial to the North, such as the development of natural gas pipelines from Russia through North to South Korea and other measures with a faster payoff that would help North Korea deal with its crippling energy crisis, its number one priority.

In shaping its policies toward the Korean peninsula, the United States should be guided by recognition of the totality of United States national interests in Northeast Asia as a whole. The importance of positive United States relations with all the countries surrounding Korea, including Russia, makes it imperative that the United States pursue policies in Korea that are compatible with its broader regional interests and goals. Moreover, the United States can only achieve its goals in Korea in close cooperation with neighboring countries that have a direct stake in what happens there.

Northeast Asia is now the epicenter of international commerce and technological innovation. Collectively, Japan, South Korea, China, Taiwan, and Hong Kong have constituted the fastest-growing economic region in the world for much of the past two decades and today account for nearly 30 percent of global GDP, far ahead of the United States, with 19 percent. The world's second- and third-largest economies in aggregate are in Northeast Asia (China and Japan). Approximately half of global foreign exchange reserves are held by Northeast Asian nations (over $700 billion). Northeast Asian economies account for nearly half of global inbound foreign direct investment ($100 billion per year). Northeast Asian economies are also becoming an increasing source of outbound foreign direct investment, which flows both within the East Asian region and to Europe and North America. The United States and the EU member states trade more with Northeast Asian economies than with each other.

In multiple and unseen ways, Americans are tied on a daily basis to the peoples of Northeast Asian societies. Many business, political, and academic leaders in Northeast Asia are graduates of American universities. American popular culture has had a profound impact on Northeast Asian societies—from baseball to fast food to music and consumer durables. Similarly, Northeast Asian goods and culture have increasingly penetrated American society.

In North Korean eyes, it is plausible that the United States, with nearly 10,000 nuclear weapons and overwhelming superiority in airpower, might stage a preemptive strike.

China's entry into the World Trade Organization is accelerating its economic relations with its neighbors along with the broader international community. Its trade with South Korea is now greater than United States trade with South Korea, and South Korean investment in China is burgeoning.

Russia and China are both rich in natural resources. Russia has 31 percent of known global natural gas reserves, and its oil reserves rank eighth in the world. The economic unification and stability of the two Koreas would be greatly enhanced by the development of either or both of two projected gas pipelines—one of which would run from eastern Siberia through China to North and South Korea, and the other from Sakhalin through the North to the South. Similarly, the projected extension of the trans-Siberian railroad through North Korea to the South would transform the peninsula, sharply illustrating the potential benefits of economic cooperation between Korea and its neighbors.

In security terms the United States has an enormous stake in stable relations with Northeast Asia, where it maintains armed forces totaling 100,000 service personnel—37,000 in South Korea, 43,000 in Japan, and another 20,000 aboard ships that patrol the sea-lanes of the region. These forces include an aircraft carrier battle group homeported in Japan as well as air fighter wings and marine and army units based in South Korea and Japan. Most important, the United States maintains a nuclear umbrella over South Korea and Japan.

China and Russia have long urged the United States to adopt a lower military profile in Korea as part of a more conciliatory approach to the North designed to promote a reduction of tensions. Both are playing the role of an honest broker between the two Koreas and would like the United States to do likewise.

If negotiations proceed on dismantling North Korea's nuclear program, the future of the United States nuclear umbrella over the South could well become the central issue. Russia has recently proposed a Korean nuclear-free zone in the context of the current nuclear crisis. In return for permanently dismantling its nuclear weapons efforts, North Korea is seeking United States security assurances. As the recommendations that follow suggest, such assurances should be given in bilateral United States–North Korean negotiations, but if agreement on such bilateral assurances is not reached, a multilateral forum could provide a face-saving way for the United States to join in binding regional security guarantees. As part of these guarantees, the task force envisages a collective pledge by the United States, South Korea, Japan, China, Russia, and the EU not to use or deploy nuclear weapons in Korea, linked to a companion pledge by North and South Korea not to manufacture, introduce, or deploy nuclear weapons, accompanied by adequate inspection safeguards in North Korea.

A proposal by Russia, China, South Korea, or Japan for a nuclear-free zone in Korea would test in a definitive fashion whether the North is serious about a settlement, and whether the United States is ready for a partnership with Northeast Asia on equal terms or remains wedded to the unilateralism asserted in the September 20 National Security Doctrine.

RESOLVING THE NUCLEAR CRISIS

The United States should pursue a three-stage bilateral negotiating strategy to achieve the verifiable dismantlement of North Korean nuclear capabilities, while supporting a multilateral diplomatic process addressed to economic as well as security issues in Korea.

A Bilateral Scenario

In the opening stage of its bilateral diplomacy, the United States should offer to negotiate directly with North Korea on all issues of concern to both sides, including the dismantlement of North Korea's nuclear weapons capabilities, its food and energy needs, and the full normalization of political and economic relations, provided that North Korea pledge not to reprocess the irradiated fuel rods that have been monitored by IAEA inspectors under the 1994 Agreed Framework and to permit the return of the recently expelled inspectors to resume their monitoring. North Korea would agree to honor this pledge for the duration of bilateral negotiations.

By prearrangement, Secretary of State Powell and North Korean Foreign Minister Paik Nam Soon would then make a joint declaration in Washington or Pyongyang. North Korea would pledge in this declaration to negotiate the verified dismantlement of all aspects of its nuclear capabilities. Both sides would pledge that they would not use force against the other during negotiations on dismantlement, and that, on the successful conclusion of dismantlement, they would categorically rule out the use of force against each other thereafter. The North would reaffirm its 1991 nonaggression commitment to the South. The United States would also pledge to respect North Korean sovereignty and not to hinder its economic development.

In the second stage, the two sides would initiate substantive negotiations in which progress toward denuclearization would be linked to United States steps that address North Korean concerns. For example, the United States could offer to resume the monthly oil shipments that were promised under the Agreed Framework and suspended last December and provide a first installment of conventional energy assistance, provided that North Korea take steps to refreeze the Yongbyon reactor, halt its uranium-enrichment program, declare where its enrichment facilities are located, invite United States inspectors to verify the freeze, and account for the material it is known to have imported for the enrichment program, especially aluminum tubing.

Critical but secondary American negotiating objectives could be a North Korean declaration detailing where it has procured its enrichment equipment and technology and a pledge to stop all foreign procurement, including dual-use items, related to enrichment. In return, the United States could expand conventional energy assistance.

In the third stage, the United States would press for the permanent dismantlement of uraniumenrichment capabilities, offering the economic incentives necessary to make this possible.

The United States should use the Agreed Framework in its existing form as a starting point in negotiating denuclearization with North Korea while at the same time renegotiating some provisions and adding new ones. For example, refreezing the Yongbyon and Taechon reactors and the resumption of oil shipments would be a reversion to existing provisions that were suspended when the uranium-enrichment program was revealed last October. So would a North Korean commitment not to reprocess the irradiated fuel rods at Yongbyon. Keeping the Agreed Framework in force retains the legitimacy of provisions advantageous to the United States, such as North Korea's commitment in Article 1, Section 3 not to reprocess the fuel rods or ship them out of the country, and to dismantle all plutonium-related facilities coincident with completion of the two light water reactors promised under the accord.

As the next recommendation spells out, Article 1, Section 1 should be renegotiated to provide for one reactor, not two, and new arrangements should be made for conventional energy assistance in place of the electricity that would have been generated by the second reactor.

Rationale: The priority given in this recommendation to stopping the reprocessing of the plutonium fuel rods reflects the fact that reprocessing would make possible the production of four to six nuclear weapons within six to eight months. Similarly, restarting the Yongbyon reactor and completing the construction of the two reactors at Taechon covered by the Agreed Framework would make possible the eventual production of 30 nuclear weapons per year. These are clearly established facts. By contrast, the CIA does not foresee an operational North Korean capability for making weapons-grade enriched uranium before "mid-decade."

There is an important precedent for making substantive negotiations conditional on a North Korea pledge not to reprocess the Yongbyon fuel rods and to readmit the IAEA inspectors to verify this pledge. In June 1994, former United States President Jimmy Carter, after obtaining Kim Il Sung's commitment to negotiate a nuclear freeze, persuaded him to initiate an immediate freeze that was to remain in effect pending formal negotiations and to permit IAEA inspectors to remain in Yongbyon to verify the freeze.

This is what gave President Bill Clinton the political cover necessary to conclude the Agreed Framework. Similarly, it should be sufficient for the Bush administration to obtain a commitment not to reprocess the fuel rods as a precondition for substantive dialogue. Insisting on the full dismantlement of North Korean nuclear capabilities as a precondition is unrealistic and could well goad North Korea into carrying out its threats to proceed with nuclear weapons development.

A Multilateral Scenario

To reinforce United States–North Korean negotiations, or as an alternative if bilateral dialogue founders, the EU, the United States, South Korea, North Korea, China, Russia, and Japan should convene in Brussels, with the EU as host, on the topic "Security and Economic Development in Korea." Such a conference would have five purposes: to give the United States a face-saving way to resume bilateral negotiations with North Korea; to impart international status to any bilateral United States–North Korean agreements; to draw North Korea into denuclearization commitments made to the participating states as a group, thus strengthening any undertakings it gives to the United States; to provide security guarantees to North Korea by the other participating states that would help make meaningful denuclearization acceptable to the North; and to plan economic aid initiatives by the other participating states that would make the benefits of denuclearization greater in North Korean eyes than the risks.

Sustained negotiations, pursued seriously, can lead to the denuclearization of North Korea.

Working groups on economic and security issues could meet in advance to develop specific proposals for consideration at the conference, such as natural gas pipelines and other energy projects urgently desired by the North and the Korean nuclear-free zone proposal mentioned earlier.

Rationale: Russia's offer to host a multilateral conference has received a cool United States reception. South Korea, as an interested party, would not be acceptable as a host to the North, and Japan, as the former colonial ruler of Korea, would be unacceptable to both the North and the South. The EU, by contrast, would be acceptable to all parties, including North Korea, which has been cultivating EU ties.

On January 29 the European Parliament called on the European Commission to convene "in the late spring or early summer seven-nation talks about the situation in the Korean peninsula, focusing on economic, security, and nuclear disarmament issues." North Korea would be likely to join such a conference only if it is preceded or accompanied by bilateral dialogue with the United States. Even then, it would be a reluctant participant, but it is likely to agree if attractive economic incentives emerge in preconference working groups.

RENEGOTIATING THE AGREED FRAMEWORK

The Agreed Framework should be renegotiated to provide for the construction of one light water reactor, not two, and the substitution of conventional energy alternatives for the electricity that would have been supplied by the second reactor.

North Korea would have to reaffirm its commitment to other existing provisions of the accord, under which it must dismantle its frozen nuclear facilities coincident with the completion of the reactor project. In addition, North Korea would have to accept new provisions that would end its effort to produce enriched uranium under adequate verification, and would have to go beyond existing provisions that require IAEA inspections to determine how much fissile material had been accumulated before 1994. The Bush administration wants these inspections to begin immediately, much sooner than the Agreed Framework requires. North Korea would be likely to accept such accelerated inspections if their schedule is linked to progress in the construction of the reactor.

In return, the United States could drop its opposition to projected gas pipelines from Siberia or Sakhalin that would go through North Korea to the South; encourage multilateral assistance for gas-fired power stations, transmission grids, and fertilizer factories along the pipeline route; and support interim energy aid from the Korean Peninsula Energy Development Organization (KEDO) to the North pending completion of the reactor and the pipeline.

Russia would be invited to join KEDO in recognition of its long collaboration with North Korea in civilian nuclear technology and its potential role as a supplier of natural gas to Korea.

Rationale: North Korea and South Korea alike oppose a revision of the 1994 accord in which both nuclear reactors would be abandoned in favor of conventional energy alternatives, for reasons discussed below. But both might well agree to reduce the KEDO commitment to one reactor, instead of two, if that would keep the nuclear agreement on track.

For the Bush administration, inducing North Korea to accept one reactor instead of two, together with strengthened nuclear inspections, could be presented in the United States as a political victory, partially vindicating Republican charges that Clinton gave North Korea too much in the 1994 accord on terms that were not tough enough.

For Pyongyang, to activate at least one of the reactors is a political imperative if only because the Agreed Framework bore the personal imprint of the late President Kim Il Sung and of Kim Jong Il. Equally important, since Japan and South Korea both have large civilian nuclear programs, North Korea regards nuclear power as a technological status symbol. Like Tokyo and Seoul, Pyongyang wants nuclear power in its energy mix to reduce dependence on petroleum.

In the case of South Korea, support for the KEDO program is possible partly because funding for the first reactor has already been secured from the National Assembly, in part from vested interests with a stake in contracts to build the reactors. The South had already spent some $800 million on the reactors by the end of 2002, and South Korean companies had lined up contracts totaling another $2.3 billion for the construction work ahead. Still, half a loaf would be better than none, and the money spent by the South so far has gone only to the infrastructure at the site and to the first reactor.

South Korea likes the KEDO project because it is confident that the reactors will someday belong to a unified Korea. By contrast Japan made its $1- billion commitment to KEDO grudgingly and has dragged its feet in meeting its obligations. In Japanese eyes, North Korea cannot be trusted to observe nuclear safety standards, and Tokyo fears another Chernobyl in Japan's backyard. Since Tokyo has already spent $400 million on the project, it is reluctant to see it scrapped entirely, but like Seoul might accept a compromise limiting the project to one reactor.

American support for a gas pipeline from Sakhalin through North Korea to the South is necessary because ExxonMobil, a United States firm, is the principal partner in the Sakhalin seabed gas concession involved and would not build the pipeline in the face of White House opposition.

RESUMING MISSILE NEGOTIATIONS

The United States should resume negotiations with North Korea to end the further development of missile capabilities that could threaten the United States and the export of its missiles, missile technology, and missile components to other states. Priority should be given first to extending the North Korean moratorium on missile testing in effect since September 1999; next, to stopping missile exports; and finally, to negotiating a permanent end to the testing, production, and deployment of all missiles with a range over an agreed threshold, with adequate verification.

In addition to multiyear United States food aid, energy aid, and other economic incentives for a missile agreement, the United States should support multilateral financial aid to develop new industries that would provide employment for the workers displaced from existing missile factories, together with United States aid drawing on the experience of the Nunn–Lugar program in Russia.

Rationale: Extending the moratorium on missile flight testing should be the most urgent United States objective in missile negotiations because the moratorium caps North Korean missile capabilities at present levels and such testing is easily verified by United States satellites.

During negotiations in 1999 and 2000, the United States made significant progress in missile negotiations with North Korea, and North Korean officials have since signaled their readiness to resume these negotiations where they left off in the context of an overall improvement in United States–North Korean negotiations.

The most hopeful progress was made in negotiations on missile exports. North Korea had offered to stop all exports of missiles, technology, and components if agreement could be reached on the amount and form of United States compensation for the losses that a cessation of exports would entail. North Korea agreed that compensation would not have to be in cash, as previously demanded, but in kind. Discussion on the amount and form were under way when negotiations were interrupted at the end of the second Clinton administration.

Hopeful progress was also made on banning the testing, production, and deployment of missiles. North Korea had proposed a ban covering all missiles with a range over 500 kilometers (300 miles). The United States had insisted on a shorter range, 300 kilometers (188 miles), combined with a 500-kilogram (1,100-pound) payload. This is the limitation specified in the Missile Technology Control Regime. Although agreement had not been reached on this issue, North Korean negotiators said that it could be resolved in a Clinton–Kim Jong Il summit. On compensation, agreement had been reached in principle that the United States would sponsor arrangements with Russia, China, and the EU for launching long-range North Korean satellites equipped solely for scientific research.

A ban on the flight testing of missiles can be verified by United States satellites. More intrusive verification procedures would be required to verify the cessation of the sale and production of missiles and components. Some of these could draw on experience under the Intermediate Nuclear Forces Treaty. The verification regime was not seriously addressed in the 1999–2000 negotiations.

Previous negotiations also did not seriously address limiting or ending the deployment of the existing Nodong and Scud missiles that are now capable of reaching Japan and South Korea.

ENDING THE KOREAN WAR

Half a century after the end of the Korean War, it is time for the United States to conclude peace agreements with the other two parties to the 1953 Armistice Agreement—North Korea and China—provided that North Korea agrees to conclude a separate agreement with South Korea, which did not sign the armistice. The United States should reconsider its position that it was not a signatory to the armistice, and South Korea should rethink its position that it does have legal status as a signatory.

Rationale: A formal end to the existing state of war is a necessary precondition for the reduction of tensions through conventional arms control negotiations. The United States position that it was not a signatory is untenable. Although General Mark Clark did identify himself in the armistice agreement as commander in chief of the UN Command, his role as head of that command was a mere extension of his position as the ranking commander of all United States forces in Korea and of the United States–South Korean Combined Forces Command. The command was from its inception multilateral in name only. As Trygvie Lie, UN secretary general during the Korean War, spelled out in his memoirs, successive United States commanders of the UN Command insisted on unfettered control over military operations, and in subsequent years even the cosmetic trappings of multilateral control have been progressively reduced.

The South Korean position that it has legal status as a signatory is based on two fallacious arguments. The first is that even though South Korean President Syngman Rhee attempted to subvert the armistice and the South refused to sign it, Rhee later agreed to abide by its provisions. This is fallacious because Rhee's commitment to honor the agreement was made only to the United States, not to North Korea. The second argument is that since General Clark, in signing the armistice, identified himself as commander in chief of the UN Command, South Korea, as one of the countries fighting under him, should thus be treated as a signatory. But 15 other countries also fought under the UN Command. In any case, General Clark's role as head of the UN Command was a mere extension of his position as the ranking commander of all United States forces in Korea and of the United States–South Korean Combined Forces Command.

The operational control that the United States can exercise over South Korean forces in time of war understandably leads North Korea to regard the United States as its main enemy, necessitating a bilateral peace agreement with the United States to bring the war to an end.

REPLACING THE ARMISTICE MACHINERY

The Military Armistice Commission set up in 1953 should be replaced with new peacekeeping machinery, together with companion steps to dissolve the UN Command. The United States should explore the October 9, 1998 North Korean proposal for the creation of a Mutual Security Assurance Commission in place of the Military Armistice Commission and the UN Command, consisting of American, South Korean, and North Korean generals. The United States should condition its participation in such a trilateral commission on North Korean agreement to activate the bilateral North–South Joint Military Commission envisaged in the 1992 North–South "Basic Agreement."

Rationale: Both the Military Armistice Commission and the UN Command are obsolete vestiges of an adversarial cold war relationship between the United States and North Korea. Their continuance would be incompatible with a peace agreement and with the normalization of relations between the two countries that the task force supports.

A trilateral commission would be appropriate because all three countries have forces on the ground in Korea. Additionally, an American general presides over the United States–South Korean Combined Forces Command and would have operational control over South Korean forces in wartime. But the United States cannot speak for South Korea. Thus, issues relating only to South Korean and North Korean forces would be addressed in the Joint North–South Military Commission. The new Mutual Security Commission would deal with all issues involving United States forces in Korea and would oversee arms control and tension-reduction proposals involving the United States and South Korea.

The dissolution of the UN Command would have no military impact, since it has had no military functions for more than two decades. In 1978, when the United States and South Korea created the Combined Forces Command, the UN Command formally transferred its authority to the new command. The same United States general commands both the Combined Forces Command and the UN Command, but he wears his UN hat only when participating in meetings of the Military Armistice Commission. The United States–South Korea Mutual Security Treaty would continue to provide an umbrella for the United States military presence when the UN Command is dismantled.

LOWERING THE UNITED STATES MILITARY PROFILE

Before opposition to the American military presence reaches serious proportions and leads to significant pressures for disengagement, the United States should defuse this opposition by lowering the United States military profile in South Korea and offering to make changes in the size, character, and location of United States deployments. Such changes could be made either through unilateral United States–South Korean action or in return for the pullback of forward-deployed North Korean forces as part of the broad process of North–South and North–United States rapprochement envisaged in the report. Unless and until a verifiable denuclearization agreement is reached with North Korea, however, the United States nuclear umbrella over South Korea should remain in force.

It is unrealistic to expect a collapse of the North Korean state, and the United States should deal with North Korea as it is.

The task force urges consideration of a structural change in the United States–South Korean military relationship designed to show greater sensitivity to South Korean sovereignty and to keep pace with progress in improving North–South and North–United States relations. In place of the tightly integrated United States–South Korean Combined Forces Command, the United States and South Korea should move toward a command struc-

> # Members of the Task Force
>
> SELIG S. HARRISON, chairman of the task force, is director of the Asia Program at the Center for International Policy and a senior scholar of the Woodrow Wilson International Center for Scholars. He is author of *Korean Endgame: A Strategy for Reunification and U.S. Disengagement* (Princeton: Princeton University Press, 2002) and has visited North Korea seven times.
>
> DAVID ALBRIGHT is president of the Institute for Science and International Security.
>
> BRADLEY O. BABSON is a senior consultant on East Asia to the World Bank.*
>
> EDWARD J. BAKER is associate director of the Harvard Yenching Institute.
>
> TED GALEN CARPENTER is vice president for defense and foreign policy studies of the Cato Institute.*
>
> ADMIRAL WILLIAM J. CROWE, JR. (USN, RET.) is a former chairman of the Joint Chiefs of Staff under President Ronald Reagan and former ambassador to the United Kingdom.
>
> ELLSWORTH CULVER is cofounder and senior vice president of Mercy Corps International.*
>
> BRUCE CUMINGS is the Norma and Edna Freehling Professor of History and director of the Korea Program, Center for East Asian Studies, University of Chicago.
>
> CARTER J. ECKERT is professor of Korean history and director of the Korea Institute at Harvard University.
>
> COL. JOHN ENDICOTT (USAF, RET.) is director of the Center for International Strategy, Technology, and Policy at the Sam Nunn School, Georgia Institute of Technology.
>
> AMBASSADOR ROBERT L. GALLUCCI, dean of the Edmund A. Walsh School of U.S. Foreign Service at Georgetown University, negotiated the 1994 Agreed Framework with North Korea.*
>
> AMBASSADOR JAMES E. GOODBY served in a variety of senior arms control positions during 37 years in the Foreign Service, including vice chairman of the United States delegation to the Strategic Arms Reduction Talks (START).*
>
> BRIGADIER GENERAL JAMES F. GRANT (USAF, Ret.), executive vice president of Zel Technologies, served as assistant chief of staff for Intelligence, United States Forces, South Korea, and deputy chief of staff for Intelligence of the United States–Republic of Korea Combined Forces Command.*
>
> AMBASSADOR DONALD P. GREGG, president of the Korea Society, served as ambassador to the Republic of Korea from 1989 to 1993.
>
> LEE H. HAMILTON, director of the Woodrow Wilson International Center for Scholars, is vice chairman of the National Commission on Terrorist Attacks upon the United States.
>
> PETER HAYES is executive director of the Nautilus Institute for Security and Sustainable Development.*
>
> AMBASSADOR JAMES T. LANEY served as ambassador to the Republic of Korea from 1993 to 1997.
>
> JOHN W. LEWIS is William Haas Professor Emeritus of Chinese Politics at Stanford University.
>
> KATHARINE H. S. MOON is the Jane Bishop Associate Professor of Political Science at Wellesley College.
>
> K. A. NAMKUNG is associate director of the Program on U.S.–D.P.R.K. Relations, University of California, Berkeley.
>
> CLYDE V. PRESTOWITZ, JR. is president of the Economic Strategy Institute.
>
> DAVID SHAMBAUGH is director of the China Policy Program and professor of political science and international affairs in the Elliot School of International Affairs at the George Washington University.*
>
> SUSAN L. SHIRK is professor of Asia-Pacific international relations at the University of California, San Diego.
>
> LEON V. SIGAL is director of the Northeast Asia Cooperative Security Project of the Social Science Research Council.
>
> JOHN D. STEINBRUNER is professor of public policy and director of the Center for International and Security Studies at the University of Maryland.
>
> DAVID WOLFF is a senior scholar of the Cold War International History Project of the Woodrow Wilson International Center for Scholars.
>
> MEREDITH WOO-CUMINGS is professor of political science at the University of Michigan.
>
> DAVID WRIGHT is codirector of the Global Security Program of the Union of Concerned Scientists.
>
> Members participated in their personal capacities, not as representatives of their organizations.
>
> * Members dissenting from portions of the report.

ture that provides South Korean forces with increasingly greater autonomy, including the eventual return of wartime operational control. Many aspects of the United States–Japan model, in which two separate operational structures are linked on a cooperative basis, could be adapted to Korea in the context of declining North–South tensions and reciprocal pullbacks from the DMZ. To make such a looser command structure workable, South Korea should commit the resources needed to modernize its command and control, intelligence, surveillance, and reconnaissance capabilities with American assistance.

The goal of the United States should be to move from its present "tripwire" role, in which United States forces are automatically drawn into any new Korean conflict, to a new role in which it would have greater flexibility in deciding whether to participate in any given conflict situation.

Rationale: South Korean military forces and defense industries have acquired increasing technological sophistication with United States help at a cumulative cost to the United States that has included $7 billion in grant military aid and $12 billion in United States–subsidized military sales. The well-trained, well-equipped South Korean forces are now capable of bearing the brunt of any North Korean attack, with United States forces in a supportive role. Faced with assuming the principal responsibility for financing and conducting its own defense, South Korea will have an increased incentive for finding a modus vivendi with the North.

Application of the United States–Japan model to the revision of the United States–South Korean command structure would not be possible in the context of the existing configuration of opposing forces at the DMZ and the attendant stress on time-sensitive and fully coordinated operations. However, a shift to this model could be studied in preparation for its introduction as tensions decline.

Former South Korean President Kim Dae Jung's national security adviser, Lim Dong Won, has proposed a 60-mile North–South "Offensive Weapon-Free Zone" in which tanks, mechanized infantry, armored troop carriers, and self-propelled artillery would be barred, including artillery using chemical or biological warfare agents. Given the fact that Seoul is closer to the DMZ than Pyongyang, North Korea would have to pull back farther than Seoul.

This proposal could be part of broader arms control negotiations that could include other tension-reduction initiatives. In negotiating a mutual pullback zone, the United States could propose that both sides be required to deploy all their artillery in the open, throughout their respective territories, to facilitate inspection and to maximize the warning time that the South would have in the event of an attack in violation of the accord.

For North and South alike, it would be costly to relocate their forces to create a mutual pullback zone. As a United States Institute of Peace working group has observed, "international financial support will be necessary to cover certain costs associated with a Korean arms reduction process, including mutual troop and equipment reductions and repositioning."

SUPPORTING NORTH KOREAN ECONOMIC DEVELOPMENT

As progress on resolving security issues continues, the United States should support economic regeneration and growth in North Korea and encourage North Korea to carry forward its recent economic reform initiatives with technical assistance from international financial institutions. Specific elements of such a policy could include:

- Support for economic cooperation between North and South Korea, including South Korean energy aid to the North;
- Revision of United States legislation and licensing regulations that would block North Korean exports to the United States, especially exports from projected North–South investment zones;
- Relaxation of the remaining United States economic sanctions imposed against North Korea during the Korean War;
- Support for exploratory dialogue between North Korea and international financial institutions designed to set the stage for discussions on the prerequisites for membership and for dialogue on North Korean reform strategy. This would be followed by an assessment of the North Korean economy by the international financial institutions and by technical assistance to help the North design an effective reform program;
- Eventual steps toward membership in the World Bank, International Monetary Fund, and the Asian Development Bank;
- Authorization for United States petroleum companies to conclude concession agreements with North Korea for seabed oil exploration and development in the Yellow Sea and to construct natural gas pipelines from Russia that would cross through North Korea to South Korea;
- Support for North Korean membership in regional economic forums;
- Support for regional arrangements embracing North Korea designed to promote expanded trade and investment links in Northeast Asia;
- Bilateral and multilateral credits that would enable United States mining companies to help modernize the North Korean mining industry, with payment in mineral products.

Rationale: Resolution of the nuclear crisis can be achieved only in the context of a broader rapprochement among North Korea, its neighbors in Northeast Asia, and the United States that includes bilateral and multilateral economic cooperation. A stable North Korea, growing economically in cooperation with its neighbors and moving toward a market economy, is necessary for stability and security in Northeast Asia. Conversely, economic stagnation and collapse in North Korea could lead to destabilizing refugee flows, enormous reconstruction costs, and war.

The economic reforms initiated by North Korea in July 2002 indicated a desire to move toward a market economy. They also revealed a lack of the technical knowledge of past experience in other countries with similar problems necessary to make the reforms effective. Technical consultations with international financial institutions, when North Korea is ready for it, would maximize the chances for a successful reform effort.

The United States has until now discouraged South Korean energy aid to the North as part of its effort to press North Korea for a more stringent nuclear-inspection regime. This policy should be ended in parallel with progress toward a verifiable denuclearization agreement with the North.

REDUCING MIGRATION TO CHINA

The United States and the international community should take urgent steps to relieve the plight of North Korean migrants into China and reduce the flow of future migration through humanitarian and economic assistance measures in North Korea. Specific steps to implement this policy could include the following:

- China and North Korea should be encouraged to expand the access of humanitarian aid organizations to monitor and as-

sist migrants in China and returnees to North Korea with food, medical aid, and other needed assistance;
- China should be urged to declare a moratorium on the forced return of North Korean migrants and asylum seekers, pending a more durable and humane solution;
- North Korea should be urged to repeal all laws that penalize citizens for leaving its territory or returning without authorization;
- South Korea, China, and North Korea should develop a joint program of targeted public and private investment in the poor border counties of North Hamgyong province in North Korea, where much of the migration originates. This program should be nonpolitical, supervised by technical experts and managed multilaterally;
- The Chinese government should grant semiresident status through a special visa to those individuals who can demonstrate that they have work and shelter. For those who are employed in seasonal agricultural work but who can demonstrate that the employer is prepared to house and feed them year-round, an annual visa could be issued;
- Beijing should consider a one-time amnesty for the relatively low numbers of North Korean migrants that remain in China.

Rationale: Mercy Corps, one of the leading United States–based humanitarian aid organizations operating in North Korea and the border areas of China, has estimated that between 50,000 and 150,000 North Korean migrants have crossed into China since 1997 in search of temporary food, shelter, and survival assistance from the local ethnic Korean population living in China's Yanbian Autonomous Prefecture. Most of these migrants cross the border illegally and are subject to arrest, detention, and deportation. On their return to North Korea, they are subject to penalties ranging from short-term detention to longer-term prison sentences.

These migrants are a highly vulnerable population in need of humanitarian aid and in many instances protection. Among the most vulnerable are unaccompanied minors, women, the elderly, the medically needy, and asylum seekers.

While most North Koreans in China are searching for temporary assistance, an increasing proportion seeks either long-term sanctuary in China or permanent resettlement in another country. Thus, the world faces a growing humanitarian tragedy on the Chinese–North Korean border that calls for urgent action to mitigate the suffering of the migrants and long-term measures to reduce the future flow as part of the broad process of engagement recommended by the task force.

NOTES

1. Interviewed on *The NewsHour with Jim Lehrer*, Public Broadcasting System, September 17, 1999.
2. Interview with Leon V. Sigal, October 31, 1996, in Leon V. Sigal, *Disarming Strangers: Nuclear Diplomacy with North Korea* (Princeton: Princeton University Press, 1998), pp. 93–94.

From *Current History*, April 2003. Abridged from "Turning Point in Korea: New Dangers and New Opportunities for the United States," a report of the Task Force on U.S. Korea Policy, cosponsored by The Center for International Policy and the Center for East Asian Studies at the University of Chicago. Copyright © 2003 by the Center for International Policy, Washington, D.C. Reprinted by permission of the author.

N. Korea Shops Stealthily for Nuclear Arms Gear

By Joby Warrick
Washington Post Staff Writer

MUNICH—The French cargo ship Ville de Virgo was already running a day late when it steamed into Hamburg harbor on April 3, its stadium-size deck stacked 50 feet high with cargo containers bound for Asia.

At the dock, harried German customs agents skimmed quickly through a fat manifest that included the usual Asia-bound staples—fertilizer, bulk chemicals, cheeses. A last-minute addition, 214 ultra-strong aluminum pipes purchased by China's Shenyang Aircraft Corp., was one of the final items cleared before the 40,000-ton ship fired its engines again and headed to Asia.

But within hours after the ship departed, the story of the manifest began to unravel. German intelligence officials discovered that the aluminum was destined not for China but for North Korea. The intended use of the pipes, they concluded, was not aircraft production, but the making of nuclear weapons.

On April 12, in a dramatic but little-noticed intervention, French and German authorities tracked the ship to the eastern Mediterranean and seized the pipes. German police arrested the owner of a small export company and uncovered a broader scheme to acquire as many as 2,000 such pipes. That much aluminum in North Korean hands, investigators concluded, could have yielded as many as 3,500 gas centrifuges for enriching uranium.

"The intentions were clearly nuclear," said a Western diplomat familiar with the investigation. "The result could have been several bombs' worth of weapons-grade uranium in a year."

The voyage and capture of the Ville de Virgo exposed one of the most ambitious attempts yet by North Korea to obtain materials for building nuclear weapons. But the episode also offers a glimpse into the shadowy world of weapons proliferation, in which missile parts and bomb materials circle the globe undetected, secreted away in cargo containers and suitcases, concealed by phony ship manifests and fictitious company names, eluding customs agents and defying international treaties.

The story of the Ville de Virgo is a case study in the workings of the gray zone, a combination of weak states, open borders, lack of controls and a ready market of buyers and sellers for weapons of mass destruction.

The attempt to import the aluminum tubes is being closely studied by intelligence agencies for possible clues about the design and origins of North Korea's uranium enrichment program. In January, North Korea announced that it was withdrawing from the international treaty that bars it from making nuclear weapons, and the country is believed by intelligence agencies to be pursuing nuclear weapons through two different routes—bombs based on uranium and those based on plutonium.

In recent months, North Korea's attempts to seek parts and technology in Europe have increased dramatically, U.S. and European intelligence officials say. Lately, they say, the attempts are becoming ever more elaborately disguised.

On April 4, just one day after the Ville de Virgo left Hamburg, a different cargo ship departed Japan's Kobe Harbor carrying three devices known as direct-current stabilizers, which also are used in uranium enrichment, according to a Japanese government account of the incident. Just as with the aluminum shipment, the electronic parts were being routed to a third country—in this case, Thailand—where the cargo would be diverted to North Korea.

In mid-May, a month after the aluminum pipes were seized, North Korea nearly succeeded in acquiring 33 tons of sodium cyanide, a chemical used in making the deadly nerve agent tabun, according to Western diplomatic sources. The chemicals were purchased legally from a German manufacturer who believed the buyer was a Singapore company. But in fact, a switch was planned that would have diverted them to Pyongyang, the North Korean capital.

Both efforts were thwarted, but intelligence officials have little doubt that others succeeded. "There are countries in the world where you can pay $2,000 to a government minister and he'll sign anything—and then confirm to you that he signed it," said Rastislav Kacer, a former Slovak deputy defense minister who helped lead an investigation into a similar attempt by North Korea to buy sophisticated radar equipment. "Documents that are fake can be made to appear very real."

In such an environment, said Kacer, now his country's ambassador in Washington, "no system is ever 100 percent leak-proof."

Special Aluminum Tubes

The French-owned Ville de Virgo is a workhorse of the modern shipping trade, a floating warehouse that moves cargo along a circuit running from Hamburg and Rotterdam to Singapore and Pusan, South Korea. At each port, goods are brought to the ship in prepacked steel containers, which are then stacked five high on the top deck. Only rarely are the containers opened and physically searched.

On the morning of April 3, the Ville de Virgo was running a day behind schedule as it took on freight and awaited paperwork in Hamburg before setting off on a nine-week, round-trip voyage to China and Korea. Local customs agents had visited the ship dozens of times in the past, and on this day, German officials say, there was nothing outwardly unusual about the ship or its cargo.

But one container on the deck held aluminum tubes, and German intelligence officials had been watching these very pipes for months.

Measuring nearly eight feet in length and nine inches in diameter, the tubes were made of a special alloy, 6061-T6, known to be both light and exceptionally strong. Similar tubes are used in a wide range of commercial products, from bicycle frames to aircraft parts. But they also are useful in the construction of machines known as gas centrifuges, which enrich uranium into the key material for nuclear weapons.

Throughout the second half of 2002, intelligence agencies in the United States and Western Europe picked up multiple signals that North Korea was attempting to acquire such tubes, along with other specialized metals used in centrifuges, U.S. and European sources say. Germany's top nonproliferation agency issued a warning in the fall that North Korean agents were known to be "obtaining sensitive goods" by using front companies or third countries as cover. Intelligence reports suggested that a large quantity of pipes—perhaps 220 tons or more—was being sought across Europe. The tubes are of a different type of aluminum than those that figured prominently in suspicions about Iraq.

Despite the increased vigilance, North Korea may have already succeeded in acquiring hundreds of such tubes, using connections and routes developed over years. "All they need is help from one company—perhaps a small company, one that may never actually see the aluminum pipes, or have them in their hands," said Eckhard Maak, a government prosecutor in Stuttgart, Germany, who helped investigate the case. "With only a phone and an Internet connection, you can send such materials across the world."

Export License Denied

The unlikely supplier of the aluminum pipes was a tiny German export company called Optronic. Its owner, Hans Werner Truppel, made a living brokering sales of optical and electronic equipment out of his house, a modest one-story dwelling in a village 85 miles northwest of Munich.

Three years ago, German law enforcement officials say, Truppel struck up a relationship with a North Korean businessman who claimed to represent an import-export company, Nam Chon Gang. At first, the North Korean company asked for help from Optronic in obtaining obscure machine parts and electronics, offering cash in payment. Truppel sold the firm vacuum pumps and machines known as angle grinders, in each case with the approval of German customs.

Then, last fall, Nam Chon Gang approached Optronic with a new wish list: Could Truppel find a supply of aluminum pipes, made of a specific alloy and cut to precise dimensions? In this case, the North Korean businessman claimed to be brokering a deal on behalf of Shenyang Aircraft Corp., one of China's top aircraft manufacturers. Later, a letter bearing Shenyang's logo vouched for the purchase, according to a law enforcement official who has seen the document. The letter said the aluminum was to be converted into airplane fuel tanks.

It all seemed legitimate, according to Truppel's Frankfurt attorney, Egon Geiss. In September, Optronic located British-made aluminum pipes at a company in nearby Ulm, Germany, and paid the equivalent of just over $80,000 for 214 of them. Truppel then began the process of securing the needed export papers.

To Truppel's surprise, the German government balked. Officials in the Trade Ministry, aware of the potential uses for such tubes, looked closely at Optronic's application and began picking it apart. The story about aircraft fuel tanks was dismissed as "not plausible," according to Maak, the prosecutor. Moreover, German officials were skeptical that a major Chinese aircraft corporation would employ an unknown North Korean firm to do its shopping.

"Why the North Korean middleman?" Maak said he wondered. "It seemed highly unusual."

The denial left Truppel baffled and financially exposed, according to Geiss. Now the businessman was stuck with 22 tons of aluminum, which he had paid for but couldn't use. Through the fall and winter, he tried to unsuccessfully sell the pipes to others at a discount. Meanwhile, the Ulm company that had sold the pipes to Truppel in September was still holding them in its warehouse and was pressuring Truppel to pick them up.

Exactly how and why the pipes ended up on the Ville de Virgo remains in dispute. Geiss said Truppel received a call from Delta-Trading, a relatively small metals production, distribution and export firm based in Hamburg. Delta offered to take the pipes and promised to secure the necessary export papers, he said. Truppel "explained to Delta in writing that he was unable to export" the pipes, Geiss added. But in the end Truppel agreed to pay Delta about $6,000—roughly half the profit he had expected to make on the deal—to take the matter off his hands.

"He assumed that Delta, because of its connections, had other legal avenues for exporting the aluminum," Geiss said of Truppel. "He understood that Delta was to take care of all the necessary arrangements." Delta declined comment.

German prosecutors say Truppel was not so naive. "He definitely knew what he was doing," Maak said. "The important thing is, Optronic was denied permission to export, and it did so anyway."

German officials were wary enough to issue a warning urging customs agents to watch for outbound shipments of aluminum pipes. Sometime after April 4 came a report that 22 tons of aluminum had moved from Ulm to Hamburg to be loaded onto the Ville de Virgo.

By the time the warning was issued, the ship and cargo were already on their way to the Mediterranean.

A Trove of Evidence

The North Korean man who drew Truppel into the aluminum scheme has never been publicly identified. But German and U.S. investigators say companies like Nam Chon Gang exist in cities throughout Europe, Japan and other regions that offer access to critical technology.

Last August, police made a rare move against such a company in Bratislava, the Slovak capital. The company, New World Trading Slovakia, was founded in March 2001 by two North Koreans who apparently were seeking a quiet location for negotiating deals with customers on three continents, Slovak officials say.

One of them, Kim Kum Jin, 51, had once served as an economic adviser at North Korea's embassy in Egypt. Kim and his partner, Sun Hui Ri, 48, quickly grew fond of their new home. They bought a Mercedes-Benz and opened shop in a luxurious high-rise in one of Bratislava's newest commercial districts, police investigators said in interviews in the Slovak capital. The couple even listed their company in the city's business registry.

But last summer, Slovak federal police, after months of surveillance, began to suspect the two were trading in weapons technology. Lacking sufficient evidence to file charges, the authorities ordered the couple to leave the country last August.

Kim and Sun left behind a trove of documents, police said, including financial records, invoices and bills of lading. The papers described multiple deals by the pair to procure materials for weapons programs, as well as millions of dollars in sales of missile technology to Egypt, Libya, Iran, Syria and Vietnam. One of their major clients, documents revealed, was an Egyptian military-industrial concern.

"They did it all by fax and computer," said an investigator with firsthand knowledge of the case, who spoke on the condition his name not be used. "None of the material ever crossed into Slovakia, which would have been a clear violation of the law. That's why they were able to operate as long as they did."

This pattern is at the heart of how governments such as North Korea manage to traffic in weapons materials. Many countries have agreed to treaties and multilateral agreements, such as the nuclear Non-Proliferation Treaty and the Missile Technology Control Regime, in an effort to restrict such dangerous transfers. But these efforts were defeated by North Korea using faxes and computers. North Korea has said it does not accept the treaties and defended its right to sell weapons abroad.

"With North Korea you have a strange mix of impressive, extensively clandestine systems and sometimes incredible naivete about how things work," said Greg Thielmann, recently retired director of the office on strategic, proliferation and military issues at the State Department's Bureau of Intelligence and Research. "But somehow they have found a way to operate in a world of export-control regimes and still buy the things they need, and still ship their missiles to other countries."

Logistical support along the way is provided by North Korea's embassies and staff, whose activities and travel are protected under the rules of diplomacy, U.S. and European intelligence officials say. Backing for complex weapons deals comes from North Korean banks, including the Vienna-based Golden Star Bank, Pyongyang's only financial institution in Europe. The imposing red stucco building near one of Vienna's busiest markets has no customers and no private accounts, yet its activities have raised alarms within Austria's Interior Ministry.

A report by the ministry's office for the protection of the constitution included a list of activities the agency had connected to the bank. It included intelligence-gathering as well as "money-laundering, the distribution of forged currency and illegal trade with radioactive substances."

Unscheduled Stop

French and German officials had little evidence in hand on April 10 when they pondered their options for dealing with the Ville de Virgo. By this time, the ship was in the eastern Mediterranean, far beyond the territorial reach of the two countries, steaming southeast toward the Suez Canal at 23 knots.

One possible solution—letting the ship proceed to an Asian port and working through the host government—was ruled out as too risky. Another option, since the ship was French-owned and technically under France's jurisdiction, was to stop the ship at sea and transfer the cargo to a French military vessel.

Instead, it was decided that the aluminum pipes simply should be removed, quickly and quietly, at the first possible port. The ship's French owner endorsed the plan.

When contacted by radio, the Ville de Virgo's captain was unaware of any controversy involving the aluminum tubes. But he agreed to a request to make an unscheduled stop in the Egyptian port of Alexandria, just outside the Suez, to remove the tubes from his ship. As the ship arrived in Alexandria on April 12, a special crew and cargo crane were waiting at the dock. Another vessel returned the tubes to Hamburg on April 28.

In Stuttgart, Truppel, the Optronic chief, was arrested for violating German export laws and was ordered held without bail. He remains imprisoned in Stuttgart awaiting trial. The company that acted as an export middleman, Delta-Trading, has not been charged. Geiss, Truppel's attorney, plans to argue that his client was tricked by Delta and North Korea.

Back at Hamburg's harbor, the watch for aluminum tubes continues. Nam Chon Gang and its mysterious North Korean entrepreneur, thwarted in one attempt to obtain the metal, might be trying again: U.S. proliferation officials said they learned from European allies of "multiple" efforts to acquire aluminum tubes in recent months.

The dimensions of the tubes suggest to nuclear experts that North Korea is attempting to build a type of gas centrifuge designed by the European consortium Urenco—a design stolen by Pakistani scientists in the 1970s. The Urenco centrifuge uses an aluminum casing that is roughly the same size as the tubes exported by Optronic, said David Albright, a physicist and president of the Washington-based Institute for Science and International Security.

But it takes more than aluminum to build a centrifuge, Albright noted. Highly specialized magnets, bearings and a metal known as maraging steel are also required. North Korea would probably have to import all those things, yet there have been no known interceptions of such materials.

"There would have to be many more shipments," Albright said. "Usually what you see is only the tip of the iceberg."

Stopping a single shipment of aluminum tubes from reaching North Korea was a setback for Pyongyang—but probably only a temporary one, he said. "You can hurt them badly," Albright said, "but in the end you can only delay them from succeeding."

Special correspondent Shannon Smiley in Berlin and researcher Robert E. Thomason in Washington contributed to this report.

From *The Washington Post*, August 15, 2003. © 2003 by The Washington Post. Reprinted with permission.

Nuclear Nightmares

Experts on terrorism and proliferation agree on one thing:
Sooner or later, an attack will happen here. When and how is what robs them of sleep.

By Bill Keller

The panic that would result from contaminating the Magic Kingdom with a modest amount of cesium would probably shut the place down for good and constitute a staggering strike at Americans' sense of innocence.

Not If But When Everybody who spends much time thinking about nuclear terrorism can give you a scenario, something diabolical and, theoretically, doable. Michael A. Levi, a researcher at the Federation of American Scientists, imagines a homemade nuclear explosive device detonated inside a truck passing through one of the tunnels into Manhattan. The blast would crater portions of the New York skyline, barbecue thousands of people instantly, condemn thousands more to a horrible death from radiation sickness and—by virtue of being underground—would vaporize many tons of concrete and dirt and river water into an enduring cloud of lethal fallout. Vladimir Shikalov, a Russian nuclear physicist who helped clean up after the 1986 Chernobyl accident, envisioned for me an attack involving highly radioactive cesium-137 loaded into some kind of homemade spraying device, and a target that sounded particularly unsettling when proposed across a Moscow kitchen table—Disneyland. In this case, the human toll would be much less ghastly, but the panic that would result from contaminating the Magic Kingdom with a modest amount of cesium—Shikalov held up his teacup to illustrate how much—would probably shut the place down for good and constitute a staggering strike at Americans' sense of innocence. Shikalov, a nuclear enthusiast who thinks most people are ridiculously squeamish about radiation, added that personally he would still be happy to visit Disneyland after the terrorists struck, although he would pack his own food and drink and destroy his clothing afterward.

Another Russian, Dmitry Borisov, a former official of his country's atomic energy ministry, conjured a suicidal pilot. (Suicidal pilots, for obvious reasons, figure frequently in these fantasies.) In Borisov's scenario, the hijacker dive-bombs an Aeroflot jetliner into the Kurchatov Institute, an atomic research center in a gentrifying neighborhood of Moscow, which I had just visited the day before our conversation. The facility contains 26 nuclear reactors of various sizes and a huge accumulation of radioactive material. The effect would probably be measured more in property values than in body bags, but some people say the same about Chernobyl. Maybe it is a way to tame a fearsome subject by Hollywoodizing it, or maybe it is a way to drive home the dreadful stakes in the arid-sounding business of nonproliferation, but in several weeks of talking to specialists here and in Russia about the threats an amateur evildoer might pose to the homeland, I found an unnerving abundance of such morbid creativity. I heard a physicist wonder whether a suicide bomber with a pacemaker would constitute an effective radiation weapon. (I'm a little ashamed to say I checked that one, and the answer is no, since pacemakers powered by plutonium have not been implanted for the past 20 years.) I have had people theorize about whether hijackers who took over a nuclear research laboratory could improvise an actual nuclear explosion on the spot. (Expert opinions differ, but it's very unlikely.) I've been instructed how to disperse plutonium into the ventilation system of an office building.

The realistic threats settle into two broad categories. The less likely but far more devastating is an actual nuclear explosion, a great hole blown in the heart of New York or Washington, followed by a toxic fog of radiation. This could be produced by a black-market nuclear warhead procured from an existing arsenal. Russia is the favorite hypothetical source, although Pakistan, which has a program built on shady middlemen and covert operations, should not be overlooked. Or the explosive could be a homemade device, lower in yield than a factory nuke but still creating great carnage.

The second category is a radiological attack, contaminating a public place with radioactive material by packing it with conventional explosives in a "dirty bomb" by dispersing it into the air or water or by sabotaging a nuclear facility. By comparison with the task of creating nuclear fission, some of these schemes would be almost childishly simple, although the consequences would be less horrifying: a panicky evacuation, a gradual increase in cancer rates, a staggeringly expensive cleanup, possibly the need to demolish whole neighborhoods. Al Qaeda has claimed to have access to dirty bombs, which is unverified but entirely plausible, given that the makings are easily gettable.

Nothing is really new about these perils. The means to inflict nuclear harm on America have been available to rogues for a long time. Serious studies of the threat of nuclear terror date back to the 1970's. American programs to keep Russian nuclear ingredients from falling into murderous hands—one of the subjects high on the agenda in President Bush's meetings in Moscow this weekend—were hatched soon after the Soviet Union disintegrated a decade ago. When terrorists get around to trying their first nuclear assault, as you can be sure they will, there will be plenty of people entitled to say I told you so.

All Sept. 11 did was turn a theoretical possibility into a felt danger. All it did was supply a credible cast of characters who hate us so much they would thrill to the prospect of actually doing it—and, most important in rethinking the probabilities, would be happy to die in the effort. All it did was give our nightmares legs.

Tom Ridge cupped his hands prayerfully and pressed his fingertips to his lips. "Nuclear," he said simply.

And of the many nightmares animated by the attacks, this is the one with pride of place in our experience and literature—and, we know from his own lips, in Osama bin Laden's aspirations. In February, Tom Ridge, the Bush administration's homeland security chief, visited The Times for a conversation, and at the end someone asked, given all the things he had to worry about—hijacked airliners, anthrax in the mail, smallpox, germs in crop-dusters—what did he worry about most? He cupped his hands prayerfully and pressed his fingertips to his lips. "Nuclear," he said simply.

My assignment here was to stare at that fear and inventory the possibilities. How afraid should we be, and what of, exactly? I'll tell you at the outset, this was not one of those exercises in which weighing the fears and assigning them probabilities laid them to rest. I'm not evacuating Manhattan, but neither am I sleeping quite as soundly. As I was writing this early one Saturday in April, the floor began to rumble and my desk lamp wobbled precariously. Although I grew up on the San Andreas Fault, the fact that New York was experiencing an earthquake was only my second thought.

The best reason for thinking it won't happen is that it hasn't happened yet, and that is terrible logic. The problem is not so much that we are not doing enough to prevent a terrorist from turning our atomic knowledge against us (although we are not). The problem is that there may be no such thing as "enough."

25,000 Warheads, and It Only Takes One My few actual encounters with the Russian nuclear arsenal are all associated with Thomas Cochran. Cochran, a physicist with a Tennessee lilt and a sense of showmanship, is the director of nuclear issues for the Natural Resources Defense Council, which promotes environmental protection and arms control. In 1989, when glasnost was in flower, Cochran persuaded the Soviet Union to open some of its most secret nuclear venues to a roadshow of American scientists and congressmen and invited along a couple of reporters. We visited a Soviet missile cruiser bobbing in the Black Sea and drank vodka with physicists and engineers in the secret city where the Soviets first produced plutonium for weapons.

Not long ago Cochran took me cruising through the Russian nuclear stockpile again, this time digitally. The days of glasnost theatrics are past, and this is now the only way an outsider can get close to the places where Russians store and deploy their nuclear weapons. On his office computer in Washington, Cochran has installed a detailed United States military map of Russia and superimposed upon it high-resolution satellite photographs. We spent part of a morning mouse-clicking from missile-launch site to submarine base, zooming in like voyeurs and contemplating the possibility that a terrorist could figure out how to steal a nuclear warhead from one of these places.

"Here are the bunkers," Cochran said, enlarging an area the size of a football stadium holding a half-dozen elongated igloos. We were hovering over a site called Zhukovka, in western Russia. We were pleased to see it did not look ripe for a hijacking.

"You see the bunkers are fenced, and then the whole thing is fenced again," Cochran said. "Just outside you can see barracks and a rifle range for the guards. These would be troops of the 12th Main Directorate. Somebody's not going to walk off the street and get a Russian weapon out of this particular storage area."

In the popular culture, nuclear terror begins with the theft of a nuclear weapon. Why build one when so many are lying around for the taking? And stealing tends to make better drama than engineering. Thus the stolen nuke has been a staple in the literature at least since 1961, when Ian Fleming published "Thunderball," in which the malevolent Spectre (the Special Executive for Counterintelligence, Terrorism, Revenge and Extortion, a strictly mercenary and more technologically sophisticated precursor to al Qaeda) pilfers a pair of atom bombs from a crashed NATO aircraft. In the movie version of Tom Clancy's thriller "The Sum of All Fears," due in theaters this week, neo-Nazis get their hands on a mislaid Israeli nuke, and viewers will get to see Baltimore blasted to oblivion.

Eight countries are known to have nuclear weapons—the United States, Russia, China, Great Britain, France, India, Pakistan and Israel. David Albright, a nuclear-weapons expert and president of the Institute for Science and International Security, points out that Pakistan's program in particular was built almost entirely through black markets and industrial espionage, aimed at circumventing Western export controls. Defeating the discipline of nuclear nonproliferation is ingrained in the culture. Disaffected individuals in Pakistan (which, remember, was intimate with the Taliban) would have no trouble finding the illicit channels or the rationalization for diverting materials, expertise—even, conceivably, a warhead.

But the mall of horrors is Russia, because it currently maintains something like 15,000 of the world's (very roughly) 25,000 nuclear warheads, ranging in destructive power from about 500 kilotons, which could kill a million people, down to the one-kiloton land mines that would be enough to make much of Manhattan uninhabitable. Russia is a country with sloppy accounting, a disgruntled military, an audacious black market and indigenous terrorists.

It's easier to take the fuel and build an entire weapon from scratch than it is to make one of these things go off.

There is anecdotal reason to worry. Gen. Igor Valynkin, commander of the 12th Main Directorate of the Russian Ministry of Defense, the Russian military sector in charge of all nuclear weapons outside the Navy, said recently that twice in the past year terrorist groups were caught casing Russian weapons-storage facilities. But it's hard to know how seriously to take this. When I made the rounds of nuclear experts in Russia earlier this year, many were skeptical of these near-miss anecdotes, saying the security forces tend to exaggerate such incidents to dramatize their own prowess (the culprits are always caught) and enhance their budgets. On the whole, Russian and American military experts sound not very alarmed about the vulnerability of Russia's nuclear warheads. They say Russia takes these weapons quite seriously, accounts for them rigorously and guards them carefully. There is no confirmed case of a warhead being lost. Strategic warheads, including the 4,000 or so that President Bush and President Vladimir Putin have agreed to retire from service, tend to be stored in hard-to-reach places, fenced and heavily guarded, and their whereabouts are not advertised. The people who guard them are better paid and more closely vetted than most Russian soldiers.

Eugene E. Habiger, the four-star general who was in charge of American strategic weapons until 1998 and then ran nuclear antiterror programs for the Energy Department, visited several Russian weapons facilities in 1996 and 1997. He may be the only American who has actually entered a Russian bunker and inspected a warhead *in situ*. Habiger said he found the overall level of security comparable to American sites, although the Russians depend more on people than on technology to protect their nukes.

The image of armed terrorist commandos storming a nuclear bunker is cinematic, but it's far more plausible to think of an inside job. No observer of the unraveling Russian military has much trouble imagining that a group of military officers, disenchanted by the humiliation of serving a spent superpower, embittered by the wretched conditions in which they spend much of their military lives or merely greedy, might find a way to divert a warhead to a terrorist for the right price. (The Chechen warlord Shamil Basayev, infamous for such ruthless exploits as taking an entire hospital hostage, once hinted that he had an opportunity to buy a nuclear warhead from the stockpile.) The anecdotal evidence of desperation in the military is plentiful and disquieting. Every year the Russian press provides stories like that of the 19-year-old sailor who went on a rampage aboard an Akula-class nuclear submarine, killing eight people and threatening to blow up the boat and its nuclear reactor; or the five soldiers at Russia's nuclear-weapons test site who killed a guard, took a hostage and tried to hijack an aircraft, or the officers who reportedly stole five assault helicopters, with their weapons pods, and tried to sell them to North Korea.

The Clinton administration found the danger of disgruntled nuclear caretakers worrisome enough that it considered building better housing for some officers in the nuclear rocket corps. Congress, noting that the United States does not build housing for its own officers, rejected the idea out of hand.

If a terrorist did get his hands on a nuclear warhead, he would still face the problem of setting it off. American warheads are rigged with multiple PAL's ("permissive action links")—codes and self-disabling devices designed to frustrate an unauthorized person from triggering the explosion. General Habiger says that when he examined Russian strategic weapons he found the level of protection comparable to our own. "You'd have to literally break the weapon apart to get into the gut," he told me. "I would submit that a more likely scenario is that there'd be an attempt to get hold of a warhead and not explode the warhead but extract the plutonium or highly enriched uranium." In other words, it's easier to take the fuel and build an entire weapon from scratch than it is to make one of these things go off.

Then again, Habiger is not an expert in physics or weapons design. Then again, the Russians would seem to have no obvious reason for misleading him about something that important. Then again, how many times have computer hackers hacked their way into encrypted computers we were assured were impregnable? Then again, how many computer hackers does al Qaeda have? This subject drives you in circles.

The most troublesome gap in the generally reassuring assessment of Russian weapons security is those tactical nuclear warheads—smaller, short-range weapons like torpedoes, depth charges, artillery shells, mines. Although their smaller size and greater number makes them ideal candidates for theft, they have gotten far less attention simply because, unlike all of our long-range weapons, they happen not to be the subject of any formal treaty. The first President Bush reached an informal understanding with President Gorbachev and then with President Yeltsin that both sides would gather and destroy thousands of tactical nukes. But the agreement included no inventories of the stockpiles, no outside monitoring, no verification of any kind. It was one of those trust-me deals that, in the hindsight of Sept. 11, amount to an enormous black hole in our security.

Did I say earlier there are about 15,000 Russian warheads? That number includes, alongside the scrupulously counted strategic warheads in bombers, missiles and submarines, the commonly used estimate of 8,000 tactical warheads. But that figure is at best an educated guess. Other educated guesses of the tactical nukes in Russia go as low as 4,000 and as high as 30,000. We just don't know. We don't even know if the Russians know, since they are famous for doing things off the books. "They'll tell you they've never lost a weapon," said Kenneth Luongo, director of a private antiproliferation group called the Russian-American Nuclear Security Advisory Council. "The fact is, they don't know. And when you're talking about warhead counting, you don't want to miss even one."

And where are they? Some are stored in reinforced concrete bunkers like the one at Zhukovka. Others are deployed. (When the submarine Kursk sank with its 118 crewmen in August 2000, the Americans' immediate fear was for its nuclear armaments. The standard load out for a submarine of that class includes a couple of nuclear torpedoes and possibly some nuclear depth charges.) Still others are supposed to be in the process of being dismantled under terms of various formal and informal arms-control agreements. Some are in transit. In short, we don't really know.

The other worrying thing about tactical nukes is that their anti-use devices are believed to be less sophisticated, because the weapons were designed to be employed in the battlefield. Some of the older systems are thought to have no permissive action links at all, so that setting one off would be about as complicated as hot-wiring a car.

Efforts to learn more about the state of tactical stockpiles have been frustrated by reluctance on both sides to let visitors in. Viktor Mikhailov, who ran the Russian Ministry of Atomic Energy until 1998 with a famous scorn for America's nonproliferation concerns, still insists that the United States programs to protect Russian nuclear weapons and material mask a secret agenda of intelligence-gathering. Americans, in turn, sometimes balk at reciprocal access, on the grounds that we are the ones paying the bills for all these safety upgrades, said the former Senator Sam Nunn, co-author of the main American program for securing Russian nukes, called Nunn-Lugar.

People in the field talk of a nuclear 'conex' bomb, using the name of those shack-size steel containers—2,000 of which enter America every hour on trains, trucks and ships. Fewer than 2 percent are cracked open for inspection.

"We have to decide if we want the Russians to be transparent—I'd call it cradle-to-grave transparency with nuclear material and inventories and so forth," Nunn told me. "Then we have to open up more ourselves. This is a big psychological breakthrough we're talking about here, both for them and for us."

The Garage Bomb One of the more interesting facts about the atom bomb dropped on Hiroshima is that it had never been tested. All of those spectral images of nuclear coronas brightening the desert of New Mexico— those were to perfect the more complicated plutonium device that was dropped on Nagasaki. "Little Boy," the Hiroshima bomb, was a rudimentary gunlike device that shot one projectile of highly enriched uranium into another, creating a critical mass that exploded. The mechanics were so simple that few doubted it would work, so the first experiment was in the sky over Japan.

The closest thing to a consensus I heard among those who study nuclear terror was this: building a nuclear bomb is easier than you think, probably easier than stealing one. In the rejuvenated effort to prevent a terrorist from striking a nuclear blow, this is where most of the attention and money are focused.

A nuclear explosion of any kind "is not a sort of high-probability thing," said a White House official who follows the subject closely. "But getting your hands on enough fissile material to build an improvised nuclear device, to my mind, is the least improbable of them all, and particularly if that material is highly enriched uranium in metallic form. Then I'm really worried. That's the one."

To build a nuclear explosive you need material capable of explosive nuclear fission, you need expertise, you need some equipment, and you need a way to deliver it.

Delivering it to the target is, by most reckoning, the simplest part. People in the field generally scoff at the mythologized suitcase bomb; instead they talk of a "conex bomb," using the name of those shack-size steel containers that bring most cargo into the United States. Two thousand containers enter America every hour, on trucks and trains and especially on ships sailing into more than 300 American ports. Fewer than 2 percent are cracked open for inspection, and the great majority never pass through an X-ray machine. Containers delivered to upriver ports like St. Louis or Chicago pass many miles of potential targets before they even reach customs.

"How do you protect against that?" mused Habiger, the former chief of our nuclear arsenal. "You can't. That's scary. That's very, very scary. You set one of those off in Philadelphia, in New York City, San Francisco, Los Angeles, and you're going to kill tens of thousands of people, if not more." Habiger's view is "It's not a matter of *if*; it's a matter of *when*"—which may explain why he now lives in San Antonio.

The Homeland Security office has installed a plan to refocus inspections, making sure the 2 percent of containers that get inspected are those without a clear, verified itinerary. Detectors will be put into place at ports and other checkpoints. This is good, but it hardly represents an ironclad defense. The detection devices are a long way from being reliable. (Inconveniently, the most feared bomb component, uranium, is one of the hardest radioactive substances to detect because it does not emit a lot of radiation prior to fission.) The best way to stop nuclear terror, therefore, is to keep the weapons out of terrorist hands in the first place.

Fabricating a nuclear weapon is not something a lone madman—even a lone genius—is likely to pull off in his hobby room.

The basic know-how of atom-bomb-building is half a century old, and adequate recipes have cropped up in physics term papers and high school science projects. The simplest design entails taking a lump of highly enriched uranium, about the size of a cantaloupe, and firing it down a big gun barrel into a second lump. Theodore Taylor, the nuclear physicist who designed both the smallest and the largest American nuclear-fission warheads before becoming a remorseful opponent of all things nuclear, told me he recently looked up "atomic bomb" in the World Book Encyclopedia in the upstate New York nursing home where he now lives, and he found enough basic information to get a careful reader started. "It's accessible all over the place," he said. "I don't mean just the basic principles. The sizes, specifications, things that work."

Most of the people who talk about the ease of assembling a nuclear weapon, of course, have never actually built one. The most authoritative assessment I found was a paper, "Can Terrorists Build Nuclear Weapons?" written in 1986 by five experienced nuke-makers from the Los Alamos weapons laboratory. I was relieved to learn that fabricating a nuclear weapon is not something a lone madman—even a lone genius— is likely to pull off in his hobby room. The paper explained that it would require a team with knowledge of "the physical, chemical and metallurgical properties of the various materials to be used, as well as characteristics affecting their fabrication; neutronic properties; radiation effects, both nuclear and biological; technology concerning high explosives and/or chemical propellants; some hydrodynamics; electrical circuitry; and others." Many of these skills are more difficult to acquire than, say, the ability to aim a jumbo jet.

The schemers would also need specialized equipment to form the uranium, which is usually in powdered form, into metal, to cast it and machine it to fit the device. That effort would entail months of preparation, increasing the risk of detection, and it would require elaborate safeguards to prevent a mishap that, as the paper dryly put it, would "bring the operation to a close."

Still, the experts concluded, the answer to the question posed in the title, while qualified, was "Yes, they can."

David Albright, who worked as a United Nations weapons inspector in Iraq, says Saddam Hussein's unsuccessful crash program to build a nuclear weapon in 1990 illustrates how a single bad decision can mean a huge setback. Iraq had extracted highly enriched uranium from research-reactor fuel and had, maybe, barely enough for a bomb. But the manager in charge of casting the metal was so afraid the stuff would spill or get contaminated that he decided to melt it in tiny batches. As a result, so much of the uranium was wasted that he ended up with too little for a bomb.

"You need good managers and organizational people to put the elements together," Albright said. "If you do a straight-line extrapolation, terrorists will all get nuclear weapons. But they make mistakes."

On the other hand, many experts underestimate the prospect of a do-it-yourself bomb because they are thinking too professionally. All of our experience with these weapons is that the people who make them (states, in other words) want them to be safe, reliable, predictable and efficient. Weapons for the American arsenal are designed to survive a trip around the globe in a missile, to be accident-proof, to produce a precisely specified blast.

But there are many corners you can cut if you are content with a big, ugly, inefficient device that would make a spectacular impression. If your bomb doesn't need to fit in a suitcase (and why should it?) or to endure the stress of a missile launch; if you don't care whether the explosive power realizes its full potential; if you're willing to accept some risk that the thing might go off at the wrong time or might not go off at all, then the job of building it is immeasurably simplified.

"As you get smarter, you realize you can get by with less," Albright said. "You can do it in facilities that look like barns, garages, with simple machine tools. You can do it with 10 to 15 people, not all Ph.D.'s, but some engineers, technicians. Our judgment is that a gun-type device is well within the capability of a terrorist organization."

All the technological challenges are greatly simplified if terrorists are in league with a country—a place with an infrastructure. A state is much better suited to hire expertise (like dispirited scientists from decommissioned nuclear installations in the old Soviet Union) or to send its own scientists for M.I.T. degrees.

Thus Tom Cochran said his greatest fear is what you might call a bespoke nuke—terrorists stealing a quantity of weapons-grade uranium and taking it to Iraq or Iran or Libya, letting the scientists and engineers there fashion it into an elementary weapon and then taking it away for a delivery that would have no return address.

That leaves one big obstacle to the terrorist nuke-maker: the fissile material itself.

To be reasonably sure of a nuclear explosion, allowing for some material being lost in the manufacturing process, you need roughly 50 kilograms—110 pounds—of highly enriched uranium. (For a weapon, more than 90 percent of the material should consist of the very unstable uranium-235 isotope.) Tom Cochran, the master of visual aids, has 15 pounds of depleted uranium that he keeps in a Coke can; an eight-pack would be plenty to build a bomb.

Only 41 percent of Russia's weapon-usable material has been secured... So the barn door is still pretty seriously ajar. We don't know whether any horses have gotten out.

The world is awash in the stuff. Frank von Hippel, a Princeton physicist and arms-control advocate, has calculated that between 1,300 and 2,100 metric tons of weapons-grade uranium exists—at the low end, enough for 26,000 rough-hewed bombs. The largest stockpile is in Russia, which Senator Joseph Biden calls "the candy store of candy stores."

Until a decade ago, Russian officials say, no one worried much about the safety of this material. Viktor Mikhailov, who ran the atomic energy ministry and now presides over an affiliated research institute, concedes there were glaring lapses.

"The safety of nuclear materials was always on our minds, but the focus was on intruders," he said. "The system had never taken account of the possibility that these carefully screened people in the nuclear sphere could themselves represent a danger. The system was not designed to prevent a danger from within."

Then came the collapse of the Soviet Union and, in the early 90's, a few frightening cases of nuclear materials popping up on the black market.

If you add up all the reported attempts to sell highly enriched uranium or plutonium, even including those that have the scent of security-agency hype and those where the material was of uncertain quality, the total amount of material still falls short of what a bomb-maker would need to construct a single explosive.

But Yuri G. Volodin, the chief of safeguards at Gosatomnadzor, the Russian nuclear regulatory agency, told me his inspectors still discover one or two instances of attempted theft a year, along with dozens of violations of the regulations for storing and securing nuclear material. And as he readily concedes: "These are the detected cases. We can't talk about the cases we don't know." Alexander Pikayev, a former aide to the Defense Committee of the Russian Duma, said: "The vast majority of installations now have fences. But you know Russians. If you walk along the perimeter, you can see a hole in the fence, because the employees want to come and go freely."

The bulk of American investment in nuclear safety goes to lock the stuff up at the source. That is clearly the right priority. Other programs are devoted to blending down the highly enriched uranium to a diluted product unsuitable for weapons but good as reactor fuel. The Nuclear Threat Initiative, financed by Ted Turner and led by Nunn, is studying ways to double the rate of this diluting process.

Still, after 10 years of American subsidies, only 41 percent of Russia's weapon-usable material has been secured, according to the United States Department of Energy. Russian officials said they can't even be sure how much exists, in part because the managers of nuclear facilities, like everyone else in the Soviet industrial complex, learned to cook their books. So the barn door is still pretty seriously ajar. We don't know whether any horses have gotten out.

And it is not the only barn. William C. Potter, director of the Center for Nonproliferation Studies at the Monterey Institute of International Studies and an expert in nuclear security in the former Soviet states, said the American focus on Russia has neglected other locations that could be tempting targets for a terrorist seeking bomb-making material. There is, for example, a bomb's worth of weapons-grade uranium at a site in Belarus, a country with an erratic president and an anti-American orientation. There is enough weapons-grade uranium for a bomb or two in Kharkiv, in Ukraine. Outside of Belgrade, in a research reactor at Vinca, sits sufficient material for a bomb—and there it sat while NATO was bombarding the area.

"We need to avoid the notion that because the most material is in Russia, that's where we should direct all of our effort," Potter said. "It's like assuming the bank robber will target Fort Knox because that's where the most gold is. The bank robber goes where the gold is most accessible."

Weapons of Mass Disruption The first and, so far, only consummated act of nuclear terrorism took place in Moscow in 1995, and it was scarcely memorable. Chechen rebels obtained a canister of cesium, possibly from a hospital they had commandeered a few months before. They hid it in a Moscow park famed for its weekend flea market and called the press. No one was hurt. Authorities treated the incident discreetly, and a surge of panic quickly passed.

The story came up in virtually every conversation I had in Russia about nuclear terror, usually to illustrate that even without splitting atoms and making mushroom clouds a terrorist could use radioactivity—and the fear of it—as a potent weapon.

The idea that you could make a fantastic weapon out of radioactive material without actually producing a nuclear bang has been around since the infancy of nuclear weaponry. During World War II, American scientists in the Manhattan Project worried that the Germans would rain radioactive material on our troops storming the beaches on D-Day. Robert S. Norris, the biographer of the Manhattan Project director, Gen. Leslie R. Groves, told me that the United States took this threat seriously enough to outfit some of the D-Day soldiers with Geiger counters.

No country today includes radiological weapons in its armories. But radiation's limitations as a military tool—its tendency to drift afield with unplanned consequences, its long-term rather than short-term lethality—would not necessarily count against it in the mind of a terrorist. If your aim is to instill fear, radiation is anthrax-plus. And unlike the fabrication of a nuclear explosive, this is terror within the means of a soloist.

If your aim is to instill fear, radiation is anthrax-plus. And unlike the fabrication of a nuclear explosive, this is terror within the means of a soloist.

That is why, if you polled the universe of people paid to worry about weapons of mass destruction (W.M.D., in the jargon), you would find a general agreement that this is probably the first thing we'll see. "If there is a W.M.D. attack in the next year, it's likely to be a radiological attack," said Rose Gottemoeller, who handled Russian nuclear safety in the Clinton administration and now follows the subject for the Carnegie Endowment. The radioactive heart of a dirty bomb could be spent fuel from a nuclear reactor or isotopes separated out in the process of refining nuclear fuel. These materials are many times more abundant and much, much less protected than the high-grade stuff suitable for bombs. Since Sept.11, Russian officials have begun lobbying hard to expand the program of American aid to include protection of these lower-grade materials, and the Bush administration has earmarked a few million dollars to study the problem. But the fact is that radioactive material suitable for terrorist attacks is so widely available that there is little hope of controlling it all.

The guts of a dirty bomb could be cobalt-60, which is readily available in hospitals for use in radiation therapy and in food processing to kill the bacteria in fruits and vegetables. It could be cesium-137, commonly used in medical gauges and radiotherapy machines. It could be americium, an isotope that behaves a lot like plutonium and is used in smoke detectors and in oil prospecting. It could be plutonium, which exists in many research laboratories in America. If you trust the security of those American labs, pause and reflect that the investigation into the great anthrax scare seems to be focused on disaffected American scientists.

Back in 1974, Theodore Taylor and Mason Willrich, in a book on the dangers of nuclear theft, examined things a terrorist might do if he got his hands on 100 grams of plutonium—a thimble-size amount. They calculated that a killer who dissolved it, made an aerosol and introduced it into the ventilation system of an office building could deliver a lethal dose to the entire floor area of a large skyscraper. But plutonium dispersed outdoors in the open air, they estimated, would be far less effective. It would blow away in a gentle wind.

The Federation of American Scientists recently mapped out for a Congressional hearing the consequences of various homemade dirty bombs detonated in New York or Washington. For example, a bomb made with a single footlong pencil of cobalt from a food irradiation plant and just 10 pounds of TNT and detonated at Union Square in a light wind would send a plume of radiation drifting across three states. Much of Manhattan would be as contaminated as the permanently closed area around the Chernobyl nuclear plant. Anyone living in Manhattan would have at least a 1-in-100 chance of dying from cancer caused by the radiation. An area reaching deep into the Hudson Valley would, under current Environmental Protection Agency standards, have to be decontaminated or destroyed.

Frank von Hippel, the Princeton physicist, has reviewed the data, and he pointed out that this is a bit less alarming than it sounds. "Your probability of dying of cancer in your lifetime is already about 20 percent," he said. "This would increase it to 20.1 percent. Would you abandon a city for that? I doubt it."

Indeed, some large portion of our fear of radiation is irrational. And yet the fact that it's all in your mind is little consolation if it's also in the minds of a large, panicky population. If the actual effect of a radiation bomb is that people clog the bridges out of town, swarm the hospitals and refuse to return to live and work in a contaminated place, then the impact is a good deal more than psychological. To this day, there is bitter debate about the actual health toll from the Chernobyl nuclear accident. There are researchers who claim that the people who evacuated are actually in worse health over all from the trauma of relocation, than those who stayed put and marinated in the residual radiation. But the fact is, large swaths of developed land around the Chernobyl site still lie abandoned, much of it bulldozed down to the subsoil. The Hart Senate Office Building was closed for three months by what was, in hindsight, our society's inclination to err on the side of alarm.

There are measures the government can take to diminish the dangers of a radiological weapon, and many of them are getting more serious consideration. The Bush administration has taken a lively new interest in radiation-detection devices that might catch dirty-bomb materials in transit. A White House official told me the administration's judgment is that protecting the raw materials of radiological terror is worth doing, but not at the expense of more catastrophic threats.

"It's all over," he said. "It's not a winning proposition to say you can just lock all that up. And then, a bomb is pretty darn easy to make. You don't have to be a rocket scientist to figure about fertilizer and diesel fuel." A big fertilizer bomb of the type Timothy McVeigh used to kill 168 people in Oklahoma City, spiced with a dose of cobalt or cesium, would not tax the skills of a determined terrorist.

"It's likely to happen, I think, in our lifetime," the official said. "And it'll be like Oklahoma City plus the Hart Office Building. Which is real bad, but it ain't the World Trade Center."

The Peril of Power Plants Every eight years or so the security guards at each of the country's 103 nuclear power stations and at national weapons labs can expect to be attacked by federal agents armed with laser-tag rifles. These mock terror exercises are played according to elaborate rules, called the "design basis threat," that in the view of skeptics favor the defense. The attack teams can include no more than three commandos. The largest vehicle they are permitted is an S.U.V. They are allowed to have an accomplice inside the plant, but only one. They are not allowed to improvise. (The mock assailants at one Department of Energy lab were ruled out of order because they commandeered a wheelbarrow to cart off a load of dummy plutonium.) The mock attacks are actually announced in advance. Even playing by these rules, the attackers manage with some regularity to penetrate to the heart of a nuclear plant and damage the core. Representative Edward J. Markey, a Massachusetts Democrat and something of a scourge of the nuclear power industry, has recently identified a number of shortcomings in the safeguards, including, apparently, lax standards for clearing workers hired at power plants.

One of the most glaring lapses, which nuclear regulators concede and have promised to fix, is that the design basis threat does not contemplate the possibility of a hijacker commandeering an airplane and diving it into a reactor. In fact, the protections currently in place don't consider the possibility that the terrorist might be willing, even eager, to die in the act. The government assumes the culprits would be caught while trying to get away.

A nuclear power plant is essentially a great inferno of decaying radioactive material, kept under control by coolant. Turning this device into a terrorist weapon would require cutting off the coolant so the atomic furnace rages out of control and, equally important, getting the radioactive matter to disperse by an explosion or fire. (At Three Mile Island, the coolant was cut off and the reactor core melted down, generating vast quantities of radiation. But the thick walls of the containment building kept the contaminant from being released, so no one died.)

One way to accomplish both goals might be to fly a large jetliner into the fortified building that holds the reactor. Some experts say a jet engine would stand a good chance of bursting the containment vessel, and the sheer force of the crash might disable the cooling system—rupturing the pipes and cutting off electricity that pumps the water through the core. Before nearby residents had begun to evacuate, you could have a meltdown that would spew a volcano of radioactive isotopes into the air, causing fatal radiation sickness for those exposed to high doses and raising lifetime cancer rates for miles around.

This sort of attack is not as easy, by a long shot, as hitting the World Trade Center. The reactor is a small, low-lying target, often nestled near the conspicuous cooling towers, which could be destroyed without great harm. The reactor is encased in reinforced concrete several feet thick, probably enough, the industry contends, to withstand a crash. The pilot would have to be quite a marksman, and somewhat lucky. A high wind would disperse the fumes before they did great damage.

Invading a plant to produce a meltdown, even given the record of those mock attacks, would be more complicated, because law enforcement from many miles around would be on the place quickly, and because breaching the containment vessel is harder from within. Either invaders or a kamikaze attacker could instead target the more poorly protected cooling ponds, where used plutonium sits, encased in great rods of zirconium alloy. This kind of sabotage would take longer to generate radiation and would be far less lethal.

Discussion of this kind of potential radiological terrorism is colored by passionate disagreements over nuclear power itself. Thus the nuclear industry and its rather tame regulators sometimes sound dismissive about the vulnerability of the plants (although less so since Sept. 11), while those who regard nuclear power as inherently evil tend to overstate the risks. It is hard to sort fact from fear-mongering.

Nuclear regulators and the industry grumpily concede that Sept. 11 requires a new estimate of their defenses, and under prodding from Congress they are redrafting the so-called design basis threat, the one plants are required to defend against. A few members of Congress have proposed installing ground-to-air missiles at nuclear plants, which most experts think is a recipe for a disastrous mishap.

"Probably the only way to protect against someone flying an aircraft into a nuclear power plant," said Steve Fetter of the University of Maryland, "is to keep hijackers out of cockpits."

Being Afraid For those who were absorbed by the subject of nuclear terror before it became fashionable, the months since the terror attacks have been, paradoxically, a time of vindication. President Bush, whose first budget cut $100 million from the programs to protect Russian weapons and material (never a popular program among conservative Republicans), has become a convert. The administration has made nuclear terror a priority, and it is getting plenty of goading to keep it one. You can argue with their priorities and their budgets, but it's hard to accuse anyone of indifference. And resistance—from scientists who don't want security measures to impede their access to nuclear research materials, from generals and counterintelligence officials uneasy about having their bunkers inspected, from nuclear regulators who worry about the cost of nuclear power, from conservatives who don't want to subsidize the Russians to do much of anything—has become harder to sustain. Intelligence gathering on nuclear material has been abysmal, but it is now being upgraded; it is a hot topic at meetings between American and foreign intelligence services, and we can expect more numerous and more sophisticated sting operations aimed at disrupting the black market for nuclear materials. Putin, too, has taken notice. Just before leaving to meet Bush in Crawford, Tex., in November, he summoned the head of the atomic energy ministry to the Kremlin on a Saturday to discuss nuclear security. The subject is now on the regular agenda when Bush and Putin talk.

These efforts can reduce the danger but they cannot neutralize the fear, particularly after we have been so vividly reminded of the hostility some of the world feels for us, and of our vulnerability.

Fear is personal. My own—in part, because it's the one I grew up with, the one that made me shiver through the Cuban missile crisis and "On the Beach"—is the horrible magic of nuclear fission. A dirty bomb or an assault on a nuclear power station, ghastly as that would be, feels to me within the range of what we have survived. As the White House official I spoke with said, it's basically Oklahoma City plus the Hart Office Building. A nuclear explosion is in a different realm of fears and would test the country in ways we can scarcely imagine.

A mushroom cloud of irradiated debris would blossom more than two miles into the air. Then highly lethal fallout would begin drifting back to earth, riding the winds into the Bronx or Queens or New Jersey.

As I neared the end of this assignment, I asked Matthew McKinzie, a staff scientist at the Natural Resources Defense Council, to run a computer model of a one-kiloton nuclear explosion in Times Square, half a block from my office, on a nice spring workday. By the standards of serious nuclear weaponry, one kiloton is a junk bomb, hardly worthy of respect, a fifteenth the power of the bomb over Hiroshima.

A couple of days later he e-mailed me the results, which I combined with estimates of office workers and tourist traffic in the area. The blast and searing heat would gut buildings for a block in every direction, incinerating pedestrians and crushing people at their desks. Let's say 20,000 dead in a matter of seconds. Beyond this, to a distance of more than a quarter mile, anyone directly exposed to the fireball would die a gruesome death from radiation sickness within a day—anyone, that is, who survived the third-degree burns. This larger circle would be populated by about a quarter million people on a workday. Half a mile from the explosion, up at Rockefeller Center and down at Macy's, unshielded onlookers would expect a slower death from radiation. A mushroom cloud of irradiated debris would blossom more than two miles into the air, and then, 40 minutes later, highly lethal fallout would begin drifting back to earth, showering injured survivors and dooming rescue workers. The poison would ride for 5 or 10 miles on the prevailing winds, deep into the Bronx or Queens or New Jersey.

A terrorist who pulls off even such a small-bore nuclear explosion will take us to a whole different territory of dread from Sept. 11. It is the event that preoccupies those who think about this for a living, a category I seem to have joined.

"I think they're going to try," said the physicist David Albright. "I'm an optimist at heart. I think we can catch them in time. If one goes off, I think we will survive. But we won't be the same. It will affect us in a fundamental way. And not for the better."

Bill Keller is a Times columnist and a senior writer for the magazine.

From the *New York Times* Magazine, May 26, 2002, pp. 22, 24-29, 51, 54-55, 57. © 2002 by Bill Keller. Distributed by The New York Times Special Features. Reprinted by permission.

Towards an internet civil defence against bioterrorism

Approaches towards the public-health prevention of bioterrorism are too little, and too late. New information-based approaches could yield better homeland protection. An internet civil defence is presented where millions of eyes could help to identify suspected cases of bioterrorism, with the internet used to report, confirm, and prevent outbreaks.

Lancet Infectious Diseases 2001; **1**: 125–127

Ronald E LaPorte, Francois Sauer, Steve Dearwater, Akira Sekikawa, Eun Ryoung Sa, Deborah Aaron, and Eugene Shubnikov

Despite the large resources put into bioterrorism preparedness,[1] there is little evidence that they work. The West Nile virus outbreak showed how vulnerable our homeland is to attacks. Confusion, error, miscommunication, and misdiagnosis reigned supreme in this near-calamity.[2] By the time the cause was discovered the epidemic was well past its tipping point. Bioterrorists would use more deadly agents, and would spread them over much greater areas, creating extraordinary havoc. As shown by this incident, public health is ill prepared to handle attacks by bioterrorists.

More than US$250 million is to be spent this year to prepare existing US public health systems for bioterrorism. The UK is also spending large amounts of money to thwart bioterrorism. However, even with the increase in funding, the West Nile virus episode and others show how poorly prepared our public health institutions are for an attack. We argue that improvement will not happen by pouring more money into established paradigms of public health, instead new and more effective information technologies need to be tested and introduced. We need to move forward towards a new approach, an internet civil defence to prevent bioterrorism.

We must face the disturbing fact that it is very difficult to predict and guard against a bioterrorist attack because there are too many targets, too many means to penetrate the targets, the occurrences are too rare, and the bioterrorists are too crafty. Pouring more money into a decaying system is like shoring up the Maginot line as the cunning enemy easily slips around our defences. We need to develop a broader, "flatter," and nimbler organisation to combat bioterrorism, with thousands more eyes. The existing hierarchical public health system is not sufficiently agile to compete with bioterrorists.

A tradition of civil defence

The military defence of the US and UK in warfare has always partly been the responsibility of their citizens. We would argue that this is also the case for the war against bioterrorism, where citizens should be the first and most important line of defence. We have the opportunity now to build an internet-based outbreak-prevention network, by connecting citizens to each other on the internet to form a civil defence.

Civil defence and home guard during World War II consisted in part of individuals who were trained to prevent battle damage within their country. Civil defence represented the identification of leaders in the community who would oversee activities during the time of air raids. These were the trusted agents who would guide people to shelters, or help to dig people out from ruined buildings. We knew when the alarm went off above the firehouse that the leaders would be a trusted source of information. During the Cold War of the 1950s and 1960s, civil defence became more coordinated; there were specific guidelines as to where to take people for protection, civilians were trained to watch over the neighbourhoods to prevent looting, and mitigation of panic was an essential component.

Towards the end of the Cold War, civil defence and home guard in our countries started to wane. A primary reason for this was that the arsenals on both sides became so destructive that a war would almost certainly yield mutual annihilation, and any attempt at civil defence would be futile.

However, with the fall of communism the prospect of complete annihilation was lessened. Focal disasters, whether they are manmade (terrorism, nuclear reactors) or infectious (bioterrorism, emerging disease), will continue to plague our citizens, and will be rare and difficult to predict.

As civil defence waned, a new form of home guard grew against an internal enemy, crime. The neighbourhood watch notion (eg, http://www.nwatch.org.uk) developed as a result of increasing incidence and fear of crime, and involves people in communities uniting to watch over each other to deter crime. Is this not what we want for bioterrorism? Why not establish the same principle, whereby neighbours watch out for each other, but do this on the internet with a global health-network neighbourhood watch to act as a deterrent against bioterrorism, and to mitigate damage in the event of an attack. The beauty of such a system is that with the internet, there is a death to distance. Thus bioterrorist experts from Russia could be brought in over the internet within minutes after the identification of an attack. The internet community leaders could provide updated information to others, to alleviate fear, reduce panic, and to guide people to safety.

The electronic matrix

Instead of building an inflexible Maginot line defence as we are now, perhaps we should consider an ever alert, flexible electronic-matrix civil defence as our first line of defence. This public-health system would collect locally information of possible bioterrorist activity, allowing suspicious behaviour to be investigated and planned bioterrorist activity to be stopped. Rather than the present 2000 public health experts worldwide on the outlook for bioterrorism, there would be 20 million brains on internet bioterrorism watch. We thus have millions of citizen watchdogs.

There are many advantages of developing a bioterrorism watch. First, we have many more "eyeballs" on the lookout for bioterrorist activity. Second, the implementation of an internet bioterrorism watch would help to reduce panic (the "War of the Worlds" effect) if a major bioterrorism event happened. What better to alleviate fear than a visit from your brother, or neighbour, or friend who says that everything is all right, because this is what had been heard from the civil-defence-network portal? We believe information from trusted agents; we do not from people we do not know. In this way, the internet civil-defence team acts as a secure information pipeline that can filter accurate information through a matrix of people we trust. Lastly, in the days and weeks following the outbreak, this network will act to bring accurate information to the people who need it the most.

The internet is being used to monitor bioterrorism and emerging diseases. There are many relevant websites; two of the most reliable are ProMed (http://www.fas.org/promed) and WHO's outbreak verification list (http://www.cdc.gov/ncidod/eid/vol6no2/grein.htm). These websites differ from the one we propose because they are either list servers or portals and can be viewed as one-to-many transmissions of bioterrorism information—eg, reports of an attack sent to a list of 10 000 people or posted for millions to see. The proposed internet civil defence is internet community-based with basically a one-to-one transmission of information, and trusted agents spreading the message to each other. Based on the idea of six degrees of separation, we can reach large numbers of people worldwide through interpersonal relationships. List servers or websites are more likely to be fraudulent than a message from your mother that can be verified with a phone call.

The idea of a neighbourhood watch can be more specific—eg, a school watch. Children are well suited to monitoring for bioterrorism, because they are some of the most susceptible targets, they transmit agents readily through the school and family, and they are among the easiest groups to monitor. In many ways they are like the "canary in the coal mine," the earliest indication of an attack. However, there is virtually no monitoring of children in either the US or UK for bioterrorism, but there is for "crime terrorism". The launch of a school watch on the internet could lead to a local-to-global watchdog system being used to guard our children.

The first step for an internet watch clearly needs to be the identification and verification of a possible bioterrorist attack. There will be false positives with this approach, but it is better to be more sensitive than specific when dealing with weapons of mass destruction. The false positive rate can be reduced, however, with certified training programmes based on the experiences of civil-defence systems from past decades.

How would we grow an internet civil-defence network? This would be quite simple, and has been done to a lesser extent with the global health network (http://www.pitt.edu/~super1/), which consists of 4600 academics who are experts in bioterrorism-prevention and the internet. The goal is to identify a core of people who have web access, and who want to protect their communities against bioterrorism. A second example would be to use existing networks of people such as Medscape (http://www.medscape.com), which has over 2 million subscribers in health, of whom 75 000 are in infectious disease. Certification systems similar to those used during World War II could evolve.

There is perhaps evidence that a civil-defence model will work more effectively than the current approaches. On the internet people are networked together looking for viruses—computer viruses—so that at first sight notification is spread worldwide. Many pairs of eyes then try to find the person who developed and unleashed the virus. This system works very well. Another example is the case of ten escaped convicts from Texas who were found as the result of people in the heart of the USA talking together through the internet.

Disinformation, where there is a variety of fraud, hacking, and malevolent cyberwar or infowar activities, presents a considerable drawback in the prevention of bioterrorism, and can happen irrespective of the system that is implemented. However, with an internet civil defence the risk can be lessened by very simple measures. The first is that we warn individuals that the only good information is that from a trusted agent from a select group of perhaps ten friends and family members. Sophisticated bioterrorists could spread misinformation using your father's name but an internet-based civil-defence system, with routine telephone call verification to a relative, would greatly reduce the probability of hacking and causing disinformation to a large percentage of the 20 million in the system.

Ideally, we would reach at least one person in each local community, as with the neighbourhood watch. This person would then set up the model for the internet neighbourhood watch by doing the exact same procedure used in traditional neighbourhood watch, but over the internet. This would create a local area civil-defence watch that, when combined with the other census tracts, and other countries, quickly branches upward to a wider area global watch.

Conclusions

Present public health approaches to the prevention of bioterrorism have too few participants, cannot cover enough contingencies to ensure that bioterrorism is thwarted, and are too costly. Little is done to address the panic that would follow an attack. Bringing the eyes, minds, and ears of citizens of the US, the UK, and the rest of the world to fight bioterrorism offers a unique opportunity to thwart and mitigate outbreaks. The civil defence/home guard model has proven to be effective in our countries. It is likely that the development of an internet civil-defence model would be very effective as first line of defence against bioterrorism. This is not to say that the internet civil-defence programme should replace current approaches, but it should be considered as a new weapon system to enhance our current homeland defence.

References

1. Khan AS, Morse S, Lillibridge S. Public-health preparedness for biological terrorism in the USA. *Lancet* 2000; **356**: 1179–82.

2. Steinhauer J, Miller J. In New York outbreak, glimpse of gaps in biological defenses. *The New York Times* 1999 Oct 11; Sect. Metropolitan desk.

From *The Lancet Infectious Diseases*, September 2001. © 2001 by Elsevier Science LTD. Reprinted by permission.

UNIT 4
North America

Unit Selections

14. **Bush's Revolution**, Ivo H. Daalder and James M. Lindsay
15. **Supremacy by Stealth: Ten Rules for Managing the World**, Robert D. Kaplan
16. **The Watchful and the Wary**, Robert Dreyfuss
17. **Economic Crossroads on the Line**, Michael Grunwald
18. **Canada Links Arrest of 19 to Possible Terrorism Ties**, Clifford Krauss

Key Points to Consider

- Explain why you agree or disagree with the shift to a preemptive doctrine in U.S. strategy, evidenced in the new National Security Strategy released in 2002.
- How long do you believe a substantial U.S. military presence will be required in Iraq to ensure a successful transition to a new political order?
- Do you agree with the thesis that the U.S. military intervention in Iraq would hasten the United States demise as a major world power due to "military overreach"? Why or why not?
- What additional measures could the U.S. and Canada take to increase security along their shared 5,500 mile border?
- If Canada continues to scale back its troop commitments and the amount of money spent on the military, what impact is this likely to have on Canada's ability to fight a war on terrorism?
- Why are many analysts predicting smoother diplomatic relations between Canada and the United States?

 **Links: www.dushkin.com/online/**
These sites are annotated in the World Wide Web pages.

U.S. Department of State
http://www.state.gov/index.cfm

The Henry L. Stimson Center—Peace Operations and Europe
http://www.stimson.org/fopo/?SN=FP20020610372

The North American Institute
http://www.northamericaninstitute.org

After a contested election, the U.S. Supreme Court declared George W. Bush the winner of the 2000 U.S. presidential election. Initially, the new president focused more on domestic priorities, including passage of a tax cut and the formulation of measures to cope with a developing recession. Foreign policy was a distant second priority. Senior Bush officials began implementing several changes in U.S. multilateral commitments that generated sharp criticism at home and abroad.

The terrorist attacks on the World Trade Center in New York and the Pentagon in Washington D.C. dramatically changed the foreign policy landscape. Americans of all political persuasions came together quickly to support the President's "War on Terrorism." For many Americans, September 11, 2001 marked the end of the post-cold war era and the start of a new "war" against global terrorism. As the War on Terrorism became the central organizing principle of America's foreign and defense policies, the Bush administration developed a new National Security Strategy in September 2002. A central principle of the new national strategy is the doctrine of preemption. The preemption principle asserts a unilateral right to intervene military in other nation-states who are pursuing covert weapons of mass destruction programs that are designed to attack Americans.

The claim that Sadam Hussein was continuing to sponsor covert biological and chemical research and development programs that were capable of being scaled up within 48 hours in order to kill thousands of Americans was the primary rationale for the U.S. military invasion of Iraq. The lack of evidence of active weapons of mass destruction programs in Iraq after the war raised new doubts about the merits of the new Bush doctrine of pre-emption. In "Bush's Revolution," Ivo H. Daalder and James M. Lindsay describe how Bush's foreign policies changed American foreign policies. His major foreign policy objectives challenge some of the most highly valued norms of the existing international order and the consequences of Bush's recent actions will be felt for years to come. While Daalder and Lindsay are critical of the United States shift to a unilateral approach under Bush, Robert D. Kaplan, in "Supremacy by Stealth: Ten Rules for Managing the World," supports Bush's national security doctrine. However, as a journalist who spends large amounts of time in developing countries, Robert D. Kaplan offers some rules and tactics of how the United States should operate at a tactical level to manage an unruly world.

While most Americans favored the military intervention into Iraq initially, President Bush saw his popularity decline in public opinion polls by the Spring of 2003 as the number of Americans killed by terrorist attacks in Iraq rose. Since last spring skepticism about the wisdom of a prolonged U.S. military occupation in Iraq has increased both at home and abroad. In response to these criticisms and changing conditions on the ground, the Bush administration changed course and announced that U.S. authorities would turn over political control to an appointed civilian Iraqi governing body by mid-summer and reduce the U.S. military presence. Swift and widespread opposition by various domestic groups in Iraq to the U.S. plan regarding appointment of Iraqi political officials as well as continued attacks on American soldiers and Iraqi police forced the U.S. to change its policy towards Iraq once again. The U.S. government has now abandoned the revised plan for transferring political power and moved to reengage the UN in the transition process. U.S. officials on the ground worked with a broader spectrum of representatives of Iraqi groups to draft a constitution and reach agreement on the contours of an interim government. Several key issues including continuing violence, growing tensions between Sunnis and Shiites, and the status of Kurds remain contentious issues. Even more contentious for many Iraqis is a U.S.-backed proposal calling for representation of women in the future political order and for gender equality under the law.

The capture of Saddam Hussein and the return of UN representation in the negotiation process in Iraq helped to deflect some criticisms of U.S. occupation in Iraq. The U.S. military occupation of Iraq stretched U.S. forces thin. U.S. military forces in Afghanistan are still attempting to maintain order in rural areas while also searching for key al Qaeda leaders, including Bin Laden, in the border area between Afghanistan and Pakistan. After reports that Bin Laden had visited Pakistan in early 2004 to meet with his top operatives, President Bush authorized whatever measures necessary to capture or kill Bin Laden.

The War on Terrorism is the primary rationale used by the Bush administration to propose an indefinite extension of provisions of the controversial National Patriot Act that gives the U.S. federal government broad new powers that may violate certain individual liberties. Robert Dreyfuss describes in "The Watchful and the Wary," how the War on Terrorism has already fundamentally eroded legal guarantees of privacy for Americans by eliminating a 30-year prohibition on the FBI and CIA from spying on Americans. Today, the FBI and CIA are building a massive intelligence network designed to spy on terrorists and on everyday Americans. Dreyfuss details how temporary actions after September 11, 2001 are now a permanent routine aspect of law enforcement. While administration officials argue that these changes are necessary to win the War on Terrorism, critics charge that such moves will erode individual liberties of Americans and thus undermine some of the most cherished rights of Americans.

By the beginning of 2004 with the Democratic primaries for the 2004 presidential elections well underway, much more criticisms were heard about many aspects of the current domestic and national security policies. Uncertainty also increased in an election year about how long the U.S. Congress and American public would be willing to support the increased defense costs associated with the U.S. occupation of Iraq and Afghanistan and the wider War on Terrorism. Budgetary surpluses at the start of the Bush administration had turned into a deficit of 4.2 billion by 2004. White House and Congressional analysts differ over the size of future projected deficits but there was a general agreement that the historical deficit would increase even more since Bush's AY04 budget request did not include a request for funds to cover military operations in Afghanistan and Iran once the extraordinary $87 billion allocations made in 2003 are spent. The budget deficit is due to many factors, including the loss of pubic revenue from a continuing recession and Bush's tax cut programs, as well as the high cost of new military activities abroad and efforts by the new De-

partment of Homeland security to increase the defense preparedness of first line defenders at the state and local level. Some critics now charge that the war on terrorism is likely to accelerate America's decline as a world power because military "overreach" will bust the country's economy.

Much like the United States, Canada immediately declared a war on terrorism in the aftermath of September 11, 2001. Canada committed military forces to the War on Terrorism after the U.S.-led air campaign against al Qaeda began. The day after U.S. and British forces launched their military assault on Afghanistan, Canada announced plans to send ships, planes, and about 2,000 service personnel to Afghanistan. The U.S. and Canadian governments undertook new security measures along the world's longest undefended border. The short-term response was to increase barriers along the 5,500-mile long frontier. However, as Michael Grunwald describes in "Economic Crossroads on the Line," the downside of the immediate security response was to create long delays and costly waits at the border, especially for Canadians. As a longer-term response, both countries agreed to explore ways to rely on technology to create a "smarter border".

Canadian officials are also working hard to identify and detain suspected terrorists within Canada's border. During 2003, Canadian security officials detained 19 individuals who were suspected of being terrorists after they were found taking flying lessons at a school near an Ontario nuclear power plant. However, the war on terrorism is also leading to tensions between the two North American neighbors. Since 2001, Canadian officials have made several official protests to the U.S. government about the current practice of detaining, without charge, Canadian citizens who are suspected of being terrorists.

How long Canada will be able to maintain a military role in the War on Terrorism remains in doubt. By the end of 2002 financial constraints on increasing the Canadian expenditures for defense raised serious questions about whether Canada would be able to continue an activist foreign policy and a long-standing commitment to contribute troops to UN international peace-keeping missions. Since 2000 the number of troops committed to peacekeeping operations has been reduced as part of cost-cutting measures. At the same time, Canada purchased new military hardware and implemented new tax cuts. Neither policy changes were enough to stop a continuing degradation of Canadian force readiness or an exodus of experienced personnel from the intelligence service. After studying the current state of the national forces, a Canadian senate committee recently recommended that the military immediately withdraw all its forces from overseas duty for two years and spend billions more to stop the armed forces from collapsing.

The limited amount of resources that Canada is willing to spend on military forces is a source of long standing tensions between the U.S. and Canada. In recent years, policy disagreements between the two countries were aggravated by the very different political ideologies of the two countries' leaders. The ideological gap between the leaders of the two countries was frequently evident in the rhetoric of senior Canadian officials. At the close of 2002, Canadian Prime Minister Jean Chretien observed that perceived Western arrogance had played a part in the September 11 attacks and warned the U.S. and other wealthy nations against "humiliating" poorer countries. However, key political officials in Canada's Liberal Party spent much of 2003 defending themselves against charges of involvement in a kickback scandal. At the end of 2003, Paul Martin was sworn in as Canada's 21st prime minister. Many analysts predict smoother relations between the U.S. and Canada under his leadership. One of Prime Minister Martin's first speeches called for a mending of relations with the United States and a new "freshness and clarity" to Canada's place in the world.

Article 14

Bush's Revolution

IVO H. DAALDER AND JAMES M. LINDSAY

"At heart, Bush is a revolutionary. Everything he has done in his first 32 months as president shows that he is committed to challenging the existing order. He has been audacious rather than cautious, proactive rather than reactive, risk-prone rather than risk-averse. In his actions as well as his doctrines, he has changed the course of American foreign policy."

George W. Bush had reason to be pleased as he peered down at Baghdad from the window of Air Force One in early June 2003. He had just completed a successful visit to Europe and the Middle East. The trip began in Warsaw, where he had the opportunity to personally thank Poland for being one of just two European countries to contribute troops to the Iraq War effort. He then traveled to Russia to celebrate the 300th birthday of St. Petersburg and to sign the papers formally ratifying a treaty committing Moscow and Washington to slash their nuclear arsenals. He flew on to Évian, a city in the French Alps, to attend a summit meeting of the heads of the world's major economies. He next stopped in Sharm el-Sheik, Egypt, for a meeting with moderate Arab leaders, before heading to Aqaba, Jordan, on the shore of the Red Sea, to discuss the road map for peace with the Israeli and Palestinian prime ministers. He made his final stop in Doha, Qatar, where troops at US Central Command greeted him with thunderous applause. Now Bush looked down on the city that American troops had seized only weeks before. As the president pointed out landmarks below to his advisers, the pilot dipped Air Force One's wings in a gesture of triumph.

Bush's seven-day, six-nation trip was in many ways a victory lap to celebrate America's win in the Iraq War—a war that many of the leaders Bush met on his trip had opposed. But in a larger sense he and his advisers saw it as a vindication of his leadership. The man from Midland, Texas, had been mocked throughout the 2000 presidential campaign as a know-nothing. He had been denounced early in his presidency for turning his back on time-tested diplomatic practices and ignoring the advice of America's friends and allies. Yet here he was traveling through Europe and the Middle East, not as a penitent making amends, but as a leader commanding respect.

As Air Force One flew over Iraq, Bush could believe that he had become an extraordinarily effective foreign policy president. He had dominated the American political scene like few others. He had been the unquestioned master of his own administration. He had gained the confidence of the American people and persuaded them to follow his lead. He had demonstrated the courage of his convictions on a host of issues—abandoning cold war treaties, fighting terrorism, overthrowing Saddam Hussein. He had spent rather than hoarded his considerable political capital, consistently confounding his critics with the audacity of his policy initiatives. He had been motivated by a determination to succeed, not paralyzed by a fear of failure. And, while he had steadfastly pursued his goals in the face of sharp criticism, he had acted pragmatically when circumstances warranted. In the process, Bush had set in motion a revolution in American foreign policy.

This revolution continues today, even in the face of growing challenges and criticism. It is a revolution not in America's foreign policy goals, but in how to achieve them. In his first 32 months in office, Bush has discarded or redefined many of the key principles governing the way the United States should act overseas. He has relied on the unilateral exercise of American power rather than on international law and institutions to get his way. He has championed a proactive doctrine of preemption and de-emphasized the reactive strategies of deterrence and containment. He has promoted forceful interdiction, preemptive strikes, and missile defenses as means to counter the proliferation of weapons of mass destruction, and he has downplayed America's traditional support for treaty-based nonproliferation regimes. He has preferred regime change to direct negotiations with countries and leaders that he loathes. He has depended on ad hoc coalitions of the willing to gain support abroad while ignoring permanent alliances. He has retreated from America's decades-long policy of backing European integration and instead exploited Europe's internal divisions. And he has tried to unite the great powers in the common cause of fighting terrorism while rejecting traditional policies that sought to balance one power against another. By rewriting the rules of America's engagement in the world, the man dismissed throughout his political career as a lightweight has left an indelible mark on politics at home and abroad.

Nevertheless, the revolution that might have seemed promising at its start has with time proved problematic. Even as he peered out the window of Air Force One to look at Baghdad, there were troubling signs of things to come. American troops in Iraq were embroiled in what had all the makings of guerrilla war. Anger had swelled overseas at what was seen as American arrogance and hypocrisy. Several close allies spoke openly about how to constrain the United States rather than how best to work with it. As the president's plane flew home, Washington was beginning to confront a new question: Were the costs of the Bush revolution about to swamp the presumed benefits?

MAKING THE WORLD SAFE FOR AMERICA

What precisely is the Bush revolution in foreign policy? At its broadest level, it rests on two beliefs. The first is that in a dangerous world the best—if not the only—way to ensure America's security is to remove the constraints imposed by friends, allies, and international institutions. Maximizing America's freedom to act is essential because the unique position of the United States makes it the most likely target for any country or group hostile to the West. Americans cannot count on others to protect them because countries inevitably ignore threats that do not involve them. Moreover, formal arrangements restrict the ability of the United States to make the most of its primacy. Gulliver must shed those constraints that he helped the Lilliputians weave.

The other belief is that an America unbound should use its strength to change the status quo in the world. Bush's foreign policy does not propose that the United States keep its powder dry while it waits for dangers to gather. Instead, the Bush philosophy turns John Quincy Adams on his head and argues that the United States should aggressively go abroad to search for monsters to destroy. That was the logic behind the Iraq war, and it animates the administration's efforts to deal with other rogue states.

These fundamental beliefs have important consequences for the practice of American foreign policy. One consequence is a decided preference for unilateral action. Unilateralism is appealing because it is often easier and more efficient, at least in the short term, than multilateralism. Contrast the Clinton administration's 1999 Kosovo war, where Bush and his advisers believed that the task of coordinating the views of all NATO members greatly complicated the war effort, with the US war in Afghanistan under Bush, in which Pentagon planners did not have to subject any of their decisions to foreign approval. This is not to say that Bush flatly rules out working with others. Rather, his preferred form of multilateralism—to be indulged when unilateral action is impossible or unwise—involves building ad hoc coalitions of the willing, or what Richard Haass, an adviser to Colin Powell, has called "à la carte multilateralism."

Second, preemption no longer is a last resort of American foreign policy. In a world in which weapons of mass destruction are spreading and terrorists and rogue states are readying to attack in unconventional ways, Bush argues that "the United States can no longer solely rely on a reactive posture as we have in the past.... We cannot let our enemies strike first." Indeed, the United States should be prepared to act not just preemptively against imminent threats, but also preventively against potential threats. Vice President Dick Cheney was emphatic on this point in justifying the overthrow of Saddam on the eve of the Iraq War. "There's no question about who is going to prevail if there is military action. And there's no question but [that] it is going to be cheaper and less costly to do now than it will be to wait a year or two years or three years until he's developed even more deadly weapons, perhaps nuclear weapons."

Bush has effectively abandoned a decades-long consensus that put deterrence and containment at the heart of American foreign policy.

Third, the United States should use its unprecedented power to carry out regime change in rogue states. The idea of regime change is not new to American foreign policy. The Eisenhower administration engineered the overthrow of Iranian Prime Minister Mohammed Mossadegh in the 1950s; the CIA trained Cuban exiles in the 1960s in a botched bid to oust Fidel Castro; Ronald Reagan channeled aid to the Nicaraguan contras in the 1980s to overthrow the Sandinista government; and Bill Clinton helped Serb opposition forces to remove Slobodan Milosevic in 2000. What is different in the Bush presidency is the willingness, even in the absence of a direct attack on the United States, to use US military forces for the express purpose of toppling other governments. This was the gist of both the Afghanistan and the Iraq wars. Unlike proponents of rollback, who never succeeded in overcoming the argument that their anti-communist policies could lead to World War III, Bush bases his policy on the belief that no one can push back.

THE TRIUMPH OF NEOCONSERVATISM?

Bush has presided over this revolution in foreign policy, but is he responsible for it? Commentators across the political spectrum say no. They give the credit (or blame) to neoconservatives within the administration, led by Deputy Secretary of Defense Paul Wolfowitz, who are determined, it is said, to use America's great power to transform despotic regimes into liberal democracies. One critic calls Bush "the callow instrument of neoconservative ideologues." Another sees a "neoconservative coup" in Washington and wonders if "George W. fully understands the grand strategy that Wolfowitz and other aides are unfolding." Pundits are not the only ones to argue that the Bush revolution represents a neoconservative triumph. "Right now, the neoconservatives in this administration are winning," Democratic Senator Joseph Biden, the ranking member of the Senate Foreign Relations Committee, said in July 2003. "They seem to have captured the heart and mind of the president, and they're controlling the foreign policy agenda."

Article 14. Bush's Revolution

This conventional wisdom is wrong on at least two counts. First, it fundamentally misunderstands the intellectual currents within the Bush administration and the Republican Party more generally. Neoconservatives—who might better be called democratic imperialists—are more prominent outside the administration, particularly in the pages of *Commentary* and the *Weekly Standard* and in the television studios of Fox News, than they are inside it. The bulk of Bush's advisers, including most notably Vice President Cheney and Defense Secretary Donald Rumsfeld, are not neocons. Nor, for that matter, is Bush. They are instead assertive nationalists—traditional hard-line conservatives willing to use American military power to defeat threats to US security but reluctant as a general rule to use American primacy to remake the world in its image.

Although neoconservatives and assertive nationalists differ on whether the United States should actively spread its values abroad, they share a deep skepticism of traditional Wilsonianism, including its commitment to the rule of law and its belief in the relevance of international institutions. They place their faith not in diplomacy and treaties, but in power and resolve. Agreement on this key point has allowed neoconservatives and assertive nationalists to form a marriage of convenience in overthrowing cold war doctrines of deterrence, even as they disagree about what kind of commitment the United States should make to rebuilding Iraq and remaking the rest of the world.

The second and more important flaw of the neoconservative coup theory is that it grossly underestimates Bush. The man from Midland is not a figurehead in someone else's revolution. He may have entered the Oval Office not knowing which general ran Pakistan, but he has been the puppeteer, not the puppet. He has governed as he said he would on the campaign trail. He has actively solicited the counsel of his seasoned advisers, and tolerated if not encouraged vigorous disagreement among them. When necessary, he has overruled them. The president has led his own revolution.

PUTTING THE REVOLUTION IN MOTION

Bush's desire to revolutionize the conduct of American foreign policy existed long before September 11, 2001. But it took the terrorist attacks in New York and Washington to move him to act boldly. Foreign policy, or more precisely, the war on terrorism, became the defining mission of his presidency. Yet, while Bush pledged within days of the attacks to go after "terrorists with global reach," he and his advisers would spend much of the next two years determining precisely what this entailed.

The full extent of Bush's war on terror became apparent when he delivered his first State of the Union address in January 2002. "A terrorist underworld—including groups like Hamas, Hezbollah, Islamic Jihad, Jaish-i-Mohammed—operates in remote jungles and deserts, and hides in the centers of large cities," Bush told Congress and the nation. But that was not all. The threat facing the United States extends beyond these terrorist groups to rogue states such as Iran, Iraq, and North Korea that are bent on acquiring weapons of mass destruction. "States like these, and their terrorist allies, constitute an axis of evil," Bush warned. "By seeking weapons of mass destruction, these regimes pose a grave and growing danger. They could provide these arms to terrorists, giving them the means to match their hatred. They could attack our allies or attempt to blackmail the United States. In any of these cases, the price of indifference would be catastrophic."

Then, using the most dire language heard in any presidential speech since John F. Kennedy's first State of the Union address four decades earlier, Bush declared that the United States could no longer afford to sit and wait until America was struck again. "Time is not on our side. I will not wait on events, while dangers gather. I will not stand by, as peril draws closer and closer. The United States of America will not permit the world's most dangerous regimes to threaten us with the world's most destructive weapons."

The importance of Bush's address lay in clearly identifying a major new threat to the United States: the combination of terrorism, tyrants, and technologies of mass destruction. This new threat is considerably broader in scope than the terrorist groups with global reach that had been the main preoccupation of the anti-terror coalition Bush assembled immediately after 9-11. Yet in highlighting the threat posed by this trinity of evil, Bush said nothing about how he proposed to defeat it. The principal elements of a strategy for dealing with the more expansive peril emerged only in the months following his State of the Union address. The key elements of this strategy, which reflect the administration's hegemonist worldview, are American power and leadership, a focus on rogue states, and the need to act preemptively. A few weeks after Bush's axis of evil speech, Vice President Cheney made clear that responsibility for meeting this threat lay squarely on America's shoulders. "America has friends and allies in this cause," the vice president told a packed gathering at the Council on Foreign Relations, "but only we can lead it. Only we can rally the world in a task of this complexity, against an enemy so elusive and so resourceful. The United States, and only the United States, can see this effort through to victory."

The concern that terrorists might acquire weapons of mass destruction led the administration to focus on states that are able and willing to help terrorists obtain these technologies. Echoing Bush's warning about the axis of evil, Cheney promised that "we will work to prevent regimes that sponsor terror from threatening America or our friends and allies with chemical, biological or nuclear weapons—or allowing them to provide those weapons to terrorists." And Rumsfeld left little doubt about how this goal is to be accomplished. "Defending against terrorism and other emerging twenty-first century threats may well require that we take the war to the enemy," he told faculty and students at the National Defense University days after the president's axis of evil speech. "The best, and in some cases, the only defense, is a good offense."

The administration pulled the main strands of its emerging strategy together in time for Bush's commencement address at West Point on June 1, 2002. Calling for new thinking to match new threats, the commander in chief told the newest generation of soldiers that the old cold war doctrines of deterrence and containment no longer provided a sufficient basis for defending America. "Deterrence—the promise of massive retaliation against nations—means nothing against shadowy terrorist networks with no nation or citizens to defend. Containment is not possible when unbalanced dictators with weapons of mass destruction can deliver those weapons on missiles or secretly provide them to terrorist allies." The United States cannot rely on treaties signed by tyrants. And while homeland and missile defense are clear priorities, "the war on terror will not be won on the defensive." Instead, Bush proclaimed that "we must take the battle to the enemy, disrupt his plans, and confront the worst threats before they emerge. In the world we have entered, the only path to safety is the path of action. And this nation will act." Ultimately, Bush concluded, the nation's "security will require all Americans to be forward-looking and resolute, to be ready for preemptive action when necessary to defend our liberty and to defend our lives."

THE NATIONAL SECURITY STRATEGY

The fullest elaboration of Bush's strategy for defeating the terrifying combination of terrorism, tyrants, and technologies of mass destruction came in the *National Security Strategy*, a document that the White House issues annually at the behest of Congress. Bush released the strategy on September 20, 2002, just as the domestic and international debate on Iraq was heating up. The document, reworked in plain English at Bush's direction so that "the boys in Lubbock" could understand it, offers the most comprehensive statement of the administration's foreign policy.

Consistent with the president's hegemonist worldview, the document puts American power at the strategy's center. This power derives from a combination of America's "unparalleled military strength" and its embodiment of freedom and democracy. "The great struggles of the twentieth century between liberty and totalitarianism ended with a decisive victory for the forces of freedom," Bush wrote in the introduction to the strategy. "In keeping with our heritage and principles, we do not use our strength to press for unilateral advantage." Instead, the goal of American power is to help make the world safe for freedom to flourish. "We will defend the peace by fighting terrorists and tyrants. We will preserve the peace by building good relations among the great powers. We will extend the peace by encouraging free and open societies on every continent." The essence of the Bush strategy, therefore, is to use America's unprecedented power to make the world more convivial to American interests.

Achieving this goal requires the removal of obstacles and threats to liberty and freedom that exist throughout the world— "to create a balance of power that favors human freedom," as Bush put it. To foster "conditions in which all nations and all societies can choose for themselves the rewards and challenges of political and economic liberty," it is necessary to create conditions that will enable people everywhere to choose democracy and free enterprise. The primary obstacle to people making that choice, the strategy document argues, "lies at the crossroads of radicalism and technology," where terrorists and tyrants are determined to acquire technologies of mass destruction. "The United States will not allow these efforts to succeed. We will build defenses against ballistic missiles and other means of delivery. We will cooperate with other nations to deny, contain, and curtail our enemies' efforts to acquire dangerous technologies. And, as a matter of common sense and self-defense, America will act against such emerging threats before they are fully formed." While Washington "will constantly strive to enlist the support of the international community, we will not hesitate to act alone, if necessary, to exercise our right of self-defense by acting preemptively" against the threat confronting the nation and, indeed, the world. "We must be prepared to stop rogue states and their terrorist clients before they are able to threaten or use weapons of mass destruction against the United States and our allies and friends."

The strategy document asserts that, after 9-11, there can be no doubt that terrorists and the rogue states that support them will stop at nothing in their attempts to strike America again. "Today, our enemies see weapons of mass destruction as weapons of choice. For rogue states these weapons are tools of intimidation and military aggression against their neighbors." Such weapons could enable them "to blackmail the United States and our allies to prevent us from deterring or repelling the aggressive behavior of rogue states." Deterrence by threatening retaliation is less likely to work "against leaders of rogue states more willing to take risks, gambling with the lives of their people, and the wealth of their nations." And, of course, "deterrence will not work against a terrorist enemy whose avowed tactics are wanton destruction and the targeting of innocents."

This is why America might have to act preemptively, the strategy argues. "The United States has long maintained the option of preemptive actions to counter a sufficient threat to our national security. The greater the threat, the greater is the risk of inaction—and the more compelling the case for taking anticipatory action to defend ourselves, even if uncertainty remains as to the time and place of the enemy's attack." Of course, force will not have to be used "in all cases to preempt emerging threats, nor should nations use preemption as a pretext for aggression. Yet in an age where the enemies of civilization openly and actively seek the world's most destructive technologies, the United States cannot remain idle while dangers gather."

> Much of the Bush rhetoric—including its justification for the Iraq War—is consistent with the notion of preventive war, not preemption.

Once the grave threat to liberty has been eliminated, it will be possible to extend the peace to every corner of the globe. This, indeed, is both a strategic and a moral imperative. Strategically, "the events of September 11, 2001, taught us that weak states, like Afghanistan, can pose as great a danger to our national interests as strong states. Poverty does not make poor people into terrorists and murderers. Yet poverty, weak institutions, and corruption can make weak states vulnerable to terrorist networks and drug cartels within their borders." Morally, the poverty that grips much of the world offends American values. "A world where some live in comfort and plenty, while half of the human race lives on less than $2 a day, is neither just nor stable."

For all its focus on ways to extend the peace, the core of the Bush strategy is defeating the enemies of freedom. Not only do terrorists and tyrants most threaten the security of America and the world, but the strategy holds that the core values of freedom, democracy, and free enterprise will triumph once these threats have been eliminated. "Americans," Bush told the nation in his 2003 State of the Union address, "are a free people, who know that freedom is the right of every person and the future of every nation. The liberty we prize is not America's gift to the world, it is God's gift to humanity." At the very core of the Bush strategy, then, lies a deeply American assumption about people all over the world: that given the chance, people everywhere will make the same choice Americans have made since gaining independence more than two centuries ago. They will embrace freedom, democracy, and free enterprise.

THE NEW SPARTA

The Bush strategy represents a profound strategic innovation—less in its goals than in the way Bush proposes to achieve them. This is why the doctrine of preemption has become the focal point of discussions about the strategy at home and abroad. After all, Bush has effectively abandoned a decades-long consensus that put deterrence and containment at the heart of American foreign policy. "After September the 11th, the doctrine of containment just doesn't hold any water, as far as I'm concerned," Bush explained in early 2003.

Critics have leveled four complaints against the preemption doctrine. First, many question why the administration decided to make a public statement about something that has long been a US policy option and, in some instances, an actual policy. "It is not clear to me what advantage there is in declaring it publicly," said Brent Scowcroft, national security adviser during the Ford and first Bush administrations. "It has been common knowledge that under some circumstances the United States would preempt. As a declaratory policy it tends to leave the door open to others who want to claim the same right. By making it public we also tend to add to the world's perception that we are arrogant and unilateral." In other words, there is much to lose and little to gain by making the doctrine public—or even turning an option into a policy.

Scowcroft's comment touched on a second objection to the preemption argument: countries may use it as a cover for settling their own national security scores. Days after the strategy's publication, Russia hinted that it might have to intervene in neighboring Georgia to go after Islamic terrorists allegedly hiding in the Pankisi Gorge. India embraced preemption as a universal doctrine. "Every nation has that right," Finance Minister Jaswant Singh said on a visit to Washington days after the strategy's publication. "It is not the prerogative of any one country. Preemption is the right of any nation to prevent injury to itself." But, as Henry Kissinger has suggested, "it cannot be in either the American national interest or the world's interest to develop principles that grant every nation an unfettered right of preemption against its own definition of threats to its security." The strategy recognizes this problem by warning nations not to "use preemption as a pretext for aggression." But the administration has not identified what separates justifiable preemption from unlawful aggression. Without a bright line that can gain widespread adherence abroad, the administration runs the risk that its words will be used to justify ends that it opposes.

Third, critics argue that the Bush strategy suffers from considerable conceptual confusion, which has real policy consequences. Most important, it conflates the notion of preemptive and preventive war. *Preemptive* wars are initiated when another country is clearly about to attack. Israel's decision to go to war in June 1967 against its Arab neighbors is the classic example. *Preventive* wars are launched by states against others before the state being attacked poses a real or imminent threat. "What made war inevitable," the ancient Greek historian Thucydides wrote about the Peloponnesian War, "was the growth in Athenian power and the fear this caused in Sparta." The purpose of initiating war in these circumstances is therefore to stop a threat before it can arise. Israel's strike against Iraq's Osirak reactor in 1981 was one example of preventive war. Cheney's argument that Iraq needed to be struck before it acquired nuclear weapons is another. Much of the Bush rhetoric—including its justification for the Iraq War—is consistent with the notion of preventive war, not preemption. The problem is that, while preemptive wars have long recognized standing in international law as a legitimate form of self-defense, preventive wars do not. Not surprisingly, a resort to preventive war in the case of Iraq has proved highly controversial.

For all the criticism of the Bush strategy's core innovation, the real debate has been a practical one about Iraq rather than a doctrinal one about preemption. In part this reflects the fact that, the rhetoric surrounding the doctrine notwithstanding, Iraq was the driving force behind its promulgation. For all the talk about an axis of evil, it was clear that the administration at least initially would focus squarely on Baghdad. Just two weeks after the president warned about the axis, Secretary of State Colin Powell told Congress, "There is no plan to start a war with these nations," referring to Iran and North Korea. "We want to see a dialogue. We want to contain North Korea's activities with respect to proliferation, and we are going to keep the pressure on them. But there is no plan to begin a war with North Korea; nor is there a plan to begin a conflict with Iran." Yet neither Powell nor any other official had anything reassuring to say about Iraq. On the contrary, Bush left no doubt that he wanted Saddam gone—and sooner rather than later. When US forces, aided by small numbers of British and Australian troops, invaded Iraq in March 2003, they did so over the objections of many key allies and without explicit United Nations authorization.

The Bush administration acts as if the world has entered a postdiplomatic age, in which making speeches or issuing ultimatums takes the place of give-and-take negotiations.

The Revolution's Results

Chinese leader Chou En-lai was once asked what he thought of the French Revolution. "It's too early to tell," he replied. The same could be said of the Bush revolution. Some may have been tempted to call it a smashing success after the impressive military victories against the Taliban and Saddam. But the ultimate question is not whether it worked in the short run, but whether it enhances the security, prosperity, and liberty of the American people in the long run. Are Americans better off with or without the Bush revolution?

The president is hardly alone in understanding that America possesses unrivaled power—especially military power. What makes him revolutionary is his willingness to use it, even over the strenuous objections of America's friends and allies. In the war on terrorism, he has used American power to set the international agenda. In the war against Iraq, he has used it to compel others to follow—or at least to accept—his chosen course. In America's policies toward the Middle East, he has used it to sideline leaders whom America preferred not to deal with, from Mullah Omar and Saddam to Yasir Arafat. In these and other instances, Bush has moved decisively to take the initiative. Rather than debate issues endlessly, he has chosen to act. And his decisions more often than not have reflected his convictions, rather than Washington's conventional wisdom.

Even so, while Bush understands that American muscle can shape events, he has overestimated what the unilateral exercise of its power can achieve. America is not omnipotent. To achieve most of its goals it still requires the cooperation of others. Washington's ability to rally allies to its side depends on identifying and pursuing common interests, not just national ones. Yet, since Bush became president, people around the world have lost trust in the United States, doubting that it has much interest in them or their problems. They fear that an America unbound has taken the tyrant's motto as its own: *Oderint dum metuant*—"Let them hate as long as they fear." And they have become more reluctant to cooperate with Washington. America suddenly faces the possibility that it will end up standing all alone, a great power unable to achieve its most important goals.

From the start, Bush has insisted that the rest of the world be measured by America's standard, not the other way around. This attitude infuses Bush's language, his polite but cursory treatment of other world leaders, and his lack of concern for their interests and their advice. Bush's approach strikes many as an arrogance born of power, not principle. And they resent it deeply.

Since September 11, Bush has painted the world in black and white, while others, particularly overseas, still paint it in shades of gray. He has distinguished between those who are "evil" and those who are "good," between those who are "for us" and those who are "against us," between those who "love freedom" and those who "hate the freedom we love." The war on terrorism is a "crusade." Osama bin Laden had to be found "dead or alive." And Bush was "sick and tired" of the games Saddam played. This rhetoric initially helped galvanize Americans to support the president's assertive and often audacious policies abroad. But it is alien to most foreigners, who, because of America's unquestioned supremacy, comprise as much a part of Bush's audience as the American people do. Not accustomed to the blunt language and locutions of west Texas, many people outside the United States see Bush's words as proof that their views do not matter.

Bush and his advisers have not tried to dispel such perceptions. Instead, they frequently express their contempt for opinions different from their own. When Gerhard Schröder used his opposition to a war against Iraq to squeeze out a narrow reelection victory in Germany, Bush refused to place the customary congratulatory phone call. National security adviser Condoleezza Rice spoke of the "poisoned" state of US-German relations.

Rumsfeld has a particular knack for twisting a knife in open wounds. He dismissed France and Germany as "old Europe" for failing to support the war against Iraq. To punish France for its opposition, he banned high-level US military participation in the annual Paris Air Show, lobbied defense industry executives not to attend the show, disinvited France from a major military exercise, and sought to exclude the chief of the French air staff from a US-hosted conference of air force commanders. He lumped Germany with Cuba and Libya as countries unwilling to help the United States in its war against Iraq. He did so even though, in addition to granting overflight and basing rights, Germany deployed hundreds of troops to Kuwait, where they manned advanced chemical and biological warfare detection vehicles. Hundreds more Germans were in Turkey as part of a NATO commitment to defend the Turks against Iraqi retaliation; thousands more protected US bases in Germany against terrorist attacks. When Rumsfeld traveled to Germany in June 2003, he thanked the Poles, Romanians, and Albanians for contributing to the wars in Iraq and Afghanistan. He said nothing about the larger and far more significant German military contribution to both these efforts.

After 9-11, Bush also made clear that only a country's support for the war on terror mattered much to the United States. Just as he has reoriented America's foreign policy agenda to focus singlemindedly on defeating terrorism, so the president expects every other country to reorient its foreign policy as well. At the same time, the Bush administration acts as if the world has entered a postdiplomatic age, in which making speeches or issuing ultimatums takes the place of give-and-take negotiations. Once America's position is clear, others are expected to follow.

The Arrogance of Power

"If we're an arrogant nation, they'll resent us," Bush observed about other countries during his second presidential debate with Al Gore. "If we're a humble nation, but strong, they'll welcome us." It was a wise observation that Bush and much of his administration somehow forgot. Resentment, not respect, best characterizes how most other countries have reacted to the Bush revolution. Early evidence came during the 2002 elections in Germany and South Korea. In both countries, the results turned on opposition to US policy.

Foreign opinion about America increasingly soured in the weeks leading up to the Iraq war. Among the eight largest European countries polled in March 2003, only in Poland did even 50 percent hold a favorable view of the United States. Britain came in a close second, with 48 percent holding a favorable view. In Italy and France, only a third did; in Germany and Russia only a quarter did; and in Spain and Turkey barely 10 percent did. By the early summer of 2003, with the Iraq debate mostly past, America's image in Europe had recovered slightly—but it still was far less favorable than it had been in 2000 or even 2002.

In much of the rest of the world few people view the United States favorably—and their numbers are dwindling. The antagonism is especially pronounced in the Arab and Islamic world. In Jordan, Indonesia, Morocco, Pakistan, and among the Palestinians, near majorities polled earlier this year believed that bin Laden would do the right thing in world affairs. By contrast, overwhelming majorities said they had no confidence in Bush's leadership.

The president and his advisers appear to worry little about America's unfavorable image abroad. On the contrary, they express surprise at foreign resentment—or even growing fear—of American power. "There were times that it appeared that American power was seen to be more dangerous than, perhaps, Saddam Hussein," Rice told European journalists months after the Iraq War. "I'll just put it very bluntly: We simply didn't understand it." Bush and his advisers are perplexed by hostile foreign reactions to their policies because of their deeply held conviction that America is a uniquely just nation and is seen abroad as being so. "I'm amazed that there is such misunderstanding of what our country is about," Bush said in October 2001. "Like most Americans, I just can't believe it. Because I know how good we are." Bush's worldview simply makes no allowance for others doubting the purity of American motives.

The Bush philosophy turns John Quincy Adams on his head and argues that the United States should aggressively go abroad to search for monsters to destroy.

Confronted with news that his policies stir anger abroad, Bush's reaction has been to insist that he was not elected to do what is popular. When asked in February 2003 about large antiwar protests in England, he responded: "First of all, you know, size of protest, it's like deciding, well, I'm going to decide policy based upon a focus group. The role of a leader is to decide policy based upon the security—in this case, the security of the people." To a point, Bush is right. He cannot run foreign policy as if it were a popularity contest. That does not mean, however, that the United States can afford to ignore how others view it from abroad. Like American presidents, foreign leaders have to take account of their own publics. When those publics oppose Bush's policies, that becomes a problem not just for their leaders, but for Washington as well.

The Iraq experience underscores that *how* America leads matters as much as *whether* it leads. Too often America under Bush has behaved like the "SUV of nations," as the journalist Mary McGrory put it. "It hogs the road and guzzles the gas and periodically has to run over something—such as another country—to get to its Middle Eastern filling station." The cumulative effect of such behavior is substantial. It has angered even America's closest allies, many of whom have come to see their role not as America's partner but as a brake on the improvident exercise of American power. It has weakened their support for American actions. And it has undermined their willingness to cooperate in dealing with those challenges that are common to them all.

Although Bush's imperious style has entailed great costs for American foreign policy, it is not the only shortcoming in his revolution. To be sure, Bush would be wiser to show what the Declaration of Independence called "a decent respect to the opinions of mankind." But more grace by itself would not be enough to allay the fears of friends and allies. The deeper problem is that the fundamental premise of the Bush revolution—that America's security rests on an America unbound—is mistaken.

This premise might be right if the unilateral exercise of American power could achieve America's major foreign policy goals. But the most important foreign policy challenges that America faces—whether defeating terrorism, reversing weapons proliferation, promoting economic prosperity, safeguarding political liberty, sustaining the global environment, or halting the spread of killer diseases—cannot be solved by Washington alone. They require the active cooperation of others.

The question is how best to secure that cooperation. Bush maintains that, far from impeding cooperation, unilateralism will foster it. If the United States leads, others will follow. They will join with America because they share its values and interests. To be sure, some countries might object to how Washington intends to lead. But Bush is convinced they will come around once the benefits of American action become clear.

The flaw in this thinking has become painfully obvious in Iraq. No doubt many countries, including all members of the UN Security Council, shared a major interest in making sure that Iraq did not possess nuclear and other horrific weapons. For most, however, that common interest did not translate into active cooperation in a war to oust Saddam from power—or even into support for such a war. A few countries actively tried to stop the march to war; many others simply sat on the sidelines.

Little has changed since the toppling of Saddam's statue in Firdos Square. Although many countries believe that stabilizing postwar Iraq is vitally important—for regional stability, international security, and their own national safety—they have not rushed to join the reconstruction effort. In September 2003, American troops constituted more than 80 percent of all forces supporting the Iraq operation—at an annual cost to the American taxpayer of more than $50 billion. Britain provides nearly half of the other forces. The remaining foreign contributions are insignificant. Hungary, for instance, is to provide 133 truck drivers. In many cases, countries have agreed to contribute troops only after Washington said it would help pay for them.

NOT FOLLOWING THE LEADER

The lesson of Iraq, then, is that sometimes when you lead, few follow. This, ultimately, constitutes the real danger of the Bush revolution. America's friends and allies might not be able to stop Washington from doing as it wishes, but neither will they necessarily be willing to come to its aid when their help is most needed. Indeed, the more that others question America's power, purpose, and priorities, the less influence America will have. If others seek to counter the United States and delegitimize its power, Washington will need to exert more effort to reach the same desired end—assuming it can reach its objective at all. If others step aside and leave Washington to tackle common problems as it sees fit, the costs will increase. This prospect risks undermining not only what the United States can achieve abroad but also domestic support for its engagement in the world. The American public, always wary of being played for a sucker, might balk at paying the price of unilateralism. Americans might rightly ask, if others are not willing to bear the burdens of meeting tough challenges, why should they? In this respect, an unbound America is a less secure America.

But Bush's way is not America's only choice. In fact, Washington has chosen differently before. When America emerged from World War II as the predominant power in the world, it could have imposed an imperium commensurate with its power—and no one could have prevented it. But Franklin Roosevelt and Harry Truman chose not to. They recognized that American power would be more acceptable and thus more effective and lasting if it were folded into alliances and multilateral institutions that served the interests and purposes of many countries. So they created the United Nations to help ensure international peace and security, set up the Bretton Woods system to help stabilize international economic interactions, and spent vast sums to help rebuild countries (including vanquished foes) that had been devastated by the war. It was not just America's victory in war, but also its magnanimity in peace, that made the twentieth century the American century.

Throughout the cold war, international institutions provided a crucial means to exert America's authority. They bound everyone else into a US-run world order. They in effect constituted what a British journalist called "America's secret empire." Bush has preferred to build his empire on American power alone rather than on the greater power that comes from working with friends and allies. His reliance on military power has proved extraordinarily effective in routing foes, but far less effective in building a lasting basis for peace and prosperity. The United States could decisively defeat the Taliban and Saddam, but rebuilding Afghanistan and Iraq would be better accomplished by working with others. The lesson is clear. Far from demonstrating the triumph of unilateral American power, Bush's wars have demonstrated the importance of basing American foreign policy on a blend of power and cooperation.

By summer 2003, many observers thought they saw signs that the Bush revolution was losing its zeal. "We want multilateral solutions," Rice reassured a European public, even while reminding them that problems like weapons proliferation have to be solved. Iran's and North Korea's nuclear ambitions, though in many ways more threatening and urgent than the dangers posed by Saddam, also are far more difficult problems to tackle unilaterally. With military force all but ruled out, a cooperative effort represents the least-bad option. In August, Bush accepted Powell's advice to seek a UN resolution encouraging other countries to contribute troops and money to the Iraq reconstruction effort. With Bush's public approval ratings falling to levels not seen since before 9-11, and with a presidential election season about to start, the conventional wisdom holds that Bush will now trim his sails rather than engage in daring gambits.

THE PERMANENT REVOLUTION?

But it is wrong to conclude that Bush is abandoning his revolution. This became clear in mid-September when he addressed the UN General Assembly. The speech was vintage Bush—clear, concise, hard-charging, with not an inch of give to his critics. Everyone has to make a choice, he said, and those that make a wrong choice (as the Taliban and Saddam did) must suffer the consequences. He proudly defended his decision to take the fight against terrorism "to the enemy" and admitted no mistakes in postwar planning. And although he called for greater UN involvement in rebuilding Iraq, he limited its role to helping Iraq write a new constitution, train civil servants, and conduct elections. For everything else—including security and returning power to Iraqis—"the coalition" (that is, the United States) would remain in control.

Bush's defiant speech reflected more than just personal pride in his decisiveness, though that is considerable. "I have not looked back on one decision I have made and wished I had made it a different way," he said. "I don't spend a lot of time theorizing or agonizing. I get things done." At heart, Bush is a revolutionary. Everything he has done in his first 32 months as president shows that he is committed to challenging the existing order. He has been audacious rather than cautious, proactive rather than reactive, risk-prone rather than risk-averse. In his actions as well as his doctrines, he has changed the course of American foreign policy. The consequences of the Bush revolution will be felt for years to come.

IVO H. DAALDER *is a senior fellow in Foreign Policy Studies at the Brookings Institution.* JAMES M. LINDSAY *is vice president and director of studies at the Council on Foreign Relations. Their most recent book is* America Unbound: The Bush Revolution in Foreign Policy *(Brookings Institution Press, 2003), from which this essay is adapted.*

Article 15

Supremacy by Stealth

It is a cliché these days to observe that the United States now possesses a global empire—different from Britain's and Rome's but an empire nonetheless. It is time to move beyond a statement of the obvious. Our recent effort in Iraq, with its large-scale mobilization of troops and immense concentration of risk, is not indicative of how we will want to act in the future. So how should we operate on a tactical level to manage an unruly world? What are the rules and what are the tools?

BY ROBERT D. KAPLAN

In the late winter of 2003, as the United States was dispatching tens of thousands of soldiers to the Middle East for an invasion of Iraq, the U.S. Army Special Operations Command was deployed in sixty-five countries. In Nepal the Special Forces were training government troops to hunt down the Maoist rebels who were terrorizing that nation. In the Philippines they were scheduled to increase in number for the fight against the Abu Sayyaf guerrillas. There was also Colombia—the third largest recipient of U.S. foreign aid, after Israel and Egypt, and the third most populous country in Latin America, after Brazil and Mexico. Jungly, disease-ridden, and chillingly violent, Colombia is the possessor of untapped oil reserves and is crucially important to American interests.

The totalitarian regimes in Iraq and North Korea, and the gargantuan difficulty of displacing them, may have been grabbing headlines of late, but the future of military conflict—and therefore of America's global responsibilities over the coming decades—may best be gauged in Colombia, where guerrilla groups, both left-wing and right-wing, have downplayed ideology in favor of decentralized baronies and franchises built on terrorism, narcotrafficking, kidnapping, counterfeiting, and the siphoning of oil-pipeline revenues from local governments. FARC (Fuerzas Armadas Revolucionarias de Colombia), for example, is Karl Marx at the top and Adam Smith all the way down the command chain. Guerrilla warfare is now all about business, and physical cruelty knows no limits. It extends to torture (fish hooks to tear up the genitals), gang rape, and the murder of children whose parents do not cooperate with the insurgents. The Colombian rebels take in hundreds of millions of dollars annually from cocaine-related profits alone, and have documented links to the Irish Republican Army and the Basque separatists (who have apparently advised them on kidnapping and car-bomb tactics). If left unmolested, they will likely establish strategic links with al Qaeda.

Arauca province, a petroleum-rich area in northeastern Colombia, near the Venezuelan border, is a pool-table-flat lesion of broadleaf thickets, scrap-iron settlements, and gravy-brown rivers. The journey from the airfield to the Colombian army base, where a few dozen Green Berets and civil-affairs officers and their support staff are bunkered behind sandbags and concertina wire, is only several hundred yards. Yet U.S. personnel make the journey in full kit, inside armored cars and Humvees with mounted MK-19 40mm grenade launchers. As I stepped off the tarmac in late February, two Colombian soldiers, badly wounded by a car bomb set off by left-wing narcoterrorists (the bomb had been coated with human feces in hopes of causing infection), were being carried on stretchers to the base infirmary, where a Special Forces medic was waiting to treat them. The day before, the Colombian police had managed to deactivate two other bombs in Arauca. The day before that there had been an assassination attempt on a local politician. And the day before that an electricity tower had been bombed, knocking out power in the region. Previous days had brought the usual roadside kidnappings, street-corner bicycle bombings, grenade strikes on police stations, and mortar attacks on Colombian soldiers—using propane cylinders packed with nails, broken glass, and feces.

As we drove through Arauca's mangy streets in a Special Forces convoy, every car and bicycle seemed potentially deadly. Yet the U.S. troops there are defiant, if frustrated. The U.S. government permits them only to train, rather than fight alongside, their Colombian counterparts, but they want the rules of engagement loosened. After a truck unexpectedly pulled out into the street, slowing our convoy and causing us to scan rooftops and parked vehicles (and causing me to sweat more than usual in the humid and fetid atmosphere), a Green Beret with experience on several continents leaned over and said, "If five firemen get killed fighting a fire, what do you do? Let the building burn? I wish people in Washington would totally get Vietnam out of their system."

Back at the base, Major Mike Oliver and Captain Carl Brosky, civil-affairs specialists who between them have served in the Balkans, Africa, and several Latin American countries, were spending the day chasing down two containers of equipment for Arauca's schools and hospital that had been held up in customs at the Venezuelan border. A week earlier, at Tolomeida, several hundred miles south, I had watched Sergeant Ivan Castro, a Puerto Rican from Hoboken, New Jersey, as he patiently taught Colombian soldiers how to sit in a 360-degree "cigar formation" while on reconnaissance, in order to rest in the field without being surprised by the enemy. Later he taught them how to peel back in retreat, without a gap in fire, after making first contact with the enemy. Castro worked twelve hours in the heat that day, speaking in a steady, nurturing tone, working with each soldier until the whole unit performed the drills perfectly.

Even as America's leaders deny that the United States has true imperial intentions, Colombia—still so remote from public consciousness—illustrates the imperial reality of America's global situation. Colombia is only one of the far-flung places in which we have an active military presence. The historian Erich S. Gruen has observed that Rome's expansion throughout the Mediterranean littoral may well have been motivated not by an appetite for conquest per se but because it was thought necessary for the security of the core homeland. The same is true for the United States worldwide, in an age of collapsed distances. This American imperium is without colonies, designed for a jet-and-information age in which mass movements of people and capital dilute the traditional meaning of sovereignty. Although we don't establish ourselves permanently on the ground in many locations, as the British did, reliance on our military equipment and the training and maintenance that go along with it (for which the international arms bazaar is no substitute) helps to bind regimes to us nonetheless. Rather than the mass conscription army that fought World War II, we now have professional armed forces, which enjoy the soldiering life for its own sake: a defining attribute of an imperial military, as the historian Byron Farwell noted in *Mr. Kipling's Army* (1981).

The Pentagon divides the earth into five theaters. For example, at the intersection of 5° latitude and 68° longitude, in the middle of the Indian Ocean, CENTCOM (the U.S. Central Command) gives way to PACOM (the Pacific Command). At the Turkish-Iranian border it gives way to EUCOM (the European Command). By the 1990s the U.S. Air Force had a presence of some sort on six of the world's continents. Long before 9/11 the Special Forces were conducting thousands of operations a year in a total of nearly 170 countries, with an average of nine "quiet professionals" (as the Army calls them) on each mission. Since 9/11 the United States and its personnel have burrowed deep into foreign intelligence agencies, armies, and police units across the globe.

Precisely because they foment dynamic change, liberal empires—like those of Venice, Great Britain, and the United States—create the conditions for their own demise. Thus they must be especially devious. The very spread of the democracy for which we struggle weakens our grip on many heretofore docile governments: behold the stubborn refusal by Turkey and Mexico to go along with U.S. policy on Iraq. Consequently, if we are to get our way, and at the same time to promote our democratic principles, we will have to operate nimbly, in the shadows and behind closed doors, using means far less obvious than the august array of power displayed in the air and ground war against Iraq. "Don't bluster, don't threaten, but quietly and severely punish bad behavior," says Eliot Cohen, a military historian at the Johns Hopkins School of Advanced International Studies, in Washington. "It's the way the Romans acted." Not just the Romans, of course: "Speak softly and carry a big stick" was Theodore Roosevelt's way of putting it.

We can take nothing for granted. A hundred years ago the British Navy looked fairly invincible for all time. A world managed by the Chinese, by a Franco-German-dominated European Union aligned with Russia, or by the United Nations (an organization that worships peace and consensus, and will therefore sacrifice any principle for their sakes) would be infinitely worse than the world we have now. And so for the time being the highest morality must be the preservation—and, wherever prudent, the accretion—of American power.

The purpose of power is not power itself; it is the fundamentally liberal purpose of sustaining the key characteristics of an orderly world. Those characteristics include basic political stability; the idea of liberty, pragmatically conceived; respect for property; economic freedom; and representative government, culturally understood. At this moment in time it is American power, and American power only, that can serve as an organizing principle for the worldwide expansion of a liberal civil society. As I will argue below, the United States has acquired this responsibility at a dangerous and chaotic moment in world history. The old Cold War system, for half a century the reigning paradigm in international affairs, is obviously defunct. Enlarging the United Nations Security Council, as some suggest, would make it even harder for that body to achieve consensus on anything remotely substantive. Powers that may one day serve as stabilizing regional influences—India and Russia, China and the European Union—are themselves still unstable or unformed or unconfident or illiberal. Hundreds of new and expanding international institutions are beginning to function effectively worldwide, but they remain fragile. Two or three decades hence conditions may be propitious for the emergence of a new international system—one with many influential actors in a regime of organically evolving interdependence. But until that time arrives, it is largely the task of the United States to maintain a modicum of order and stability. We are an ephemeral imperial power, and if we are smart, we will recognize that basic fact.

The "American Empire" has been discussed ad nauseam of late, but practical ways of managing it have not. Even so, the management techniques are emerging. While realists and idealists argue "nation-building" and other general principles in Washington and New York seminars, young majors, lieutenant colonels, and other middle-ranking officers are regularly making decisions in the field about how best to train Colombia's army, which Afghan tribal chiefs to support, what kind of coast guard and special forces the Yemeni government requires, how the Mongolians can preserve their sovereignty against Chinese and Russian infiltration, how to transform the Romanian military into a smaller service along flexible Western command lines, and so forth. The fact is that we trust these people on the ground to be keepers of our values and agents of our imperium, and to act without specific instructions. A rulebook that does not make sense to them is no rulebook at all.

The following rules represent a distillation of my own experience and conversations with diplomats and military officers I have met in recent travels on four continents, and on military bases around the United States.

RULE NO. 1
PRODUCE MORE JOPPOLOS

When I asked Major Paul S. Warren, at Fort Bragg, North Carolina, home of the Army's Special Operations Command, what serves as the model for a civil-affairs officer within the Special Operations forces, he said, "Read John Hersey's *A Bell for Adano*—it's all there." The hero of Hersey's World War II novel is Army Major Victor Joppolo, an Italian-American civil-affairs officer appointed to govern the recently liberated Sicilian town of Adano. Joppolo is full of resourcefulness. He arranges for the U.S. Navy to show local fishermen which parts of the harbor are free of mines, so that they can use their boats to feed the town. He finds a bell from an old Navy destroyer to replace the one that the Fascists took from the local church and melted down for bullets. He countermands his own general's order outlawing the use of horse-drawn carts, which the town needs to transport food and water. He goes to the back of a line to buy bread, to show Adano's citizens that although he is in charge, he is their servant, not their master. He is the first ruler in the town's history who doesn't represent a brute force of nature. In Hersey's words,

> [Men like Joppolo are] *our future in the world. Neither the eloquence of Churchill nor the humanness of Roosevelt, no Charter, no four freedoms or fourteen points, no dreamer's diagram so symmetrical and so faultless on paper, no plan, no hope, no treaty—none of these things can guarantee anything. Only men can guarantee, only the behavior of men under pressure, only our Joppolos.*

One good man is worth a thousand wonks. As *The Times of India* wrote on July 7, 1893, the mind of a sharp political agent should not be "crowded with fusty learning." Ian Copland, a historian of the British Raj, wrote that "extroverts and sporting types, sensitive to the cultural milieu," were always necessary to win the confidence of local rulers. In Yemen recently I observed a retired Special Forces officer cementing friendships with local sheikhs and military men by handing out foot-long bowie knives as gifts. In a world of tribes and thugs manliness still goes a long way.

The right men or women, no matter how few, will find the right hinge in a given situation to change history. The Spartans turned the tide of battle in Sicily by dispatching only a small mission, headed by Gylippus. His arrival in 414 B.C. kept the Syracusans from surrendering to the Athenians. It broke the Athenian land blockade of Syracuse, rallied other Sicilian city-states to the cause, and was crucial to the defeat of the Athenian fleet the following year. The United States sent a similarly small mission to El Salvador in the 1980s: never more than fifty-five Special Forces trainers at one time. But that was enough to teach the Salvadoran military to confront more effectively the communist guerrillas while beginning to transform itself from an ill-disciplined constabulary force into something much closer to a professional army.

"You produce a product and let him loose," explains Sidney Shachnow, a retired Army major general. "The Special Forces that dropped in to help [the Afghan warlord Abdul Rashid] Dostum, the guys who grew beards, got on horses, and dressed up like Afghans, were not ordered to do so by Tommy Franks. These were decisions they made in the field."

Shachnow himself is a perfect example of the kind of man he describes. Hard and chiseled, he calls to mind Ligustinus, a Roman centurion who spent nearly half his life in the Army—in Spain, Macedonia, and Greece—and was cited for bravery thirty-four times. Shachnow is a Holocaust survivor. Born in 1933, in Lithuania, he endured a Nazi concentration camp as a boy; emigrated to Salem, Massachusetts; joined the Army as a private out of high school; after reaching the rank of sergeant first class attended officers' training school; and served two combat tours in Vietnam, where he was wounded twice. He rose to be a two-star general and a guiding light of the Special Forces. His success resulted from decisions made on instinct and impulse, and from an ability to take advantage of cultural settings in which he did not naturally fit—exactly the ability that U.S. trainers and commandos in El Salvador, Afghanistan, and so many other places have had to possess.

"A Special Forces guy," Shachnow told me, "has to be a lethal killer one moment and a humanitarian the next. He has to know how to get strangers who speak another language to do things for him. He has to go from knowing enough Russian to knowing enough Arabic in a few weeks, depending on the deployment. We need people who are cultural quick studies." Shachnow was talking about a knack for dealing with people, almost a form of charisma. The right man will know how to behave in a given situation—will know how to find things out and act on them.

RULE NO. 2
STAY ON THE MOVE

Xenophon's Greek army cut through the Persian Empire in 401 B.C., with the troops freely debating each step. We should be mobile in the same way—get bogged down militarily nowhere, but make sure we have military access everywhere. Because we have to manage a world in which—as always—old regimes periodically crumble, disaster lies in becoming too deeply implanted in more than a handful of countries at once.

Here our provincialism helps. As Hayward S. Florer, a retired Special Forces colonel, told me, "Even our Special Ops people are insular. Sure, we like the adventure with other cultures, learning the history and language. But at heart many of us are farm boys who can't wait to get home. In this way we're not like the British and French. Our insularity protects us from becoming colonials."

Colonialism is in part an outgrowth of cosmopolitanism, the intellectual craving to experience different cultures and locales; it leads, inexorably, to an intense personal involvement in their fate. "We want an empire not of colonies or protectorates but of personal relationships," a Marine lieutenant colonel at Camp Pendleton, in California, told me. "We back into deployments. There doesn't need to be a policy directive from the Pentagon—half the time we don't know what the policy is. We get a message from a Kenyan or Nigerian officer who studied here that his unit needs training. We try to do it. We help decide, based on our needs in a region, who we want to help out." The U.S. military is constantly doing favors for other militaries, favors we call in when we need to. This is how we sometimes get access to places. The formal base rights that we have in forty countries may in the future be less significant than the number of friendships maintained between U.S. officers and their foreign counterparts. With that in mind, the military needs to establish a formal data system for tracking such relationships. At present the method of keeping abreast of these crucial ties is largely anecdotal.

The best tools of access are the so-called "iron majors," a term that really refers to all mid-level officers, from noncommissioned master sergeants and chief warrant officers to colonels. In a sense majors run our military establishment, regardless of who the Secretary of Defense happens to be. Up through the rank of captain an officer hasn't closed the door on other career options. But becoming a major means you've "bought into the corporation," explains Special Forces Major Roger D. Carstens. "We're the ones who are up at four A.M. answering the general's e-mails, making sure all the systems are go."

The United States has set up military missions throughout the formerly communist world, creating situations in which U.S. majors, lieutenant colonels, and full colonels are often advising foreign generals and chiefs of staff. Make no mistake: these officers are policymakers by another name. A Romanian-speaking expert on the Balkans, Army Lieutenant Colonel Charles van Bebber, has become well known in top military circles in Bucharest for helping to start the reform process that led to Romania's integration with NATO. Such small-scale but vital relationships give America an edge there over its Western European allies. One of the reasons that countries like Romania and Bulgaria supported the U.S. invasion of Iraq is that they now see their primary military relationship as being with America rather than with NATO as such.

In formerly communist Mongolia, U.S. Army Colonel Tom Wilhelm, a fluent speaker of Russian who studied at Leningrad State University, is an adviser to the local military. With Wilhelm's help, Mongolia has reoriented its defense strategy toward international peacekeeping—as a means of gaining allies in global forums against its rapacious neighbors, Russia and China. The planned dispatch of a Mongolian contingent to help patrol postwar Iraq was the result of what one good man—in this case, Wilhelm—was able to accomplish on the ground. I recently followed him around on an inspection tour of Mongolia's Gobi Desert border with China. We slept in local military outposts, rode Bactrian camels, and spent hours in conversation with mid-level Mongolian officers over meals of horsemeat and camel's milk. It is through such activities that relationships are built and allies are gained in an era when anyplace can turn out to be strategic.

RULE NO. 3
EMULATE SECOND-CENTURY ROME

Provincialism is the aspect of our national character that will keep the United States from overextending itself in too many causes. But owing to the wave of immigration from Asia, the Middle East, and Latin America that began in the 1970s, the United States is an international society comparable to Rome in the second century A.D., when the empire reached its territorial zenith under Trajan and, more important, was granting citizenship to elites in the Balkans, the Middle East, and North Africa. (Trajan and Hadrian, in fact, were both from Spain.) Our military, intelligence, and diplomatic communities must now turn to our Iranian-, Arab-, and other hyphenated Americans—our potential Joppolos. At a time when we desperately need more language specialists, it is shameful that we are seeking out so few of the many native speakers at our disposal. The financial incentives we offer them are simply insufficient, and the waiting period for security clearance has become farcically long. This situation has been changing of late for the better: it needs to continue to do so.

Trained area specialists are likewise indispensable. In 1976 Secretary of State Henry Kissinger entrusted the eminent Arabist and diplomat Talcott Seelye, in Lebanon, to carry out two discreet evacuations of American citizens from that war-torn country with the help of the Palestine Liberation Organization—which we did not recognize at the time. Seelye, who was born in Beirut, may not have wholly agreed with Kissinger's foreign policy—but that didn't matter. He knew how to get the job done. The fact that Arabists and other area specialists may be emotionally involved, through marriage or friendship, with host countries—often causing them to dislike the policies that Washington orders them to execute—can actually be of benefit, because it gives them credibility with like-minded locals. In any case, such tensions between policymakers and agents in the field are typical of imperial systems. We should not be overly concerned about them.

True, comparison is the beginning of all serious scholarship, and area experts are ignorant of much outside their favored patch of ground. Their knowledge of the current reality in a given country is so prodigious that they often cannot imagine a different reality. That is why area experts can say what is going on in a place, but cannot always say what it means. Still, it is impossible to implement any policy without them, as Kissinger and others learned.

Colonel Robert Warburton, the Anglo-Afghan who established the Khyber Rifles regiment on the Northwest Frontier of British India in 1879, was one kind of person needed to manage our interests in distant corners of the world. Warburton spoke fluent Pashto and Persian, and was at home among both aristocratic Englishmen and Afridi tribesmen. The normally cruel and perfidious Afridis held him in such high esteem that he did not need to go armed among them. Warburton was less a cosmopolitan than a nuts-and-bolts journeyman, whose linguistic skills came from birth and circumstance more than from intellectual curiosity. The American equivalents of Warburton can be found among Arab-Americans posted to Central Command and Latino-Americans posted to Southern Command—people who fit into places like Yemen and Colombia, but who want only to return to their suburban American homes afterward.

Southern Command, in particular, is full of Spanish-speaking noncommissioned officers: ethnic Mexicans, Dominicans, Cubans, and Puerto Ricans. The relative shortage of speakers of Arabic and other languages in the rest of the military indicates that in the Special Forces, at least, languages may soon have to be recognized as an "occupational skill"—like weaponry, communications, battlefield medicine, engineering, and intelligence, one in which every noncommissioned officer must spend a year specializing. If each Special Forces unit had a couple of officers who were fluent in several languages spoken in the theater command (Arabic, Persian, and Turkish in CENTCOM, for example), our ability to project power would dramatically increase.

The forward basing of area commands is another strategy that would encourage area expertise and language skills. In the years to come we should consider moving Central Command headquarters from Tampa, Florida, to the Middle East, and Southern Command headquarters from Miami back to Panama, where it was until 1997. There is simply no substitute for being in the region when it comes to absorbing language and culture. As a journalist, I have found that in my profession people on location always have better instincts for the local situation than people back in the United States, even if they don't always draw the proper conclusions. Many a mid-level officer has told me that the same holds true in the military.

RULE NO. 4
USE THE MILITARY TO PROMOTE DEMOCRACY

In an age of expanding democracy, military and intelligence contacts are more important than ever. Civilian politicians in weak and fledgling parliamentary systems come and go. But leading military and security men remain as behind-the-scenes props, sometimes even getting themselves elected to high office—as has happened in Nigeria, Venezuela, and Russia. "Whoever the President of Kenya is, the same group of guys run their special forces and the President's bodyguards," one Army Special Operations officer told me. "We've trained them. That translates into diplomatic leverage."

The U.S. military's bilateral relationships with foreign armies and their officer corps play a substantial role in safeguarding democratic transitions. Militaries have been the pillars of so many Third World societies for so long that the advent of elections can scarcely make them politically irrelevant, especially in Africa and Latin America. In some places, such as Turkey and Pakistan, the military and security services have at times actually enjoyed a reputation for greater liberalism than the civilian authorities. In Colombia in the mid-1990s the civilian government was tainted by drug money; the military police, who were seen to be less corrupt, helped to save our bilateral relationship.

U.S. security-assistance programs also professionalize foreign militaries, thus helping to prevent coups and to improve the human-rights climate. In the 1980s in El Salvador, Colonel J. S. Roach, a member of the operational planning team there, observed that "the Salvadoran military understood they weren't supposed to violate human rights, but they believed they were driven to extreme measures by extreme circumstances." One can debate what members of El Salvador's military "understood," but Roach's team and others pounded home the point that violating human rights almost never makes sense from a pragmatic perspective, because it costs the military the civilian support so necessary to rooting out guerrilla insurgents. "Human rights wasn't a separate one-hour block at the beginning of the day," Roach said. "You had to find a way to couch it in the training so that it wasn't just a moralistic approach." Human-rights abuses didn't come to an end in El Salvador, but observers agree that they were sharply curbed.

The world is a gritty, messy place, and there are no perfect solutions. But the fact is that Third World military men are more likely to listen to American officers who brief them about human rights as a tool of counterinsurgency than to civilians who talk about universal principles of justice. At any rate, it isn't only civilians who talk about universal principles: mid-level officers from around the world are regularly sent to Fort Benning, Georgia, for training in the history and necessity of protecting human rights. (The protestors who perennially chain themselves to the gates of Fort Benning, calling its previously named School of the Americas the "School of Torturers," are implicitly championing the worst possible strategy if they want Latin armies to take human rights seriously—a strategy of isolation, which cuts foreign officers off from American society and values.)

In fact, in places where democracy is especially weak (Peru and Indonesia are obvious examples), a phone call from a U.S. general to a local officer will often advance diplomacy (and also civil society) more effectively than a phone call from the ambassador. Particularly in previously hostile areas, such as the ex-Soviet Caucasus and Central Asia, new diplomatic relationships are being eased by the U.S. military's training of border guards and security services. In other places, as in Chile a decade ago, the resumption of a bilateral military relationship with the United States cements a successful democratic transition.

The much larger truth is that the very distinction between our civilian and military operations overseas is eroding. In 1994 two Special Forces officers helped the Paraguayan government to craft new laws just after Paraguay's constitution was adopted. The U.S. military will increasingly churn out such chameleons: operatives who combine the traits of soldier, intelligence agent, diplomat, civilian aid worker, and academic. And at the same time that our uniformed officers are acting more like diplomats, our diplomats, particularly our ambassadors, are acting more like generals. It is under the State Department's auspices, not the Pentagon's, that helicopters are leased to the Colombian army to fight narcoterrorists and that a campaign is waged to track small planes suspected of transporting cocaine in the Colombia-Peru-Ecuador region. America's war against narcoterrorists in Colombia has two overseers: General James T. Hill, head of Southern Command, and Anne Patterson, the ambassador to Colombia.

The model for our future diplomats might be Deane Hinton, who oversaw the counterinsurgency operation as the ambassador to El Salvador in the early 1980s and then oversaw U.S. efforts to arm Afghan guerrillas as the ambassador to Pakistan from 1983 to 1987. In both those cases a military strategy would have been unavailing in the absence of a successful "interagency" strategy, which backed diplomatic initiatives and humanitarian aid packages with the power of a cocked gun. The same will be true in Colombia and in al Qaeda-infested Yemen. At the moment "interagency" is a dirty word among many in the field, connoting overlapping bureaucracies with conflicting agendas. But a supple and flexible civilian-military chain of command is an immensely useful tool.

Of course, in violent and chaotic parts of the world such as Afghanistan and Yemen, it is only natural that the soldier will at first be more conspicuous than the Peace Corps worker. Because parts of Yemen have become too dangerous for American civilians, the U.S. military is training the Yemeni military to better project power in the tribal badlands, so that, among other things, our foreign-aid personnel can return there. In Central and South America the U.S. military regularly vaccinates farm animals and treats them for diseases, and the villagers are not less grateful than they would be if the help came from civilians. The same was true with Mongolians treated by a four-person Air Force dental mission dispatched recently by Pacific Command to the Mongolian-Chinese border. The Air Force officers treated eighty-five local inhabitants the day I was there, and also handed out toys to the children. It is the efficacy of a humanitarian mission that morally sanctifies it; not whether it is carried out by civilians or soldiers. And if it serves U.S. interests as this one did—so much the better.

RULE NO. 5
BE LIGHT AND LETHAL

Economy of force—doing the most with the least—has been an imperative of the U.S. military, diplomatic, and intelligence communities since the beginning of the Cold War. It will become even more important as our resources are stretched. Here we can learn a great deal from the history of U.S. policy in Latin America over the past several decades: although many journalists and intellectuals have regarded this policy as something to be ashamed of, the far more significant, operational truth is that it exemplifies how we should act worldwide in the foreseeable future.

With Europe the principal Cold War battleground, and Asia the secondary front because of the threat posed by Communist China and North Korea, Latin America took a back seat for decades. The U.S. military had to make do with limited resources while operating in a vast continent. It succeeded thanks to unconventional warfare, which helped the host governments do the real work. In practice that meant aggressive intelligence operations and Special Forces training of local units, combined with domineering diplomacy.

The results were not always pretty and, frankly, not always moral—consider what occurred in Chile in the aftermath of the 1973 coup against Salvador Allende Gossens. Yet for a relatively small investment of money and manpower the United States defeated a belligerent Soviet and Cuban campaign at its back door while paving the way for the democratic transitions and market liberalizations of the 1980s and 1990s. Our "quiet professionals" helped to hunt down and kill the hemispheric agitator Ernesto "Che" Guevara in Bolivia in 1967. Fifty-five Special Forces trainers in El Salvador accomplished more than did 550,000 soldiers in Vietnam. A four-member Special Forces "mobile training team" convinced the Salvadoran police that rather than shooting leftist demonstrators at rallies, they should provide escape routes for the protesters to run away. That turned out to be the most effective kind of human-rights policy.

Economy of force in Latin America produced regimes that in almost every case were better than what the Cubans and the Russians offered. Even in Chile, despite the iniquities of the dictator Augusto Pinochet Ugarte, who took power following Allende's overthrow, the military regime lowered the infant mortality rate from seventy-nine to eleven per 1,000 births and reduced the poverty rate from 30 percent to 11 percent. Privatization gave post-Allende Chile Latin America's only economy comparable to those of the "Asian tigers." America's no-frills molding of political reality in the Western Hemisphere did not require the approval of the UN Security Council, and it did not run the risk of quagmire. There were usually few Americans on the ground in any one Latin country.

Economy of force offers a logic appropriate to an intractable world. Becoming implanted in more than a handful of countries at once spells disaster. And everyone—humanitarian interventionists included—now admits that nation-building, whether in Bosnia, Afghanistan, or Colombia, is fraught with danger, difficulty, and great expense. We shouldn't try to fix a whole society; rather, we should identify a few key elements in it, and fix them.

For example: Because a national army is essentially unreformable without wholesale social and cultural change, we should work to improve only its elite units, using trainers from the U.S. military elite. When it comes to military operations, specialized units should concentrate on the most critical targets; in Colombia, for instance, these would be the 150 or so hydrochloride laboratories throughout Colombia that refine cocaine into its final form. And because individual leaders affect history as much as large social forces do, our efforts should be invested primarily where current leaders seem particularly talented and determined. (Alvaro Uribe Vélez, the President of Colombia, is by all accounts a dynamic workaholic, however embattled, committed both to protecting human rights and to eliminating rogue forces. Were Andrés Pastrana, Uribe's less forceful predecessor, still Colombia's leader, it is doubtful that the United States would be making quite the effort it is in a place like Arauca.)

The most obvious tool to carry out an economy-of-force strategy is the Special Forces, which, as Lieutenant Colonel Kevin A. Christie told me, can perform the military equivalent of "arthroscopic surgery." Relatively small numbers of Special Forces and Marines can maximize U.S. influence in a large number of countries without risking what the Yale historian Paul Kennedy has called imperial "overstretch." Nevertheless, we shouldn't get carried away. A big increase in the number and use of Special Forces could make them less special, and therefore less effective.

A less obvious resource is the Coast Guard, which handles most anti-terrorism and drug-interdiction efforts at sea. Even in the jet-and-information age 70 percent of intercontinental cargo travels by ship, making the seas as strategic as ever. The U.S. Coast Guard, with 38,000 in its active ranks, is the world's seventh largest navy. In Colombia, which has more miles of navigable river than of passable roadway, the Coast Guard has been essential in drug-patrol training. In Yemen, Bob Innes, a retired Coast Guard captain who worked for many years in Colombia, is building a coast guard to prevent more al Qaeda attacks on oil tankers. Our strategy in Colombia and Yemen is unspoken but simple: establish not a totally reformed military but a self-sustaining structure of a few specialized units. That's the best we will be able to do, and it will not require a heavy American military presence.

The ultimate in economy of force is the "one-man mission," in which a single officer is attached to a foreign army, often at a remote base, to train and advise it. Because there are usually no other Americans around, the officer cannot escape from the local environment, even when he is off duty. Thus he rapidly acquires a hands-on knowledge of the terrain and its inhabitants, making him an intelligence asset for years to come. The military should consider making more use of such missions.

RULE NO. 6
BRING BACK THE OLD RULES

I refer to the pre-Vietnam War rules by which small groups of quiet professionals would be used to help stabilize or destabilize a regime, depending on the circumstances and our needs. Covert means are more discreet and cheaper than declared war and large-scale mobilization, and in an age when an industrial economy is no longer necessary for the production of weapons of mass destruction, the American public, burdened with large government deficits, will demand an extraordinary degree of protection for as few tax dollars as possible. Impending technologies, such as bullets that can be directed at specific targets the way larger warheads are today, and satellites that can track the neurobiological signatures of individuals, will make assassinations far more feasible, enabling the United States to kill rulers like Saddam Hussein without having to harm their subject populations through conventional combat.

As for international law, it has meaning only when war is a distinct and separate condition from peace. As war grows more unconventional, more often undeclared, and more asymmetrical, with the element of surprise becoming the dominant variable, there will be less and less time for democratic consultation, whether with Congress or with the UN. Instead civilian-military elites in Washington and elsewhere will need to make lightning-quick decisions. In such circumstances the sanction of the so-called international community may gradually lose relevance, even if everyone soberly declares otherwise.

Bringing back the old rules would help to circumvent the UN Security Council, which in any case represents an antiquated power arrangement unreflective of the latest wave of U.S. military modernization in both tactics and weaponry. In the future we should attempt to manage most problems long before they get to the Security Council, by increasingly emphasizing Special Forces and an intelligence service bolstered by its own military wing—an emphasis we applied successfully in Afghanistan. Of course, the CIA's military wing will never be large enough to do everything. Thus the CIA and the Special Forces need to coordinate their efforts more closely, under "black," or super-clandestine, rules of engagement. Not only should the CIA be *greener* (that is, have a larger uniformed military wing), but the Special Forces should be *blacker*.

To be sure, such clandestine methods might not be enough to change a regime like Iraq's. But that kind of regime is exceedingly rare; the diplomatic farce at the UN a few months back, with France and Germany working indefatigably to contain the power of a democratic United States rather than that of a Stalinist, weapons-hungry Iraq, need not be repeated.

As shocking as some of the above may sound, much of what I advocate is already taking place. The old rules, with their accent on discretion, were on the way back even before 9/11. Witness the increasing use of security-consulting firms and defense contractors that employ—in places as diverse as South America, the Caucasus, and West Africa—retired members of the U.S. military to conduct aerial surveillance, to train local armies, and to help struggling friendly regimes. Consider Military Professional Resources, Inc. (MPRI), of northern Virginia, which during the mid-1990s restructured and modernized the Croatian military. Shortly afterward Croatian battlefield success against the Serbs forced Belgrade to the peace table.

Encouraging an overall moral outcome to the Yugoslav conflict involved methods that were not always defensible in narrowly moral terms; the Croats, too, were murderers. And moral ambiguity is even greater in protracted wars, such as the Cold War and the war on terrorism, in which deals will always have to be struck with bad people and bad regimes for the sake of a larger good. The war on terrorism will not be successful if every aspect of its execution must be disclosed and justified—in terms of universal principles—to the satisfaction of the world media and world public opinion. The old rules are good rules because, as the ancient Chinese philosophers well knew, deception and occasional dirty work are morally preferable to launching a war.

RULE NO. 7
REMEMBER THE PHILIPPINES

The first large-scale encounter between the U.S. military and a guerrilla insurgency came as the United States tried to consolidate control over the Philippine archipelago, a former colony of Spain, after our victory in the Spanish-American War of 1898. Unfortunately, many of the lessons our military learned from that encounter were for a long time ignored, because the military's performance in one dimension was overshadowed by allegations involving another. As Brian McAllister Linn wrote in his dense and masterly book *The U.S. Army and Counterinsurgency in the Philippine War, 1899-1902* (1989), some charges of American brutality against Filipino civilians were certainly justified, and without question the brutality drew press attention that colors the episode still. Brutality is always inexcusable—but in this instance it was hardly the whole story. Max Boot concludes in *The Savage Wars of Peace* (2002) that U.S. actions in the Philippines constitute "one of the most successful counter-insurgencies waged by a Western army in modern times." Given the challenges ahead, our experience a century ago in the anarchic Philippines may be more relevant than our recent experience in Iraq.

Modern communications, which seem to unify the world to some degree, often foster the illusion that policies can be one-size-fits-all. Mid-level commanders in the Philippines, however, lacking helicopters and radios, were forced to become policymakers in their own patch of jungle. That was a good thing, and it promoted skills that we need more of: in a rugged topography given over to anarchy—which describes much of the world today—the political, military, and cultural situation is going to vary from micro-region to micro-region. The commanders in the Philippines who were particularly successful emphasized small, mobile units; developed native intelligence sources; and gained information by interrogating captured guerrillas. In some parts of the archipelago the United States was able to exploit ethnic divisions; in other parts it was foolish even to try. In some parts a purely military strategy was called for; in others a civil-affairs and humanitarian-aid component was an absolute necessity. Nevertheless, as Linn observed,

> It was only when the Army could separate the guerrillas from the civilians and prevent the guerrillas from disrupting civil organization that social reform was possible. Officers in the Philippines, no matter how benevolent their intentions, realized that the military objective, the defeat of the guerrillas, was the most essential of their tasks.

In other words, in areas still not pacified by our troops, it is perfectly appropriate to see more soldiers than aid workers. But those soldiers, as William Howard Taft (then the head of the Philippine Commission) and Brigadier General Frederick Funston both observed, should be led by field officers of exceptional character, with hands-on area expertise.

RULE NO. 8
THE MISSION IS EVERYTHING

No mission should ever be compromised by diplomatic punctilio. That sounds obvious, and at the same time is often impossible to implement. But here is what happens when this rule is broken.

In the late 1990s Nigerian soldiers deputized by the international community were in Sierra Leone, not only to keep the peace but also, if truth be told, in some cases to steal alluvial diamonds. Like other African peacekeeping contingents in Sierra Leone, the Nigerians weren't always paid by their own government, even though the government was getting money from the international community to provide peacekeeping. Some of these contingents were openly incompetent; the Zambians, for instance, were a battalion of mechanics, cooks, and clerks. But the United Nations said little about any of this; instead it officially accepted the obvious falsehood that all national armies are roughly equal. Diplomatic nicety had completely compromised the mission. The result: the peacekeeping effort nearly collapsed as demoralized and incompetent peacekeepers surrendered without a fight to murderous teenage paramilitaries, who closed in on the capital of Freetown. Order was restored only after the British government dispatched commandos to Sierra Leone. Mounted on rooftops at the airport, a contingent of those commandos shot and killed any rebel who emerged from the bush. For the British, only the mission mattered.

When Hans Blix, the chief UN weapons inspector, demonstrated little enthusiasm for bringing Iraqi scientists and their families out of Iraq (even though other Iraqi scientists, once outside their country, had in the past provided valuable intelligence to the West) he revealed that for the UN, yet again, the mission was not everything.

Unfortunately, for the United States the mission is not always everything either. It is often hamstrung by diplomacy and domestic public opinion. The Special Forces are allowed to train and advise local counterparts, but because of restrictions imposed by the United States and, often, the host country as well, they typically have to wave good-bye when local troops take to the field to fight. This can be demoralizing to our elite units, whose members are not draftees serving out their time but professional warriors prepared daily to take measured risks—risks that may seem incredible to timid politicians and other outsiders. And when host-country soldiers are wounded, we should not be prohibited from helping them get to our field clinics, as is sometimes the case. Our elite units should be allowed to provide air cover for local allies, and to help direct operations on the ground. There are no such limits in Afghanistan; ideally that would be the case everywhere. Successful imperial militaries have traditionally fought alongside indigenous troops.

Moreover, arbitrary troop limits set by Congress, known as "force caps," which have restricted Green Beret trainers to fifty-five in El Salvador and our troops in Colombia to 400, should be more flexible. Also, embassy Marines and Army support staff should not be part of the calculation; force caps should apply only to the advisers and training teams in the field. Every one of our Green Berets is a force multiplier, to the extent that an extra ten or twenty of them could make an exponential difference in the success of a mission. If a cap needs stretching a bit, the U.S. ambassador and the U.S. military commander in the host country should be able to stretch it on their own. To think that any of this would risk another Vietnam is alarmist.

Compromising the mission, moreover, can mean needlessly compromising our soldiers' safety. Since the destruction of the Marine barracks in Beirut in 1983, and of military apartments at Khobar Towers, in Saudi Arabia, in 1996, our generals and politicians have needed a commandment: Thou shalt not be sitting ducks. U.S. troops should never be concentrated in a place where they cannot aggressively patrol the surrounding area. Yet that was the situation I observed near the town of Saravena, in northeastern Colombia: because the rules of engagement set by our policymakers and the Colombian government did not allow for aggressive patrolling, a few dozen Green Berets and their support staff were concentrated there in barracks vulnerable to a possible attack by cylinder bombs. That may be politically sound, but it is tactically dumb. And it is morally wrong, because it denies our warriors the means of self-defense. In Saravena the mission was not everything.

RULE NO. 9
FIGHT ON EVERY FRONT

In their recent article "An Emerging Synthesis for a New Way of War," published in the *Georgetown Journal of International Affairs*, Air Force Colonels James Callard and Peter Faber describe what they call "combination warfare"—a concept derived from a 1999 Chinese text by two colonels in the People's Liberation Army, Qiao Liang and Wang Xiangsui. In the twenty-first century a single conflict may include not only traditional military activity but also financial warfare, trade warfare, resource warfare, legal warfare, and so on. The authors explain that it may eventually involve even ecological warfare (the manipulation of the heretofore "natural" world, altering the climate). Because combination warfare draws on all spheres of human activity, it is the ultimate in total war. It "seeks to overwhelm others by assaulting them in as many domains … as possible," Callard and Faber write. "It creates sustained, and possibly shifting, pressure that is hard to anticipate."

Combination warfare has already begun, though it has yet to be codified in military doctrine. The most important front, in a way, may be the media. Like the priests of ancient Egypt, the rhetoricians of ancient Greece and Rome, and the theologians of medieval Europe, the media constitute a burgeoning class of bright and ambitious people whose social and economic stature can have the effect of undermining political authority. The media increasingly, and dramatically, affect policy yet bear no responsibility for the outcome.

In terms of U.S. national interests, media attitudes have gotten both worse and better in recent years. American leaders deal less and less with strictly American media and more and more with global ones, as elite U.S. news organs increasingly make use of foreign nationals and global cosmopolitans with multiple passports. The new, global media think in terms of abstract universal principles—the traditional weapon of the weak seeking to restrain the strong—even as the primary responsibility of our policymakers must be to maintain our strength vis-à-vis China, Russia, and the rest of the world. On the other hand, it is impossible to ignore the resurgence of patriotism among American journalists; the political divide between Europe and the United States in the buildup to the war in Iraq, and during this war itself, was mirrored by a divide between the European and the U.S. media. Still, this trend may be ephemeral.

Because the consequences of attack by weapons of mass destruction are so catastrophic, the United States will periodically have no choice but to act pre-emptively on limited evidence, exposing our actions to challenge by journalists, to say nothing of millions of protesters who are increasingly able to coordinate their demonstrations worldwide. The enormous anti-war demonstrations on several continents last February revealed that life inside the post-industrial cocoon of Western democracy has made people incapable of imagining life inside a totalitarian system. With affluence often comes not only the loss of imagination but also the loss of historical memory. Thus global economic growth in the twenty-first century can be expected to create mass societies even more deluded than the ones we have now—the very actions necessary to protect human rights and democracy will become increasingly hard to explain to those who have never been deprived of them. The masses "show no concern for the causes and reasons" behind their own well-being, observed the Spanish philosopher José Ortega y Gasset in *The Revolt of the Masses* (1929), a book that was equally prescient about the Fascist rallies of the 1930s and the youth rebellion of the 1960s. Indeed, the peace demonstrators last February appeared to have no idea whatsoever that their very freedom to demonstrate had been won by war and conquest in the service of liberty—precisely what the U.S. and British governments were proposing to do in Iraq. Of course, the masses are uninterested, as Ortega noted. "Since they do not see, behind the benefits of civilization, … they imagine that their role is limited to demanding these benefits peremptorily, as if they were natural rights."

A nation whose businesses can regularly sell products that people neither want nor need should be able to market a foreign policy better than it usually does. Just as leading companies harvest the best former government officials, our government will have to find the budget and the will to hire away the best communicators for this marketing effort. We also need diplomats who are fluent in local languages and dialects and whose sole job is to appear on foreign talk shows (in the Middle East and elsewhere) and be available to local journalists for interviews, so as to better represent our point of view. This occurs too infrequently at the moment. Here, too, we desperately need more area experts; and we need more hyphenated Americans and language specialists inside government. Moreover, it is now a strategic imperative that the United States Information Agency, gutted by the Clinton Administration under pressure from Senator Jesse Helms, be reinvented.

Some may argue that an effective information strategy is largely a matter of telling and spinning the truth. But the truth needs lots of help in societies marked by mass illiteracy, where rumors and conspiracy theories are the rule rather than the exception. That is because where few of the men and almost none of the women can read, news can be communicated only orally; thus it is even more quickly subject to distortion. In the context of mass illiteracy, the growing array of CNN-like networks in Arabic and other languages creates the conditions for a tidal wave of hysteria to be generated by a single inaccurate news report. Destructive rumors and conspiracy theories need to be countered quickly.

Indeed, the best information strategy is to avoid attention-getting confrontations in the first place and to keep the public's attention as divided as possible. We can dominate the world only quietly: off camera, so to speak. The moment the public focuses on a single crisis like the one in Iraq, that crisis is no longer analyzed on its merits: instead it becomes a rallying point around which lonely and alienated people in a global mass society can define themselves through an uplifting group identity, be it European, Muslim, anti-war intellectual, or whatever.

Nevertheless, although media coverage of the war in Iraq was unprecedented, many wars will continue to be fought with few journalists in sight, and consequently with little public awareness. Look at the Congo, where more than three million people have died in conflict since the late 1990s without any significant peace protests in the West. Military conflicts in Colombia, the Philippines, Nepal, and other places may as well be happening in secret. Our intelligence officers, backed by commando detachments, should in the future be given as much leeway as they require to get the job done, so that problems won't fester to the point where we have to act in front of a battery of television cameras.

RULE NO. 10
SPEAK VICTORIAN, THINK PAGAN

As noted, imperialism in antiquity was in many respects a strain of isolationism: the demand for absolute security at home led powers to try to dominate the world around them. That pagan-Roman model of imperialism contrasts sharply with the altruistic Victorian one, exemplified by Prime Minister William Ewart Gladstone in his comment about protecting "the sanctity of life in the hill villages of Afghanistan." Americans are truly idealistic by nature, but even if we weren't, our historical and geographical circumstances would necessitate that U.S. foreign policy be robed in idealism, so as to garner public support and ultimately be effective. And yet security concerns necessarily make our foreign policy more pagan. The idealistic shorthand of "democracy," "economic development," and "human rights," by means of which the media make sense of events in distant parts of the world, conceals many harsh and complicated ground-level truths. Remember that even Gladstone's vision was more effectively implemented by the realpolitik of statesmen such as Lord Palmerston, Benjamin Disraeli, and the Marquess of Salisbury, who kept illiberal empires like Germany and Russia at bay, sometimes through sheer deviousness, and also arranged for the retaking of Sudan from Islamic extremists.

By sustaining ourselves first, we will be able to do the world the most good. Some 200 countries, plus thousands of nongovernmental organizations, represent a chaos of interests. Without the organizing force of a great and self-interested liberal power, they are unable to advance the interests of humanity as a whole.

And there is this coda: Just as, following the explorations of Portuguese and other mariners, the oceans became a new arena for great power struggles, so will outer space. We have recognized this by creating a U.S. Space Command, which is now a part of the U.S. Strategic Command. The only question now is whether the United States will invest enough in the military technology required to dominate space. If a less liberal power such as China does so instead, then American dominance will be particularly short-lived, no matter how successful the war on terrorism.

No doubt there are some who see an American empire as the natural order of things for all time. That is not a wise outlook. The task ahead for the United States has an end point, and in all probability the end point lies not beyond the conceptual horizon but in the middle distance—a few decades from now. For a limited period the United States has the power to write the terms for international society, in hopes that when the country's imperial hour has passed, new international institutions and stable regional powers will have begun to flourish, creating a kind of civil society for the world. The historian E. H. Carr once observed that "every approach in the past to a world society has been the product of the ascendancy of a single Power." Such ascendancy allows all manner of worldwide connections—economic, cultural, institutional—to be made in a context of order and stability. There will be nothing approaching a true world government, but we may be able to nurture a loose set of global arrangements that have arisen organically among responsible and like-minded states.

If this era of reluctant imperium is to leave a lasting global mark, we must know what we are up to; we must have a sense that supremacy is bent toward a purpose and is not simply an end in itself. In many ways the few decades immediately ahead will be the trickiest ones that our policymakers have ever faced: they are charged with the job of running an empire that looks forward to its own obsolescence.

Winston Churchill saw in the United States a worthy successor to the British Empire, one that would carry on Britain's liberalizing mission. We cannot rest until something emerges that is just as estimable and concrete as what Churchill saw when he gazed across the Atlantic.

Robert D. Kaplan is an Atlantic Monthly *correspondent and the author of* Warrior Politics: Why Leadership Demands a Pagan Ethos *(2001),* Eastward to Tartary: Travels in the Balkans, the Middle East, and the Caucasus *(2000),* The Arabists: The Romance of an American Elite *(1993), and other books.*

Article 16

The Watchful & the Wary

From FBI and CIA headquarters to small-town police departments, the government is building a massive intelligence network designed to spy on terrorists—and on everyday Americans.

BY ROBERT DREYFUSS

IF THE FEDERAL BUREAU of Investigation is spending a lot of money tracking terrorists, not much of it is being squandered on frills, judging from the looks of the FBI's ramshackle field office in the federal building high above Chicago's Loop. Workstations in the sprawling suite occupied by the FBI's Joint Terrorism Task Force are shoved together chockablock, and the place looks, well, lived in. Files are strewn over battered desks, where rows of agents sit gazing at computer screens and talking quietly on the phone. "We call it the trailer park, because it's such a mess," says John Raucci, a 16-year FBI veteran who runs the task force.

But the modest appearance is deceiving. Though seemingly cobbled together, the task force is the business end of an ambitious anti-terrorist and domestic spying apparatus that is rapidly being assembled by the federal government. It's a system that seeks to merge the powers of domestic law enforcement, including the FBI, with those of the Central Intelligence Agency, the Pentagon, and the Department of Homeland Security. Originally envisioned to catch Al Qaeda-style terrorists, it has been given the authority to vacuum up vast amounts of information and to analyze and file intelligence on individuals and organizations, including political and religious groups and citizens not suspected of any crime.

Authorities can now more easily gain access to phone and email logs, travel information, credit files, library records, and reams of other personal data. Using new law-enforcement powers created by the 2001 Patriot Act and an order by Attorney General John Ashcroft easing restrictions on FBI investigations, they can find out which websites people visit and which classes they're taking, infiltrate political meetings and worship services, tap phone lines, monitor postings in online chat rooms, even secretly search homes and businesses. Since 9/11, there has been a dramatic increase in the use of surveillance warrants—which allow government agencies to spy on individuals and organizations without judicial review—as well as far-reaching subpoenas that require businesses to disclose information about their customers. And the government is working to combine information on individuals that is now scattered in countless public and private computer files into one huge database that could be used to look for patterns in people's financial or travel habits.

Most of these efforts are taking place in secret, making it nearly impossible to find out exactly how extensively the new intelligence-gathering network is being used. But a few specifics have emerged. At the University of Massachusetts, for example, the FBI, working with a campus police officer recruited to the local terrorism task force, questioned faculty and students about their political views; nationwide, the bureau now has broad access to student records. Some public libraries have begun destroying their records after being contacted by the FBI about the reading habits of patrons. According to one University of Illinois survey, 550 libraries report having received such requests; the FBI says that its agents have contacted no more than 50 libraries. And last year, the owners of a Beverly Hills dive shop went public about an FBI effort to collect the names of everyone who had ever taken scuba lessons at diving schools or YMCA pools.

The intelligence network is growing fast. "We've doubled, even tripled here since 9/11," says Raucci, whose Chicago terrorism task force—one of 66 around the nation—consists of 125 FBI officers as well as representatives of the CIA and the Internal Revenue Service; postal, customs, and immigration inspectors; the Bureau of Alcohol, Tobacco, and Firearms; and city, county, and state police. Of the task force's five squads, two are dedicated to investigating Al Qaeda and other Islamist networks, while another exclusively targets domestic groups—"Posse Comitatus, the Aryan Nation, the Church of the Creator, and extremists like the animal-liberation groups," says Raucci; in addition, the FBI is providing intelligence training for thousands of officers at 200 police departments in the Chicago suburbs. The CIA has won new authority to engage in domestic spying, something it was long prohibited from doing; new domestic intelligence departments have been created at the Pentagon's Northern Command and the Department of Homeland Security, and key lawmakers are debating the creation of a massive "National Intelligence Agency" along the lines of Britain's MI5.

Together, the Bush administration's measures are unraveling the web of safeguards placed on law enforcement in the 1970s, after it was revealed that the FBI, the CIA, and the U.S. Army had conducted covert operations against campus radicals, anti-war groups, and civil rights organizations. And, critics warn, they stand to make government surveillance of citizens—once considered a temporary, post-September 11 phenomenon—a permanent, routine aspect of law enforcement. "A combination of lightning-fast technological innovations and the erosion of privacy protections," warned an American Civil Liberties Union report in January, "threatens to transform Big Brother from an oft-cited but remote threat into a very real part of American life."

ON THE STREETS of Chicago, the power of the new intelligence apparatus is already being felt—and not by terrorists. "Everything has changed since 9/11," says Emile Schepers, a longtime local activist and head of the Chicago Committee to Defend the Bill of Rights.

Schepers looks like a protester out of central casting. On a blustery cold and snowy day in early April, his bulky frame is sheathed in a tattered raincoat and he's unshaven, with a shaggy mustache and long, unkempt hair. "Last month," he says, "we organized a large anti-war demonstration, at which a police photographer came up to me and, right in my face, snapped a picture." At the protest, police made more than 800 arrests. "What suggests heightened intelligence and surveillance," Schepers says, "is that the police went into the crowd and selected activists and leaders, and arrested them."

Until recently, the Chicago Police Department was prohibited from photographing or videotaping political demonstrations. But in 2001, in a move that cleared the way for closer cooperation between police and the FBI, the department won a decades-long effort to rid itself of a court-imposed consent decree that had restricted its ability to spy on political and religious groups. The prohibition dated back to 1982, when a court forced Chicago to abandon its political spying unit, the heir to the old Red Squad that had harassed left-wing, civil rights, and anti-war activists from the 1950s to the '70s. Now, Chicago police are free to amass files on organizations, keep track of their members, infiltrate meetings, and record demonstrations, all on nothing more than the suspicion that a group might be inclined toward violence.

Critics warn that the administration's measures stand to make government surveillance of citizens—once considered a temporary, post-September 11 phenomenon— a permanent, routine aspect of law enforcement.

Larry Rosenthal, an attorney for the city of Chicago, insists that police will respect Chicagoans' First Amendment rights. Officers, he says, are still prohibited from "intelligence gathering not motivated by a law-enforcement purpose or which is intended to punish, harass, or retaliate." But he pointedly notes that under the new rules the police can keep track of organizations like ACT UP, the militant AIDS group, or anti-abortion activists. "Take a mainline anti-abortion group with fringe members who plant bombs," he says. "If a bomb goes off, at least you know who the members are, because they're in an intelligence database." Rosenthal says the Chicago police hope to integrate their criminal and intelligence databases more closely with those of the FBI and police departments in other cities, meaning that information collected anywhere in the United States will be instantly accessible to a long list of federal and local agencies. "Take the case of a demonstration against the World Trade Organization," he says. "The responsible thing to do is to find out if any of the people coming [to Chicago] caused trouble in other cities."

It won't just be unruly demonstrators, however, who end up in intelligence files. Police around the nation, from large cities to small towns, are being trained in how to gather intelligence on everyday activity in their communities. In Joliet (population 106,221), 35 miles southwest of Chicago, officers have been through more than 4,000 hours of anti-terrorism training with the FBI since 9/11, says Fred Hayes, the city's deputy chief of police. Joliet's intelligence unit has also set up a liaison to the police intelligence division in New York City, which is led by a former CIA chief of covert operations.

Bill Fitzgerald, a former Joliet police chief, runs the Institute for Public Safety Partnerships, a branch of the Justice Department that conducts training for local police. Under the new system, Fitzgerald explains, someone in Joliet might report a neighbor to the police for suspicious behavior. A police officer may then visit the person, fill out a "field interview card," and pass it to the intelligence officer in his department who would send it to the FBI task force. The process could begin with something as innocuous as a jittery citizen reporting a man in a turban loading boxes into a car—the kind of tip that, in fact, led to the arrests of numerous Arabs and Muslims in the immediate aftermath of 9/11. If that man is seen a week later at, say, a protest in Des Moines, police could access the FBI's database and call up the Joliet report. "At least they tell us it's supposed to work that way," says Fitzgerald.

Which is exactly what worries Harvey Grossman, legal director for the ACLU of Illinois. Slight and soft-spoken, the graying civil libertarian sits in his cluttered office in the warrenlike ACLU headquarters, poring over legal papers. "If you're involved in a street demonstration, you don't have any expectation that you are going to be part of a government file," he says. Even the mere threat of such a file, he notes, can put a chill on political dissent: Government workers, people who might want security clearances or whose background might be investigated for a future job, or people who simply want to remain anonymous while participating in protests all have reason to fear the inclusion of their name in a police file, especially since such files are increasingly being shared among government agencies.

The ACLU is especially concerned about the FBI and police gathering information on the political views of people who are not suspected of any wrongdoing. After the 1991 Gulf War, the FBI interviewed Arab American leaders about terrorism and, in the process, asked their views about Iraq's invasion of Kuwait, about Middle East politics, and about a Palestinian homeland. In the settlement of a lawsuit brought by the ACLU, a judge later ordered the records destroyed. Now, worries Grossman, the FBI may be using its new powers to compile similar dossiers on countless Americans—and, given the intense secrecy that surrounds the war on terrorism, it will be a lot harder to uncover them.

Already, there are indications that police in some cities have sought to collect information on individuals' political beliefs. In New York this past February, police intelligence officers interrogated anti-war protesters about their political affiliations and past protest activities, then entered the information into a database. Under pressure from the ACLU, the program was halted—as was a more extensive spying program, begun in the late 1990s by the Denver Police Department, in which police amassed files on 3,200 individuals and 208 organizations, including Amnesty International and the American Friends Service Committee, many of which were characterized as "criminal extremist."

A Justice Department training manual directs local police to pay attention to citizens' political affiliations. "In the 1950s we said we were after communist sympathizers," explains its author. "Now, we say terrorist sympathizers."

Under a Justice Department training curriculum, police are taught to pay attention to citizens' political affiliations and to look out for "enemies in our own backyard." Police are directed to collect information on the "structure, philosophy, number of members, [and] locations" of groups including "the Green Movement," which is defined as "environmental activism that is aimed at political and social reform with the explicit attempt to develop environmentally friendly policy, law, and behavior."

David Carter, a professor of criminal justice at Michigan State University who developed the Justice Department curriculum, says it's important for police to investigate even groups that so far have not engaged in any violent acts. Violence or terrorism could emerge from a wide range of movements, he says, including "the groups involved in the [World Economic Forum] protests in New York, or the World Trade Organization protesters," along with other "social-extremist groups," black separatists, and militias. "In the 1950s we said we were after communist sympathizers," Carter explains. "Now, we say terrorist sympathizers."

FOR THE FBI, putting together a national intelligence network is more than a new mandate: It's the bureau's chance to redeem itself after being widely castigated for missing clues in advance of the September 11 terrorist attacks. Leading members of Congress, think tanks, and government commissions have called for the establishment of a national intelligence agency to take over the bureau's anti-terrorism efforts; President Bush's national security team has discussed that option, but so far has not made a decision. "We're either going to create a working, effective, substantial domestic intelligence unit in the FBI or create a new agency," says Senator Richard Shelby (R-Ala.), who until recently chaired the Senate Intelligence Committee.

In response to such pressure, the FBI is scrambling to beef up its intelligence division. "We've shifted intelligence analysts from the criminal side," says Raucci, of the Chicago Joint Terrorism Task Force. "The lion's share work on the terrorism side now, and we are making strides to recruit more trained intelligence people, including from the military services." At headquarters in Washington, FBI Director Robert Mueller has established a "super squad" of terrorism specialists, opened a new Office of Intelligence, and asked Congress for a dramatic budget increase—a total of $4.6 billion for next year, nearly 50 percent more than the $3.1 billion the FBI spent in 2000. The new focus on terrorism has led the FBI to nearly abandon the war on drugs, leaving that mission to the Drug Enforcement Administration, and to significantly scale back efforts against organized crime, white-collar crime, and gangs.

The FBI and other branches of the Justice Department are also making ample use of the new powers granted them under the Patriot Act. They have filed at least 18,000 anti-terrorism subpoenas and search warrants since September 11, 2001, and conducted more than 1,000 surveillance operations with permission from the Foreign Intelligence Surveillance Act court—a special court that can grant warrants without evidence that the person being investigated is engaged in criminal activity.

During the same period, Ashcroft requested 170 emergency spying warrants, which allow the FBI to investigate individuals without judicial review; in the 23 years since the court had been established, a total of only 47 emergency warrants had been issued. In addition, the FBI has stepped up its use of national-security letters—subpoenas that can be used to compel businesses such as banks, Internet service providers, and airlines to turn over large amounts of customer data. When the ACLU sued to force the bureau to reveal how often it has used such letters, the FBI released a list that filled five pages, but with most of the information blacked out.

So far, all of this has netted the government few, if any, coups in the fight against terrorism. Data compiled by the Transactional Records Access Clearinghouse, based on Department of Justice numbers, show that in 2002, federal prosecutors nationwide charged 1,208 individuals with terrorism-related crimes. But the average sentence for those convicted on such charges was just two months—signaling that the crimes were mostly minor. Indeed, most of them reportedly involved offenses such as violating immigration rules, using a fake Social Security number, even getting drunk on an airplane.

Still, the Bush administration has launched a campaign to further extend federal spying powers by asking Congress to make the Patriot Act—parts of which are scheduled to expire in 2005—permanent, and by drawing up plans for a measure dubbed "Patriot II." The proposal would further lower the standards for court approval of FBI searches and surveillance, and it would allow agents to subpoena library, Internet, and bookstore records without going to court at all; it would also create a DNA database with samples from anyone suspected of being connected to a terrorist group. In addition, the association that represents state motor vehicle departments nationwide is creating the equivalent of a national ID card by combining driver's license records from all 50 states; besides names, addresses, and photos, the database could include a biometric identifier such as a fingerprint or retinal scan.

The CIA, meanwhile, is establishing a much closer relationship with U.S. law enforcement. CIA officials are leading the new federal agency in charge of coordinating anti-terrorist activity, the Terrorist Threat Integration Center, where FBI agents work side by side with CIA analysts. According to John Rizzo, senior deputy general counsel for the CIA, the agency has also deployed officers to many of the FBI's terrorism task forces, where they have begun working with local police, helping to set investigative priorities and identify targets for surveillance.

"Previously, this kind of activity would have been considered unpalatable," says Rizzo, because of the rules keeping law enforcement and intelligence apart. "The CIA has never dealt with state and local officials," he adds, "but we are going to have to learn to do so."

Even the federal government's sophisticated spy satellites and electronic eavesdropping equipment can be used against U.S. targets and U.S. citizens. Under an existing, though little-known, law, such high-tech gear can be directed against Americans as long as it is temporarily placed under the control of the FBI or another law-enforcement agency, according to a former high-ranking CIA official. "If the intelligence community has a technology that law enforcement thinks is of use to it," the official says, "it can just be seconded to the law-enforcement agency."

During the sniper attacks in the Washington, D.C., area last fall, the Pentagon and the National Security Agency met to consider whether spy satellites might be useful in trying to find the snipers. Richard Schiffrin, the acting general counsel for the Defense Intelligence Agency, took part in those meetings. Participants, he recalls, kept harking back to the thriller starring Will Smith as an attorney under surveillance by a government agency that uses everything from infrared heat-seeking devices to listening gadgets to reconnaissance satellites. "*Enemy of the State*," he says, "came up in every discussion."

Article 17

Wednesday, December 26, 2001

Economic Crossroads on the Line

Security Fears Have U.S. and Canada Rethinking Life at 49th Parallel

BY MICHAEL GRUNWALD
Washington Post Staff Writer

WINDSOR, Ontario

Long lines of trucks are always inching across the Ambassador Bridge, hauling cargo between the United States and Canada like freight trains of infinite length. This is the busiest border crossing in North America—more goods pass between Detroit and Windsor than between the United States and the entire European Union—but David Jolly, the bridge's manager, now watches the relentless march of commerce with two unpleasant thoughts.

The first is: *Hurry up!* The border between Detroit, synonymous with the Big Three automakers, and Windsor, the "Automotive Capital of Canada," basically bisects a huge assembly line. The flow of traffic and trade has slowly recovered since Sept. 11—a day of 20-mile backups, 14-hour waits and multimillion-dollar factory shutdowns—but volume remains down, security hassles up and delays a bit unpredictable. "See that line of trucks just sitting on the bridge?" Jolly asks. "That's bad. That's money."

His second thought is: *Uh-oh*. That's because trucks that enter the bridge are not checked until they reach the other side. So if terrorists decide to blow up the bridge with a truck bomb, Jolly doesn't see any way to stop them. "Everyone on that bridge would be fresh meat," he says. "It's stupid, but it's reality right now."

Reality may change, though: The events of Sept. 11 have left the U.S.-Canadian border in flux. The short-term response has been to increase barriers along the 5,500-mile frontier. But the dramatic costs of doing so have helped energize long-term commitments from both nations to use technology to create a "smarter border," decreasing barriers yet increasing security, reordering life at the 49th parallel for the forseeable future.

The terrorist attacks quickly inspired new security procedures and binational agreements, while redirecting money and manpower to long-neglected border agencies. Before Sept. 11, for example, half the border's 126 official crossings were unguarded at night; now the orange cones that used to block U.S. entry roads have been supplemented by people. Customs agents in both countries are asking far more questions and searching far more vehicles and asking to see passports or birth certificates as well as driver's licenses. Programs designed to help frequent border-crossers zip through designated customs lanes were suspended for fear that terrorists might already have the required electronic passes.

The Sept. 11 attacks have left the Ambassador Bridge between Detroit and Windsor—and the rest of the U.S.-Canadian border—in flux.

Now, the U.S. Coast Guard stops all boats that cross the border in the Great Lakes and provides escorts for all oil and gas tankers. The Border Patrol shifted 100 agents here from the Mexican border; the National Guard was assigned here temporarily, then indefinitely. The agency that oversees the Ambassador Bridge is spending an unsustainable $50,000 a week on private security. Traffic is no longer allowed to pile up inside the nearby Detroit-Windsor Tunnel, reducing the potential casualty count of a rush-hour bombing but creating gridlock on nearby city streets.

The result is that the world's longest undefended border is slightly more defended, and more complex to cross. So fewer people are coming across, and businesses that depend on easy cross-border traffic are bleeding money. Now U.S. and Canadian leaders eager to preserve the world's largest trading partnership are shifting focus, dusting off a bevy of reports and accords that were supposed to stop the strangling of trade at the border long ago.

Article 17. Economic Crossroads on the Line

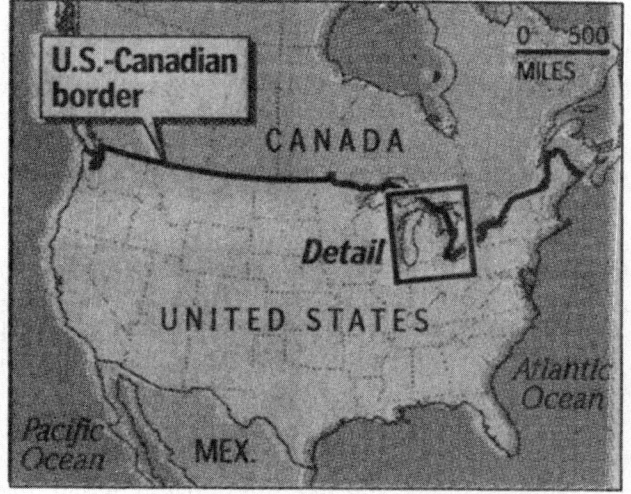

THE WASHINGTON POST

"Closing this border is not the answer," said U.S. Ambassador Paul Cellucci, a former Massachusetts governor. "We need that border open for business. More open than ever."

For some officials, the ultimate goal is a border so seamless it would barely be noticed. They rattle off staggering statistics from the post-NAFTA era: 200 million vehicle crossings a year, $1.4 billion in cross-border trade a day, one truck over the Ambassador Bridge every six seconds. The consequences of delays and disruptions, they say, can be equally staggering.

"Now everyone realizes that we can't just tweak and tinker," said Perrin Beatty, a former foreign minister and defense minister of Canada who is now president of the Canadian Manufacturers and Exporters. "For years, we've been running this border in the dumbest possible way. Now we need a fundamental rethinking of what this border means."

'Dying on the Vine'

Windsor is just south of Detroit, just across the Detroit River. The Big Three all have factories here, too. But this suburban-style city of 200,000 people will never be confused with Detroit. "It's that chummy, everybody-knows-everybody flavor that we take pride in most," explains Windsor's Web site. "From minor-league hockey teams to charity groups, we're all one big happy family."

In normal times, Windsor hosts 9 million visitors a year, most of them American day-trippers. That's more Americans than visit Montreal or Toronto or Vancouver. But since Sept. 11, in the words of Detroit Regional Chamber of Commerce President Richard Blouse, Windsor has been "dying on the vine." Jim Yanchula, Windsor's downtown revitalization chief, says the city is "hanging on by our fingernails."

> **Dave Jolly, the Ambassador Bridge manager, says, "Everyone on that bridge would be fresh meat" if terrorists decided to blow it up." It's stupid, but it's reality."**

The problem, beyond the general economic slowdown, is that American leisure travelers who used to pop over the border to take advantage of Canada's weaker dollar, lower drinking age or ritzier casinos are staying home. Traffic at the Detroit-Windsor Tunnel is down 40 percent; one-fourth of its workers have been laid off. The Ambassador Bridge, which opened a few weeks after the stock market crash of 1929, has not struggled this badly since it went bankrupt during the Great Depression. At first, Windsor entrepreneurs thought the American bridge and tunnel crowd—especially Arab Americans, who are heavily concentrated around Detroit—just didn't feel like going out. Now it's clear the fear is border-specific.

So Freed's of Windsor, a huge clothing store, is offering customers free alterations while they wait, and a free meal at the restaurant across the street, but its suits have been staying on the racks. Windsor's Raceway has offered 2-for-1 dinner specials, but standardbred racing revenue is down 20 percent and slot machine revenue has fallen 30 percent. At Cheetah's, a downtown Windsor strip club, attendance is down 80 percent; some nights there are more naked women than clothed men. Even the Windsor casino, which covers six city blocks and was one of the world's most profitable gaming operations before Sept.

11, laid off 762 employees, mostly dealers and money-counters.

"There wasn't as much money to count," explained casino spokesman Jim Mundy.

U.S. customs agents search vehicles entering the United States in Detroit via the 80-year-old Ambassador Bridge. Security checks at the crossing, the busiest in North America, sometimes create hours-long traffic backups.

The terrorist attacks also disrupted the daily routines of 30,000 commuters who cross the border every morning, including about 1,600 Canadian nurses who work in Detroit-area hospitals. After Sept. 11, at least two dozen nurses quit, and many others were forced to start their commutes as early as 3 a.m., not an easy time to get a babysitter.

In recent weeks, the overall downturn in traffic volume—along with the swift upturn in border personnel, which means more lanes open at major crossings—has actually reduced delays at the border. Still, border gridlock was an unpredictable problem here before the attacks, and now the fear of it is spreading. Remo Mancini, another Ambassador Bridge official, frets that if officials cannot restore confidence in the border, "vast regions of North America will be deindustrialized."

It was easy to see why one recent morning in the Volvo 18-wheeler that Rick Baker drives for a Canadian firm. Baker was hauling a 7,200-pound load of auto seat parts from Tillsonburg, Ontario, to Plymouth, Mich. That meant a trip across the four-lane Ambassador Bridge, a trip he sometimes makes four times a day.

At 10:19 a.m., Baker had cruised almost all the way to the toll plaza on the Canadian side, with only four trucks ahead of him in line. But at 10:29, there were still four trucks ahead of him, and an orange Coast Guard helicopter was flying ominously overhead.

"Uh-oh," Baker said. "We're stuck."

Stuck is bad for business. During the 1980s, most manufacturers adopted a "just-in-time" delivery system that moves parts directly from trucks to assembly lines. Since Sept. 11, some suppliers have been forced to maintain a "safety stock" in case of a temporary border shutdown, increasing warehousing costs. Truckers have budgeted more time on the road to make their strict "windows," which means steeper costs all around. A headline in Crain's Detroit Business warned: "Just-in-time could become just-in-case."

Baker started moving across the 1.5-mile-long bridge at 10:37; the truck lanes had been shut down for repainting. With 28,000 vehicles crossing the bridge—and occasionally breaking down on it—every day, shutdowns are not unusual these days. Two bomb threats shut down the bridge for over an hour. Another half-dozen times, Jolly closed lanes to move a single truck across the bridge after auto executives called to say they would have to shutter an assembly line—a potential $1 million-an-hour loss—if their shipment did not arrive soon.

In any case, Baker was not done with his morning crossing. He still had to get through customs and to the highway, which required an elaborate series of turns through clogged back roads. "No hats off to the design guys," he grumbled. A customs agent waved him through without asking him or his visitor for ID, but he still didn't reach the highway until 10:56.

For all the hype about security crackdowns, Baker's trailer had been inspected only three times since Sept. 11. But the border was still the time-consuming variable in his daily journeys.

"Thirty-seven minutes and they didn't even check us out," he said. "Don't you feel secure knowing you have such a secure border?"

A Tale of Two Nations

"You seem familiar, yet somehow strange," a man tells his date over martinis in a recent New Yorker cartoon. "Are you by any chance Canadian?"

It would be hard to find two allies closer than the United States and Canada, but the cultural gap between the nations extends beyond socialized medicine, hockey mania, the metric system and "eh"—not to mention trade disputes over potatoes, tomatoes and softwood lumber. It certainly extends to the border, which has always been viewed differently on either side.

For one thing, 90 percent of all Canadians live within 100 miles of the border, so Canada has paid much closer attention to it, while the United States has focused more on the flow of drugs and illegal immigrants from Mexico. The U.S. Border Patrol, for example, had 9,091 agents at the southern border before Sept. 11, and only 334 agents on the northern line, which is more than twice as long.

There was also an attitude gap. Canada sends 87 percent of its exports to the United States, so its emphasis has always been on facilitating trade. The United States sends 25 percent of its exports to Canada, the top trading partner of 38 states, but its economy is not so reliant on cross-border trade. Its emphasis has been on law enforcement, with remote cameras and sensors deployed all along the border. So while Canada has automated and revamped its customs system, for example, the United States has not even integrated the computers of its border agencies.

The gap is especially obvious at the Detroit-Windsor border, where the two countries could not even coordinate on a single electronic pass for frequent travelers. Before Sept. 11, convenience-minded Canada offered its CanPass program free to just about anyone with basic identification. The security-conscious United States fingerprinted and interviewed all applicants for its $129 SENTRI passes and required a slew of extra documents.

Still, when James W. Ziglar, commissioner of the Immigration and Naturalization Service, testified before Congress recently about the northern border, he conceded that it has been "accurately described as porous." The Justice Department's inspector general reported last year that one border sector had identified 65 smuggling corridors but had only 36 sensors to monitor them. The report noted that another sector was "6 to 7 percent effective" in stopping illegal crossings. The border is just too long—and in most areas, too rural—to make sure no one can sneak across. Last year, the Border Patrol caught only 11,000 illegal immigrants up here, vs. more than 1 million down south.

This was not a comforting thought after Sept. 11—not when Canada's intelligence service has warned that nearly every known terrorist organization has a Canadian branch, and when Canada has much more liberal asylum policies than the United States. Americans may remember a Palestinian refugee named Abu Mezer, who was caught three times sneaking into the United States from Canada before he was arrested in 1997 for plotting an attack on New York's subways. Ahmed Ressam, an Algerian who had forged a birth certificate to obtain a Canadian passport, was arrested in 1999 after taking a ferry into Washington state with explosives, which he planned to use to blow up Los Angeles International Airport during millennium celebrations.

In the days after Sept. 11, several U.S. politicians arguing for tighter borders cited rumors that some of the hijackers had crossed from Canada, but the rumors proved to be untrue. And as time has passed, the mood has shifted, with such officials as Treasury Secretary Paul H. O'Neill and Attorney General John D. Ashcroft joining their Canadian counterparts to push for a more cooperative border control system that could thwart terrorists without strangling trade. The two nations have begun taking steps to shift their overall enforcement focus to North American "perimeter security," to stop dangerous foreigners "where the water hits the rocks."

"Nobody's pointing fingers," Cellucci said. "There's just a real sense of urgency. We all recognize that if this border becomes an impediment, that's one more way for the terrorists to win."

Giving Fred a Break

The way to stop the terrorists from winning, Jolly says, is to give Fred a break.

Fred is an American guy, or maybe a Canadian guy. He crosses the border every day to work at, say, General Motors. He's no danger to anyone, but he has to wait in customs lines every morning and evening and answer the same rote questions every morning and evening.

"Look, the guy you've never seen before, knock yourself out, take his car apart, make sure he's not a bad guy," Jolly says. "But leave Fred alone. Hassling Fred is not the solution."

One solution is giving electronic passes—with fingerprints or other biometric identifiers to make sure they're not stolen—to frequent travelers and truckers who pass background checks. Tom Ridge, the Bush administration's homeland security director, recently announced that one of the array of suspended electronic-pass programs—a rare binational program limited to the Port Huron-Sarnia crossing—would be reinstated and expanded. Canada is also launching a Customs Self-Assessment program to use transponders and FedEx-style tracking technology to let importers pay duties electronically, with only occasional random inspections at the border to make sure they are reporting cargo honestly. That should help Fred. And if stopping enemies at the border is like finding needles in a haystack, these programs should dramatically shrink the haystack.

Bus driver Kiely Smith, who ferries gamblers between Detroit and Windsor, says border agents have even searched her bus's engine compartment.

Meanwhile, the two countries have agreed to share more intelligence and to beef up seaport and airport security to help keep terrorists out of North America. Canada has begun to tighten some of its asylum policies, and Ridge has floated a plan to streamline the alphabet soup of U.S. border agencies. A Windsor city effort to post bridge and tunnel waiting times at *www.bordernow.com* has helped reduce uncertainty about delays.

But there are still snags. The Bush administration has scaled back congressional proposals to triple U.S. border personnel. And Jolly's *uh-oh* problem, for now, remains a problem: U.S. officials have proposed posting American customs agents in Canada and vice versa so they could detect an explosives-laden truck *before* it reached the Ambassador Bridge or the nearby tunnel, but Canadians protective of their sovereignty have resisted. The National Guard has begun spot-checking a few vehicles as they leave Detroit, but the bridge and tunnel operators say they are still losing sleep over their vulnerability.

Jolly's *hurry up* problem is not really solved, either. Even if increased manpower helps keep more lanes open, even if the electronic passes divert most vehicles into designated lanes, neither Detroit-Windsor crossing is well-equipped to deal with modern traffic streams—and planners expect commercial traffic to triple over the next two decades. The bridge and tunnel, both 80 years old, empty into narrow and jammed city streets. Plans to link them to highways, or add customs booths, or even build a third crossing had gained little traction even before war and recession began squeezing budgets.

"The system was already broken before September 11th, and then it collapsed entirely," Beatty said. "Now we have to resolve the issues that were blocking trade before September 11th."

While they're at it, Elsie Ciaglia mused one recent afternoon, they might want to resolve the issues blocking the teal-green 12:30 p.m. Commuter Express bus to the Windsor casino. Ridership was way down for two months after Sept. 11, but recently the bus was full again—in part because all riders got $10 from the casino plus a $15 meal voucher with their $5 trip. Ciaglia, a retired secretary, was riding it once a week from her local Kmart to the nickel slots, but she was sick of the holdup at the bridge. "I bet this takes 20 minutes," she griped as the bus approached customs at 1:45 p.m.

The 55 elderly passengers all had to disembark and produce documents—the new standard procedure. There was much jostling and griping. "What a mess!" growled Angelo Pederiva, 80, a retired bricklayer with an Italian accent as thick as Chico Marx's. "Do we look like terrorists?" asked Lillian Dobbins, who was out of her house for the first time after spending the two months after Sept. 11 watching Tom Brokaw. "This isn't so bad," replied the driver, Kiely Smith. "One time they even made me open the engine compartment."

At 2:07, the bus was back on the bridge. Ciaglia was right: 22 minutes to check out the old folks. "I should have bet money!" she crowed. "I won't win in Canada, that's for sure."

From *The Washington Post*, December 26, 2001. © 2001 by The Washington Post. Reprinted with permission.

Canada Links Arrest of 19 to Possible Terrorism Ties

CLIFFORD KRAUSS

TORONTO, Aug. 23—A document filed at a detention hearing this week for 19 students and other immigrants from Pakistan detained by Canadian security officials for possible ties to terrorism cited a "pattern of fraudulent document use to obtain or maintain immigrant status" by the men, ages 18 to 33.

The men were detained on Aug. 14 after an investigation found that one of them was taking flying lessons at a school near an Ontario nuclear power plant.

Officials would disclose little about the investigation, but the four-page document sketched a picture of a mysterious group of men living in apartments with only computers and mattresses. The men appeared interested in explosives and in the Pickering Nuclear Generating Station outside Toronto, according to the document.

There had been unexplained fires in at least two of the men's apartments, and in police monitoring, two of the men had been seen walking outside the gates of the Pickering plant at 4:15 a.m. on a day in April 2002. The men said they wanted to take a walk on a beach.

One man was training to fly at a school whose flight paths cross over the Pickering plant, the document said. It said the men were in contact with unidentified sources who "have access to nuclear gauges" that contain small amounts of the isotope cesium 137, which can be used for making crude nuclear explosives.

"Based upon the structure of this group, their associations and connected events, there is a reasonable suspicion that these persons pose a threat to national security," the document said.

There seems little likelihood that the group was anywhere near to carrying out an attack. Government spokesmen played down the threat to security.

But the court document said a man who had lived with one suspect had worked for a charity group named Global Relief Foundation, which the United Nations has linked to supporters of Al Qaeda and other terrorist groups.

Some men are being held on immigration violations and others without charges. Under new antiterrorism laws, landed immigrants and foreign citizens can be detained several days on suspicion of threatening national security. A closed detention hearing is expected next week.

An investigation into at least some of the men by the Royal Canadian Mounted Police and other security forces has apparently been going on for more than a year. A police hot line received the first tips about the group shortly after the Sept. 11, 2001, attacks.

A Canadian immigration officer in Mexico City became suspicious in February of an application by one of the men for permanent residency to attend an Ottawa business college. The man had no apparent source of income, but showed a bank balance of $40,000.

The school turned out to be a fraudulent operation. Investigators found that 31 Pakistanis had used the school to enter Canada.

Mohammed Syed, a Toronto lawyer representing two suspects, said the police action "smacks of racism because they happen to be from Pakistan and are Muslim."

From the *New York Times*, August 24, 2003. © 2003 by The New York Times Company. Reprinted by permission.

UNIT 5
Latin America

Unit Selections
19. **Free Trade on Trial**, The Economist
20. **Latin America's New Political Leaders: Walking on a Wire**, Michael Shifter

Key Points to Consider
- What are some of the most serious threats to democracy in Latin America today?
- Which groups benefited and which groups lost out as a result of the North American Free Trade Agreement (NAFTA)? What position are these same groups likely to have on the proposed Free Trade Agreement of the Americas?
- What are some of the costs and benefits of changing the status of illegal immigrants from Mexico and other Latin American states to legal guest workers?
- What should the United States, Canada, and European countries do to ease the economic crisis in Argentina?
- Do you believe the United States should intervene in Haiti? Why or why not?

 **Links: www.dushkin.com/online/**
These sites are annotated in the World Wide Web pages.

Inter-American Dialogue
http://www.iadialog.org

For nearly two centuries, the United States viewed Latin America as being within its exclusive sphere of influence. Over the past two decades, most countries in the region shifted from military dictatorships to democracies. Civilian-led governments adopted neo-liberal economic reforms and followed the recommendations of the World Bank and the International Monetary Fund: privatize state-owned corporations, reduce public payrolls and subsidies, promote free trade and direct foreign investment. Political and economic change occurred during a period of prosperity but few economic benefits trickled down as most citizens remained poor.

Increased numbers of citizens in most countries are now demanding a greater share of the wealth created in recent years. Democratic development and economic reform are now firmly established as the overarching norms in the hemisphere but the global economic slowdown, currency and political crises, and continuing insurgencies in several countries are creating challenges. During the first years of the Twenty-first Century, difficult economic conditions led to a series of leftist parties gaining power in several countries, including the hemisphere's largest country, Brazil. Most observers interpreted the peaceful political shifts to the left as evidence that Latin America's democracies, although fragile, are likely to survive. Even Argentina, a country where the economic wealth of most middle class families was wiped out in a short period after dramatic currency devaluations degraded the value of frozen assets, experienced a peaceful change of leaders in 2003. However, real problems persist. Many analysts worry that economic difficulties in Mexico may signal wider economic hardships throughout the region in the coming years.

Few Americans appreciate that North American Free Trade Association (NAFTA) countries purchase 40 percent of U.S. exports. The state of the Canadian and Mexican economies is as important for the continued health of the U.S. economy as are economic trends in large states such as Texas. NAFTA has been in existence for nearly a decade. Economists estimate that about 350,000 manufacturing jobs were lost in the U.S. due to the pact while 2 million better paying jobs were created. The Bush administration supports creating a larger free-trade zone for all the Americas that would extend from Canada to the tip of South America. The article "Free Trade on Trial," describes why, on the tenth anniversary of the NAFTA agreement, it remains unpopular in the United States, Canada, and Mexico. While intra-area trade and foreign investment expanded greatly, economic growth in Mexico has been dismal, manufacturing jobs have declined in the United States, and environmental problems associated with the Maquiladora cluster have worsened. Most analysts now agree that the benefits of NAFTA were overstated.

Despite the mixed results of NAFTA, the leaders of the three countries continue to support the concept of a Free Trade Area of the Americas that would include 34 countries and 800 million people. When and if such a large free trade zone comes into existence remains uncertain. Opponents in the U.S. fear further job losses, and increased risks associated with growing economic interdependence throughout the Western Hemisphere may mean that the U.S. will be more vulnerable to the problems of their closest neighbors in the future. Recent poll data also indicate strong opposition in Mexico where many farmers believe their livelihood is threatened by tough competition from U.S. farmers who receive large subsides from the U.S. government. Other critics charge that free trade policies—as long as agricultural subsidies continue—will kill any hope for developing countries to compete with the United States.

Prior to the September 11 terrorist attacks, President Bush signaled his desire for a "special relationship" with Mexico by visiting during his first foreign visit. Fox and Bush were making substantial progress towards an agreement that would redefine the status of millions of Mexicans living in the U.S., from illegal immigrants to legal guest workers. The historic agreement and many other inter-American issues were shoved to the back burner by the terrorist attacks in 2001. The fact that Vicente Fox did not support the UN resolution calling for a U.S. military intervention in Iraq put an additional strain on relations. However, by the end of 2003 President Bush returned to immigration reform proposals that had been shelved for two years. During a visit to Mexico, Bush proposed to make it easier for immigrants to work legally in the United States while stepping up security and enforcement procedures along the border to better control illegal immigration. An estimated 8 million immigrants now live in the United States. Bush's new proposal allows long-time illegal immigrants to live legally in the United States but precludes them from pursuing U.S. citizenship. Although immigration policies are Mexico's top priority, President Fox reacted cautiously to the new U.S. proposal. Whether the U.S. will be able to meet the practical challenges of working with its neighbors is one of the defining questions for inter-American relations in the early part of the next century.

Financial volatility and the global economic downturn have created hardships for large numbers of citizens in Latin American countries and led to increased demands for political changes. Some of the demands have been peaceful, others have not. In Bolivia, President de Lozada was forced to resign in 2003 after he lost the support of key political allies during a revolt by indigenous workers and students protesting the planned sale of natural gas supplies through Chile to the United States and Mexico. In 2002 a leftist opposition leader, Luiz Inacio Lula da Silva was elected President and assumed power in the first peaceful presidential transition in Brazil in over 40 years. Luiz da Silva promised a new style of government and a crusade against hunger, injustice, and want but quickly lost a great deal of his initial popular support. Another populist politician, President Hugo Chavez of Venezuela, weathered a prolonged nation-wide strike by workers and middle class opponents who brought the country to a near standstill in 2002 and again in 2003. Despite coup rumors and calls by President Bush to hold early elections, President Chavez refused to step down and continues to defy demands for a referendum designed to oust him from power. In "Latin America's New Political Leaders: Walking on a Wire," Michael Shifter describes how the current generation of leaders must find ways to produce results to everyday problems. This

poses a formidable problem since most members of the electorate in Latin American countries have lost faith in the formulas of socialism, popular in the 1970s, or in the 'neoliberlism' formulas of the 1990s.

The shocks associated with the Asian economic crisis and the reduction in world demand for commodity exports as well as international confidence in the national currencies has declined in part because all countries in the region have payment deficits. After years of borrowing from foreign investors to support a currency system that pegged Argentina's currency to the U.S. dollar, the centrist government of President de la Rua and its successor government composed of Peronist politicians were forced to resign as popular resistance boiled over in violent protests against de la Rua's austerity measures and failures by the new government as well. Even though Argentina's debt represents 25 percent of all emerging market debt, foreign investors were well positioned to weather the losses incurred by Argentina's financial crisis.

In contrast, millions of unemployed and middle class Argentines and elites in neighboring states have had to cope with additional adverse effects generated by the downward spiral of the Argentine economy. The conditions are the worst most young Argentineans can remember. The new president, Nestor Kirchner, elected in April 2003 as a traditional Peronist, formed a government of veterans and newcomers. The new government will find it difficult to turn things around. The International Monetary Fund, after threatening to cut off further lending to Argentina, gave the government a "transitional loan" to prevent an official default. The new loan, combined with outstanding ones, means that young Argentineans are further in debt today than they were last year. The disastrous results of the de la Rua economic program provide a cautionary tale of how market reforms can miss their mark and produce devastating results.

Today, millions of Latin Americans are concerned about the prospect of political unrest and economic collapse. Poor citizens in several societies appear willing to give up some measure of democracy and accept authoritarian governments that they believe can solve their problems. Also in 2003, an official commission in Peru confirmed that 69,000 people were killed or disappeared from 1980 through 2000, a period of brutal fighting between the Shining Path rebel group and military forces, known as "a time of national shame."

At the start of 2004, political opponents and rebels were poised to invade the capital of Haiti. They, along with senior U.S. officials, demanded the resignation of President Aristide to prevent Haiti from sliding into full fledge political chaos. Throughout the crisis, France called for an international humanitarian police force to intervene to prevent widespread unrest. However, after Aristide's resignation the Bush administration, rather than the UN, immediately sent military troops to restore peace to the island.

Until September 11, 2001, stopping the flow of illegal drugs was the primary concern of U.S. government officials working in Latin America. Today, new U.S. programs designed to combat terrorism and money laundering schemes in the region by terrorists have been added to an already ambitious set of U.S. programs designed to stem the flow of illegal drugs from war-torn countries such as Colombia, and to promote democracy and human rights throughout the Western hemisphere. Since 2002 the United States has deployed troops selectively to such countries as Colombia to help local troops protect a U.S.-company owned pipeline and to relieve U.S. troops guarding suspected terrorists held without charge on the U.S. military base at Guatanamo Bay, Cuba. Thus, as U.S. foreign policy priority shifted to support the war on terrorism operations, funding for the war on drugs and other longer term programs became secondary priorities.

Article 19

Free trade on trial

The North American Free-Trade Agreement is ten years old this week. It has proved a success, though not in the way its advocates promised

FROM the start, the North American Free-Trade Agreement was bitterly controversial in all three of the countries taking part—the United States, Canada and Mexico. Its terms, which went into effect on January 1st 1994, were argued over line by line: despite its name, the agreement fell far short of scrapping all trade restrictions, and the fine print of the various exemptions and exclusions gave rise to heated argument. More than this, the agreement was attacked as bad in principle. Everybody recognised that NAFTA was an extraordinarily bold attempt to accelerate economic integration—or, as critics put it, an experiment in reckless globalisation. As such, they said, it would destroy jobs, make the poor worse off and start an environmental race to the bottom.

Equally, advocates of the agreement made some bold claims about the good it would bring. Far from destroying jobs, it would create lots of new and better ones; incomes would rise and the poor would benefit proportionately; growth would accelerate and, to the extent that this posed environmental challenges, extra resources would be available to meet them.

Unsurprisingly, a mere ten years' experience has settled few of these quarrels. Today, most trade economists read the evidence as saying that NAFTA has worked: intra-area trade and foreign investment have expanded greatly. Trade sceptics and anti-globalists look at the same history and feel no less vindicated. Look at Mexico's growth since 1994, they say—dismal for much of the period. Look at the contraction of manufacturing employment in the United States. As for the environment, go to the places south of the border where the *maquiladoras* cluster, and take a deep breath.

Politically, the sceptics, ten years on, can fairly claim victory. NAFTA is unpopular in all three countries. In Mexico, which stood to gain most from freer trade (since its barriers were so much higher at the outset) and which has indeed benefited greatly according to most economic appraisals, the agreement is widely regarded as having been useless or worse. In a poll conducted at the end of 2002 by Ipsos-Reid for the Woodrow Wilson Centre in Washington, only 29% of Mexicans interviewed said that NAFTA has benefited Mexico; 33% thought that it had hurt the country and 33% said that it had made no difference. In all three countries, the perceived results of NAFTA seem to have eroded support for further trade liberalisation.

NAFTA's champions are partly to blame for this: they oversold their case. It was never plausible, for instance, to expect that NAFTA would be a net creator of jobs. Trade policy is not a driver of overall employment; it affects the pattern of jobs, rather than the total number. To the extent that NAFTA succeeds in stimulating trade and cross-border flows of investment, jobs in each member country are created in some industries and destroyed in others. This was bound to be a painful process for some, even if it succeeded in making the member countries' economies more efficient overall, and hence in raising average incomes. Here was another instance of false advertising: NAFTA was never going to be, as some enthusiasts claimed, a win-win proposition for all of North America's citizens, even if all three countries could hope to gain in the aggregate.

Yes, it worked

So far as its economic effects are concerned, the right question to ask of NAFTA is simply whether it indeed succeeded in stimulating trade and investment. The answer is clear: it did. In 1990 the United States' exports to, and imports from, Canada and Mexico accounted for about a quarter of its trade; now they account for about a third. That is a dramatic switch, especially when one notes that the United States' non-NAFTA trade has itself grown strongly over the period. There is plenty of economic evidence to suggest that expanded trade, as a rule, raises incomes and future rates of growth. So it is pretty clear that NAFTA achieved as much as one could sensibly have expected it to achieve.

Why then is the agreement so widely regarded by non-economists as a failure? The answer lies partly in the interplay of politics and economics, and accordingly is different in each of the member countries. But one theme is common to all three: a tendency to blame NAFTA in particular, and international integration in general, for every economic disappointment of the past ten years, however tenuous the connection may be.

Debate in the United States has been preoccupied by fears over loss of jobs—by the "giant sucking sound" of work moving south, in Ross Perot's phrase from the early 1990s. A variety of estimates of NAFTA's direct effect on American labour have been made—with job losses running as high, according to one disputed study, as 110,000 a year between 1994 and 2000.

But, as already noted, direct losses do not tell the whole story: changing the pattern of employment is after all one of the reasons for promoting trade. So long as lost jobs are balanced by new ones, the overall effect on employment will be small. As Gary Hufbauer and Jeffrey Schott of the Institute for International Economics point out, between 1994 and 2000 the United States economy created more than 2m new jobs a year. Manufacturing employment has dwindled (with NAFTA as one relatively minor cause among many); jobs in other industries have more than made up the losses. And since the mid-1990s, at any rate, the great majority of new jobs created have paid above-median wages.

Against this background, even NAFTA's highest estimated direct losses can hardly be regarded as crippling. America's evident disenchantment with liberal trade has less to do with the economic depredations of the 1990s—when the economy boomed, in fact—than with a political failure to make the case for free trade against its increasingly vocal and well-organised opponents.

In Canada, initial concerns were less to do with the flight of low-skilled manufacturing jobs, because trade with Mexico seemed a less pressing issue than it was for the United States, and more to do with other sorts of international competition. As it turned out, Canadian unemployment fell markedly during the 1990s (from 11% of the labour force in 1993 to 7% in 2000). The main fear, instead, was that closer integration with the American economy would threaten Canada's European-style social-welfare model, either by leading certain practices and policies (such as the generous minimum wage) to be regarded as directly uncompetitive, or else by pressing down on the country's base of corporate and personal taxes, thereby starving public-spending programmes of resources.

Canadian public spending was indeed squeezed somewhat during the 1990s—not because NAFTA eroded the tax base, but because public borrowing had reached an unsustainable level of 8% of GDP in the early 1990s. The problem was successfully addressed: Canada has lately run a budget surplus. Despite the fiscal retrenchment, and despite NAFTA, its social-welfare model stands intact, and in sharp contrast with that of the United States. The fact is, most Canadians are willing to pay the higher taxes that are required to finance generous public services (including universal health care). As long as this remains true, NAFTA poses no threat to the Canadian way of life.

Down south

What about Mexico? The very point of NAFTA, to listen to some of its advocates, was to destroy the Mexican way of life—and replace it with something better. The overall verdict on NAFTA rests heavily on whether the pact proved a success for the country it was bound to affect most. NAFTA was never going to have much impact on the huge economy of the United States. But as recently as the mid-1980s Mexico was still an almost completely closed economy. For Mexico, NAFTA promised to be revolutionary.

Unfortunately, soon after NAFTA came into effect, the country was overwhelmed by a largely unrelated economic shock, the Tequila crisis of 1994-95. Huge capital inflows into the country in the early 1990s were followed by rapid outflows towards the end of 1994, causing the peso to plunge. The authorities were forced to float the currency on December 20th of that year, and before long it had lost nearly half of its value against the dollar.

The financial system collapsed, with many banks going under as years of bad loans were exposed. In the end, at huge cost, the government had to bail out the banks. The repercussions of the Tequila crisis for Mexico were immense. The banks, for instance, have still not fully recovered, and the subsequent lack of credit and financial services does much to explain the anaemic performance of Mexico's domestic economy over the past decade. All this makes judging the effects of NAFTA very difficult.

Take real wages. Although Mexican workers have managed impressive gains in productivity over the past ten years to compete with America and Canada, real wages have not kept pace. This allows NAFTA's critics to argue that the typical Mexican has not benefited from the treaty as he should have done, and that big business has creamed off most of the profits.

The truth is different. The Tequila crisis led to an immediate fall of about 20% in Mexican wages (more in dollar terms), while productivity kept going up. So although real wages have been rising ever since the country began to recover in 1996, they are only just reaching their levels of before the crisis. The lasting influence of higher productivity on wages may not be clear for another decade, when the effects of the Tequila crisis have fully faded away. That said, the country recovered much more quickly from the Tequila crisis than from its previous financial crises in 1982 and 1986—and this was indeed mainly due to NAFTA. Speedily arranged help from Bill Clinton's administration spurred the strong recovery. That aid sprang from America's desire not to let its new partner go under.

The closeness of the link to America, the destination of almost 90% of Mexican exports, is of course a disadvantage when America goes into recession, as it did in 2001. Mexico lost thousands of export jobs in that downswing. On the other hand, NAFTA has insulated Mexico against the financial instability that swept through Argentina, Brazil and other parts of South America in the first years of the new century. It has given Mexico an investment-grade credit rating, and allowed it to issue—almost uniquely in Latin America—very long-term local-currency bonds and mortgage-backed securities. Investors now think of Mexico more as a North American than a Latin American country.

Former President Carlos Salinas de Gortari embraced NAFTA mostly to attract more foreign investment and to boost the *maquiladora* manufacturers (set up in 1965 to allow tariff-free import of materials for assembly and re-export to the United States). Mexico's trade has surged, especially with the United States. In 2002 it totalled $250 billion, and the country's traditional deficit with its northern neighbour has been converted into a surplus in every year of NAFTA membership.

After 1994 foreign direct investment also shot up. NAFTA was designed to make investors feel more legally secure, and foreign companies duly poured in to take advantage of Mexico's closeness to the world's wealthiest market. The rise in export manufacturing also greatly reduced the country's dependence on the volatile price of oil. Moreover, NAFTA jobs in export businesses have usually been good ones, paying on average substantially more than jobs in the rest of the economy.

It hardly needs saying, however, that Mexico has no shortage of problems that NAFTA has so far failed to solve. One is the challenge of providing decently paid work for all those who need it. The chief symptom of the failure to do that, of course, is the continuing outflow of migrants.

The biggest pressure on emigration, in turn, is the crisis in the countryside. The traditional Mexican farmer had about eight hectares of his own land and some communal land for livestock. This made his family self-sufficient in everything from maize and beans to meat and milk. Even before NAFTA this traditional rural economy was disappearing, as demographic pressure caused the land to be subdivided, and many *campesinos* now eke out a living year by year, ever on the edge of disaster. "If the weather does not help us, we are completely lost," says Dionisio Garcia, who farms a smallholding in the southern state of Tlaxcala.

Most of Mr Garcia's colleagues have simply given up. He estimates that up to 90% of the heads of families in his area now spend at least six months of the year working in Canada or the United States. "What they earn there in four months, we don't earn here in a year," he says. They are part of an estimated 1.3m people who have left the land since 1994. The young, besides, are no longer much interested in making a living from the land; they are going off to drive taxis in the city, or to sell air-conditioners.

Mr Garcia says that he can no longer sell his surplus maize to Mexican wholesalers because he has been undercut by cheaper and better American imports. For him, NAFTA and free trade have been "totally bad". And yet trade in Mexico's two staples, maize and beans, is still not free; the last tariffs will remain until 2008.

The flood of corn from America's midwest is the most hated aspect of NAFTA for Mexicans. The government argues that it has to import so much because Mexico's small farmers cannot feed all Mexicans, let alone turn a profit. But critics allege that Americans are selling so cheap that they are, in effect, dumping the stuff. Besides, they receive vast subsidies from their government. NAFTA explicitly pledges to eliminate these, but it has not done so yet.

Some Mexican farmers have shown that they can make a good living under NAFTA. Export earnings from horticulture have tripled since 1994, to over $3.5 billion; exports of fresh vegetables have risen by 80% and fresh fruit by 90%. If farmers can exploit local conditions and invest in a crop that can be exported during American or European winters, they can make money. The star performer is the Hass avocado from the state of Michoacán, in the west of Mexico, where the climate is mild and the soil fertile. Before NAFTA, the United States banned it because of infestation by insects. After a clean-up and monitoring operation, supervised under NAFTA rules, avocados from Michoacán were accepted into most states of America in 1997. Exports have increased from 6,000 tonnes to 30,000 tonnes a year.

Overall, though, Mexico continues to rely on low-cost assembly, and the advantage of preferential entry into the American market. Increasingly, other countries offer cheaper labour. With China's accession to the World Trade Organisation, Mexico has already lost much of the advantage that NAFTA gave it. Many Mexicans still think that a reviving American economy, by itself, can buoy their own. But in the next upswing, as America deepens its trade links with other states, this may prove untrue.

NAFTA alone has not been enough to modernise the country or guarantee prosperity. It was never reasonable to suppose that it would be—though that did not stop many of its advocates saying so. NAFTA has spurred trade for all its members. That is a good thing. But trade can do only so much. Sadly, successive Mexican governments have failed to deal with the problems—corruption, poor education, red tape, crumbling infrastructure, lack of credit and a puny tax base—that have prevented Mexicans and foreign investors alike from exploiting the openings which freer trade afforded. Don't blame NAFTA for that.

From *The Economist*, January 3, 2004, pp. 13-14, 16. © 2004 by The Economist Newspaper, Ltd. Reprinted with permission. Further reproduction prohibited. www.economist.com

Latin America's New Political Leaders: Walking on a Wire

"Today's underlying political currents in Latin America are less about ideology and more about a public desire to find leaders who can effectively address everyday problems, and who do so honestly. The formulas of the past—whether 'socialism' in the 1970s or 'neoliberalism' in the 1990s—have been widely questioned, and largely dismissed. With traditional ideas and structures breaking apart, new leaders are being called on to produce results."

MICHAEL SHIFTER

In 1970, Salvador Allende finally ascended to the Chilean presidency on his fourth attempt. The Socialist candidate with a radical agenda won the prize that had long eluded him—only to be overthrown in a generation-defining military coup led by General Augusto Pinochet on September 11, 1973.

Nearly three decades after the Chilean coup, new faces of political leadership are emerging in Latin America, giving form and definition to the next generation. The sharply drawn left–right battles and schisms intrinsic to the cold war no longer drive politics. Rather, today's underlying political currents in Latin America are less about ideology and more about a public desire to find leaders who can effectively address everyday problems, and who do so honestly. The formulas of the past—whether "socialism" in the 1970s or "neoliberalism" in the 1990s—have been widely questioned, and largely dismissed. With traditional ideas and structures breaking apart, new leaders are being called on to produce results.

In 2003, Luiz Inácio da Silva, widely known as "Lula," is the principal focus of regional attention. Lula was resoundingly elected Brazil's president after he had failed, like Allende, in three previous attempts. Predictably, as the leader of the Workers' Party took the reins of Latin America's largest and most significant country, all eyes have turned to the region's new, charismatic leader—and the most promising experiment for social renewal.

Will Lula balance Brazil's mounting demands for greater social justice and equality with the formidable constraints and pressures imposed by international financial institutions? Will the United States not only recognize the need for but also help support Lula's alternative social policies? In short, will Lula succeed in charting a new path?

Brazil's passing of the guard from President Fernando Henrique Cardoso to Lula epitomizes the shift in Latin America in the political winds, and in political leadership, over the course of the past decade. Cardoso, not long ago considered the quintessential leftist intellectual, served two terms as president of a center-right government whose main achievements included democratic continuity and economic stability. At the same time, Cardoso's government accumulated a huge public debt, failed to achieve sustained growth, and despite improvements in many social indicators, could not make a significant dent in Brazil's vast disparities in wealth. Brazilians—from the poor to a sizeable share of the business community—in the end clamored for political change.

A SOUR MOOD

For Latin America, Lula's election to the Brazilian presidency comes at a critical moment. Stories abound in publications such as *The Economist*, *Financial Times*, and *Newsweek* about a backlash against neoliberalism, a resurgent populism, and a decided turn to the left in Latin America. The high hopes and expectations that accompanied the early 1990s, when free trade and spreading democracy were widely touted, are a fading memory.

The regional economic picture is especially gloomy and dramatic. According to the World Bank, Latin America in 2002 suffered its most disastrous economic performance in nearly two decades, with a negative growth rate of 1.1 percent. In Argentina and Venezuela the outlook is notably grim, marked by stunning economic declines and social disintegration. Experts argue that even under the most sanguine scenarios, any substantial economic rebound in these countries will take years, if not decades.

Moreover, since 1998, for the region as a whole, per capita income has dropped some 0.3 percent per year. Although projections for 2003 are slightly more upbeat, with the regional growth rate possibly reaching 2 percent, it is difficult to imagine that such a modest performance will bring much relief to Latin America's acute social conditions.

Public opinion survey data also offer little to cheer about. According to the respected Latino-barómetro, which has conducted a comparative survey annually since 1995, public dissatisfaction with government performance has steadily risen because of stubbornly poor and disappointing economic results. The inability of governments in many countries to reduce mounting unemployment and crime has also fueled enormous citizen discontent. Although the survey findings reveal that most Latin Americans continue to see democracy as the preferred political model and that they have not given up on market-oriented prescriptions for economic difficulties, these views are largely arrived at by default and are far from a ringing endorsement of the core ideas that were expected to deliver improvements in citizens' lives. Frustration with economic policies that have yielded few tangible benefits runs deep throughout the region.

Despite so many profound problems, elections remain the accepted way to select leaders. Viewed in wider historical perspective, this adherence to the democratic norm is remarkable and cause for optimism. But the signs suggest voters are groping for fresh answers and alternatives to the status quo. The profile of a Latin American leader—perhaps exemplified by Cardoso—who is an internationalist and blends technical or academic prowess and sophistication with a knack for politics—appears to be receding. "Populist" or "leftist" may be the most convenient and common terms employed to characterize new leaders who deviate from this model, and who represent the promise of a more thoroughgoing and committed social agenda. But there are scant resources available to fund meaningful social programs. And at a time that has seen the questioning and dissolution of traditional political structures and ideas, these labels obscure more than they illuminate.

BRAZIL'S "AMERICAN DREAM"

Lula differs fundamentally from many of the other political leaders who have recently emerged on Latin America's political stage. What sets him apart is his longstanding dedication to building the Workers' Party (PT), widely regarded not only as the most solid and coherent political party in Brazil, but as among the most effective parties in Latin America. Lula's two-decade project—born under a Brazilian dictatorship that began in the 1960s and only came to an end in 1985—is especially noteworthy in a region where political parties are in deep crisis, broadly discredited, and perceived as ossified and corrupt.

Although the PT has yet to be tested at the national level, it has had ample—and successful—governing experience at the state and especially municipal levels in Brazil. The PT is far from the amorphous movements and inchoate groupings that often pose as political parties in other Latin American countries. Yet it is not monolithic but composed of factions that range from decidedly hard-line to pragmatic.

Lula himself has undergone a remarkable political evolution. Over the years his positions on a variety of issues have moderated. In the 2002 election campaign he was referred to as "Lula lite" because of the alliances he made with more conservative factions, a move that would have been difficult to imagine just a few years ago. Moreover, his personal story is extraordinary, embodying the "American dream" with his rise from poverty through the ranks of the powerful metalworkers' union and eventually to party leader. It also testifies to the vitality of Brazil's democracy and the gradual loosening of its rigid social structures.

Lula's call for a "social pact"—agreements involving government, business and labor unions aimed at assuring economic and political stability—signals a shift to an alternative political arrangement, perhaps more in line with European concepts. Still, it will not be easy for Lula to contend with pressures from members of his own base, who see the PT in power as "their turn." Demands made by the international financial community will similarly test Lula's political talents. His administration's success may hinge less on the ideological direction Lula decides to pursue than on the competence of his team and the ability to devise sound and coherent policies that deliver growth and some social progress while maintaining economic stability.

CHÁVEZ'S VENEZUELA: REVOLUTION OR CHAOS?

Nearly four years before Brazilians elected Lula as president, the majority of Venezuelans put their confidence in Hugo Chávez. Chávez, a former paratrooper who led a failed military coup in February 1992, was the product and beneficiary of a traditional political party system that had collapsed from decades of corruption and mismanagement. Since the early 1980s, no Latin American country has suffered the precipitous economic decline and social decay that has struck Venezuela, made all the more egregious when viewed against the country's substantial oil wealth. A succession of failed administrations and a protracted period of party unresponsiveness led to a sharp public repudiation. Chávez, a superb orator who issued a scathing indictment of the old order, took advantage of the political vacuum.

While there is still some question about which of the "two" Lulas will ultimately leave his mark on Brazil's presidency, there is little doubt that Chávez's authoritarian instincts are stronger than his democratic tendencies. He has often defied and violated the constitution and has constantly lashed out at key elements of Venezuelan civil society, such as the media, the Roman Catholic Church, and unions. Gabriel García Marquez, Colombia's Nobel Prize–winning author, wrote an especially prescient profile of Chávez just before he took office in February 1999: "I was overwhelmed by the feeling that I had just been traveling and chatting pleasantly with two opposing men. One to whom the caprices of fate had given an opportunity to save his country. The other, an illusionist, who could pass into the history books as just another despot."

It is clear that honesty and effectiveness— not rigid formulas or sweeping blueprints— are what the current generation is looking for.

Although Chávez has been variously described as a "populist," "revolutionary," and "leftist," he has been so in rhetoric only. He seems, rather, a throwback, a figure who more closely resembles Argentina's legendary Juan Domingo Perón of the 1950s and 1960s. His actions and politics have been erratic and inept, with detrimental results for Venezuela. (The poorest Venezuelans, Chávez's core constituency, have suffered most from the country's prolonged drift and chaos.) Moreover, his confrontational style, with the charged rhetoric about Venezuela's "rancid oligarchy," has only exacerbated social tensions.

In 2002, Venezuela endured unprecedented political and social polarization, along with extraordinary bitterness and mistrust. The government and its supporters jockeyed for position with opposition forces—made up of an array of nongovernmental and professional groups and diverse political figures organized under the umbrella Democratic Coordinator—that mounted a strike to press for Chávez's ouster, either through his resignation or through new elections. With each side digging in, the prospects for any reconciliation seemed negligible. In the unfolding crisis, critical factors included Chávez's ability to figure out how to keep the oil flowing. For the world's fifth-largest supplier of oil, the fate of the petroleum sector and the state-run oil company Petróleos de Venezuela, SA, is decisive.

Another key factor is the Venezuelan military, which has also been badly divided between forces that support and those that oppose Chávez. As the violence has escalated and threatens to erupt into something that could conceivably resemble a civil war, many observers have tried to anticipate how, and at what point, the armed forces might react.

When this crisis is resolved. Venezuela will need considerable time to heal the deep wounds that have for many years run through society and have been aggravated under Chávez. But an opposition that lacks effective political leadership, a clear strategic focus, and any vision for a post-Chávez Venezuela—especially ideas about how to unite the country, including Chávez's mostly poor constituency—does not augur well for a swift reconciliation. Chávez's ability to hold on to a core base of support in the context of such a disastrous performance underscores the depth of rage and mistrust many Venezuelans feel toward an old order in need of deep-seated reform.

ECUADOR'S CONSUMMATE OUTSIDER

Ecuador also has seen a former coup plotter elected to the presidency. Like Chávez, Lucio Gutiérrez was a former army lieutenant colonel who tried to topple a democratically elected government. Joining with Ecuador's powerful indigenous movement and other military officials, Gutiérrez, unlike Chávez, succeeded in his January 2000 attempt. Although short lived, the coup catapulted Gutiérrez onto Ecuador's political stage and, less than three years later he was, to the surprise of many, overwhelmingly elected president.

Although Gutiérrez expressed admiration for Chávez at the time of the coup, he has since sought to distance himself from the controversial Venezuelan president. The initial signs suggest that, unlike Chávez, Gutiérrez will attempt to reach out and engage with his country's broad-based civil society and managerial and entrepreneurial sectors. The new Ecuadoran president's announcement of his intention to seek an agreement with the International Monetary Fund, his acceptance of the country's "dollarization" scheme (under which the United States dollar became the official currency in January 2000), and his emphasis on fiscal discipline do not signal a shift to the left or the beginning of a resurgent populism.

Yet Gutiérrez is, in one key sense, similar to Chávez. He too is the product of the wholesale rejection of his country's traditional political class, which has generally failed to deliver tangible benefits to vast sectors of the population. That Gutiérrez—as well as his rival in the second round of voting, banana magnate Álvaro Noboa—are the consummate political "outsiders" shows that most Ecuadorans have little appetite for "more of the same" (Ecuador's politics has been notably unsettled; Gutiérrez is the country's fifth president in six years). To be politically successful, he will have to produce results for the country's most marginal group, especially the poorest sectors and a sizable indigenous population.

At the same time, Gutiérrez faces formidable challenges. He inherits an enormously difficult economic situation. Although he can count on the important support of the Confederation of Indigenous Peoples of Ecuador, Latin America's strongest and most coherent indigenous confederation, to pursue his reform agenda, Gutiérrez will nonetheless have to deal with a broad array of parties represented in a remarkably fragmented Congress. In a country distinguished by notoriously sharp geographic, ethnic, and social divisions, Gutiérrez, a political neophyte, will find it difficult to forge alliances and build coalitions to accomplish his goals of reducing corruption, making the Ecuadoran economy more competitive, and working toward greater social equality.

THE EXPECTATIONS GAME

In taking on such a mammoth political task, Gutiérrez may find the experiences of other current Latin American presidents instructive. With their parties failing to enjoy a majority in their respective legislatures, Mexican President Vicente Fox (in office since December 2000) and Pe-

Article 20. Latin America's New Political Leaders: Walking on a Wire

ruvian President Alejandro Toledo (elected to the post in June 2001) have had some difficulty constructing coalition governments to implement their agendas.

After more than seven decades of single-party authoritarian rule, Mexico expressed its profound desire for change by voting for a former Coca-Cola executive and National Action Party governor from Guadalajara. But Fox has failed to satisfy the high—perhaps unrealistic—expectations that were generated by his administration. To be sure, Mexico's political system has opened up considerably—indeed, irreversibly. And the Fox government has escaped the many serious charges of corruption that had been leveled against previous Institutional Revolutionary Party (PRI) administrations. After more than two years in office, his approval rating is still roughly 50 percent.

Criticism has revolved around Fox's inability to "get things done," most notably in fiscal reform as well as other areas. The Mexican president has struggled to strike deals and work effectively with a Congress still dominated by the PRI. Fox's cabinet and advisers are capable, but have not come together as an effective team. Still open to question is Fox's ability to articulate a clear program for the country, devise a political strategy, and instill the necessary discipline to achieve his main policy aims.

Peru's Toledo came to office following an eight-month transition government in Peru that had been preceded by the highly corrupt and decade-long authoritarian regime of President Alberto Fujimori and his national security chief, Vladimiro Montesinos. Unlike Fox, Toledo—born in poverty and proudly indigenous—had virtually no political background, apart from leading the opposition against the previous government, which he did with great courage. In addition, his political party, Péru Posible, is very heterogeneous and does not have programmatic coherence or discipline.

Toledo's lack of political experience and struggle to gain some traction have exacted a heavy toll. After his first year in office, Toledo's approval rating had dropped from a high of nearly 70 percent to below 20 percent. Toledo promised considerably more—"thousands of jobs"—than he has been able to deliver, frustrating many Peruvians. He has had difficulty making decisions and outlining a coherent policy course. In June 2002, Toledo's failure to consult with local officials about the privatization of electrical companies in Peru's second-largest city of Arequipa helped spark widespread protests, with huge political costs. Toledo's self-inflicted difficulties have been compounded by the ferocity of opposition politics and signs that remnants of the Fujimori regime may be trying to sabotage the current administration.

Toledo's slide—and the mounting protests and unrest in the streets—are all the more remarkable in light of the country's 4 percent growth rate in 2002, the region's highest. In addition, the Toledo government has vigorously pursued human rights abuses and corruption that occurred under previous governments. After resolving a paternity suit and recognizing his daughter in October 2002, Toledo, who had mishandled the controversy, recovered some public support. The change did not, however, translate into an electoral victory for his party in Peru's regional elections in November.

The ability to master coalition politics and shape a governing consensus will also be the crucial question in Bolivia, under the administration of Gonzalo Sánchez de Lozada, who assumed office in August 2002. Sánchez de Lozada, who previously served as president from 1993 to 1997 and presided over wide-ranging reforms, including innovative capitalization and privatization schemes, faces an even more monumental task than he did in his first term. His relationship with his coalition partner, Jaime Paz Zamora of the Movement of the Revolutionary Left, is strained. Most significantly, Sánchez de Lozada will have to learn how to work with Evo Morales, the popular indigenous figure and leader of the numerous and increasingly mobilized coca growers. Morales, who is pressing for major social changes and could upset Bolivia's political system, registered stunningly strong support in the last election. As in Ecuador, the indigenous population in Bolivia is an important actor in shaping the country's politics.

Also in August 2002, Colombia, the only Latin American country in the midst of a civil conflict, elected a new leadership when Álvaro Uribe succeeded Andrés Pastrana as president. Uribe won a resounding victory and secured an impressive mandate in the May 2002 elections. He promised to implement his vision of "democratic security," applying a firm hand to the country's armed actors. With the collapse of peace talks in February 2002 and the hardening of domestic and international public opinion, Colombians have rallied around their new president's plan to make the country's security forces more effective. Five months into his term, Uribe enjoyed a 75 percent approval rating, the highest of any elected leader in Latin America.

Although Uribe successfully obtained public backing for his agenda, it is uncertain whether that backing will be fleeting or enduring. Much, of course, will depend on his administration's ability to reassert government authority and control over the country and protect Colombians against widespread violence. Colombia's fiscal imbalance is also serious, and it will be hard to adopt the familiar formula for greater spending restraint within a "war economy." Sharp cutbacks could well increase the risk of greater social polarization. Still, Uribe's leadership style—he is indefatigable and has built a highly disciplined cabinet and team of advisers—has broad appeal in Columbia, and he appears to have taken welcome initiative and generated some momentum. To the delight of most Colombians, Uribe is genuinely in charge.

Frustration with economic policies that have yielded few tangible benefits runs deep throughout the region.

Fundamental strategic concerns loom, however. Even under the most optimistic scenario, it will take Colombia years to substantially reverse its long-term deterioration in public order, much less to pursue meaningful social reform and reconciliation. Uribe will have to manage high public expectations and will need to think through—beyond adopting a series of new, albeit risky, security measures—how to end the conflict and successfully incorporate the country's varied armed actors into Colombian society. Whether he will be able to exhibit the necessary flexibility, vision, and commitment to human rights standards under such extraordinarily difficult conditions remains a major question.

WHAT HAPPENED TO POLITICAL RENEWAL?

Rarely has Latin America witnessed the sort of implosion of a major country like that which occurred in Argentina in December 2001. As a result of a series of converging factors both internal and external, the government led by President Fernando de la Rúa collapsed, followed by a foreign debt default of some $140 billion. The country lived through a rapid succession of presidents and tremendous uncertainty until Congress finally settled on Eduardo Duhalde, a Peronist, as Argentina's leader.

Duhalde's political skills have received mixed reviews as he has struggled to gain the necessary authority to govern the country effectively. According to the current timetable, Argentines will select Duhalde's successor in April 2003. Yet few analysts believe the new president will be able to restore the severely damaged legitimacy and credibility of the country's traditional political class; that is a longer-term challenge. Today, political parties and leaders are held responsible for Argentina's meltdown, an attitude reflected in the plummeting levels of public confidence in survey after survey. It is instructive that Carlos Menem, Argentina's wily two-term president who brought the country economic stability in the 1990s but has been accused of massive corruption and disregard for democratic institutions, is a serious contender for his old job. His support suggests that, for at least some Argentines, a proven ability to get the job done may trump deep ethical qualms.

Perhaps even more striking is the prospect that Alan Garcia, who presided over Peru's unremitting economic chaos and political violence in the late 1980s, has a good chance of returning to his old job in 2006. In 2001 Garcia, a crafty politician and superb communicator, received more than 48 percent of the vote in the second round runoff election against Toledo. And in November, his American Popular Revolutionary Alliance was the big winner in Peru's regional elections. Further, the headline of a *New York Times* article in December 22, 2002—"Peru's Former President Plots His Return to Power"—suggests a possible comeback as well by Fujimori, who resigned in disgrace and is in exile in Japan. For those who have stood for reform and renovation, few developments would be more dispiriting. While this prospect is hardly reassuring, it highlights the continuing search for political leadership in Latin America.

IS ANYONE PAYING ATTENTION?

Latin America's new cast of political leaders clearly reflects profound changes within the region's complex societies. The landscape is evolving in fundamental ways, conditioned by globalization—its benefits and downsides—and substantially influenced as well by the United States. And the emerging strategic concept, developed in Washington in response to the terrorist attacks on September 11, 2001 and manifested in shifting priorities and resources, has a direct bearing on Latin America's political dynamics. The region's prevailing political uncertainties pose a crucial test for United States policy in the hemisphere.

The key question is whether the United States can be positively and wisely engaged in Latin American affairs so that it can help support leaders committed to democratic politics and economic policies that are market oriented and emphasize a strong social agenda. Such leaders understandably resist the failed formulas for economic reform championed by Washington but are nonetheless generally interested in a cooperative relationship with the United States.

The initial signs from the United States in the post–September 11 period are not encouraging. The principal effect appears to have been distraction, as senior government officials have been consumed by urgent actions and decisions regarding the Middle East and the global effort to combat terrorism. Latin America has seldom been a high priority for the United States, but the lack of high-level sustained attention and engagement has coincided with unprecedented turbulence and deepening crises in many key countries. In the context of globalization, more is at stake for the United States in the region than in the past.

Illustrations of this indifference, or neglect, abound. Perhaps the most dramatic case is Argentina, where the United States treated the country's crisis largely as a fiscal matter, overlooking critical political and foreign policy implications. Former Treasury Secretary Paul O'Neill's remarks were seen as particularly callous. (In an August 2001 interview with CNN, for example, O'Neill said: "We're working to find a way to create a sustainable Argentina, not just one that continues to consume the money of the plumbers and carpenters in the United States who make $50,000 a year and wonder what in the world we're doing with their money.")

It is not surprising that, according to the Latinobarómetro, anti-American sentiment—that is, criticism of United States policies—is appreciably higher in Argentina than anywhere else in the region. Although it would be a stretch to attribute Argentina's political precariousness to the passivity of the United States government at a critical moment, greater political engagement might have reinforced American backing of democratic leaders.

In Venezuela, too, the absence of any strategic engagement is striking, especially in view of a possible war with Iraq and the fact that Venezuela accounts for roughly 15 percent of crude oil imports to the United States. American ineptitude was evident in its handling of both the botched coup in April 2002 and the public call for "early elections" in December 2002. The tacit approval of an unconstitutional act in the former case—the State Department initially failed to express any concern about Chávez's forced ouster—and the public association with opposition forces in the latter have done little to enhance United States credibility on the democracy question in Latin America. (The image of Venezuelan generals standing behind a business leader they had tried to install as president had especially chilling echoes of previous episodes of military takeovers in Latin America, including the American-backed Chilean coup of 1973.) In the current crisis, Washington has also failed to take full advantage of regional institutions to press for a constitutional resolution.

In Mexico it is difficult to separate Fox's political fortunes from his administration's relationship with the United States. This is especially so in light of the expectations raised by the personal relationship between Fox and Bush and the excitement that accompanied the state dinner for Fox held in Washington just days before the terrorist attacks. Without assigning any responsibility, it is undeniable that Mexico's lower place on the American foreign policy agenda in the post–September 11 period—and growing friction between the two countries—have created some political fallout for Fox. Although bilateral cooperation on myriad issues remains significant, Mexico's domestic politics are unusually sensitive to American actions and decisions.

In country after country, it is clear that what the United States does—or fails to do—is germane to the national political scene, and often affects the prospects of particular leaders. In his first year in office, for example, Toledo went up in opinion polls only on two occasions: first, when President George W. Bush visited Lima and second, when the United Sates Congress approved Andean trade legislation extending preferences and giving products from Peru, Colombia, Bolivia, and Ecuador greater access to United States markets. The long-anticipated free trade agreement between the United States and Chile, signed in December 2002, gave a political boost to Chile's president, Ricardo Lagos. The Bush administration's bailouts of Uruguay ($1.5 billion) and Brazil ($30 billion), which sought to contain contagion from Argentina's collapse and buttress Brazil at a moment of political uncertainty, also yielded political dividends in those countries. And in a telling twist, a critical statement made by the United States ambassador in Bolivia regarding the coca growers' candidate, Evo Morales, nearly succeeded in getting Morales elected! ("I want to remind the Bolivian electorate that if they vote for those who want Bolivia to return to exporting cocaine, that will seriously jeopardize any future aid to Bolivia from the U.S.")

Some in Washington, taking advantage of the transformed climate following September 11, have referred to a new "axis of evil" in Latin America, with Lula and Gutiérrez joining Chávez and, of course, the stalwart Fidel Castro. Such hysteria even found its way into a disturbing October 24 letter to President Bush from Republican Representative Henry Hyde, the chairman of the House International Relations Committee, who warned that Lula was a dangerous "pro-Castro radical who for electoral purposes had posed as a moderate." If the United States government were to fashion policies based on such simplistic, Manichaean terms, the Latin American "axis of evil" would risk becoming a self-fulfilling prophecy.

A more apt phrase that captures the political dynamic in much of Latin America is "axis of upheaval." The region's turmoil offers an opportunity for the United States to avoid the sloppiness that often results from neglect—or the hard-line unilateralism that tends to flow from a "good guy versus bad guy" mentality. Positive, sustained, high-level engagement could translate into greater sensitivity to the problems facing Latin American countries. What is required from Washington is increased flexibility that gives struggling leaders in the midst of enormously difficult circumstances more room to maneuver and undertake new social and economic policies. How the Bush administration deals with the Lula presidency is likely to be a major test in this regard.

But the chief responsibility for shaping Latin America's political future rests squarely with a fresh set of political leaders. Their task will be far from easy. Latin America's old problems, such as endemic poverty and fragile institutions, persist and are only aggravated by heightened expectations held by ever-expanding segments of the region's population. Globalization's liberating forces have made this possible. But globalization's dark side is also keenly evidence in the region, with a widening schism between intensifying social demands and political institutions in disarray.

In seeking to navigate their way through these complex challenges, Latin America's new political leaders will be put to a severe test. It is clear that honesty and effectiveness—not rigid formulas or sweeping blueprints—are what the current generation is looking for.

MICHAEL SHIFTER, *a Current History contributing editor, is vice president for policy at the Inter-American Dialogue and an adjunct professor at Georgetown University.*

UNIT 6
Europe

Unit Selections

21. **Europe Enlarged, America Detached?**, Simon Serfaty
22. **America as European Hegemon**, Christopher Layne
23. **Forget Asylum-Seekers: It's the People Inside who Count**, The Economist
24. **How the Armies of Europe Let Their Guard Down**, Philip Shiskin
25. **A Nervous New Arrival on the European Union's Block**, The Economist

Key Points to Consider

- Explain why you do or do not believe an enlarged European Union will lead to more integration within NATO.

- Who should pay for NATO's activities "out-of-area"?

- Given the differences between European countries and the U.S. over Iraq, do you agree with Christopher Layne that an enlarged Europe, France, Germany, Russia, and sometimes China are developing new habits of diplomatic cooperation to oppose the United States?

- Do you agree that one of the most challenging tasks facing European countries is how to integrate the minorities they already have into their national and regional societies?

- Is the difference between the United States and European national military merely a difference of spending more or less on their military?

- After gaining membership in the European Community, is Poland likely to become a major factor within the European Community in the future?

 **Links: www.dushkin.com/online/**
These sites are annotated in the World Wide Web pages.

Central Europe Online
http://www.centraleurope.com

Europa: European Union
http://europa.eu.int

NATO Integrated Data Service
http://www.nato.int/structur/nids/nids.htm

Social Science Information Gateway
http://sosig.esrc.bris.ac.uk

After the September 11, 2001 attacks in the United States, European investigators were surprised to learn that an extensive interlocking set of terrorist cells in Italy, Germany, Spain, Britain, France, Belgium, and numerous other countries remained in place. Many of the members of these cells were indoctrinated with combat videos from Chechnya, absorbed into al Qaeda by Osama Bin Laden's agents in Europe, and trained in Afghanistan for operations against the West. As authorities continue to identify hidden cell members within their own societies, they coordinate their efforts to rout al Qaeda fighters who had infiltrated the ranks of ethnic Albanian guerrilla forces in Macedonia, Croatia, Bosnia, and Kosovo. Two of the suicide hijackers involved in the September 11 attacks have been traced to a training camp in Bosnia.

The year of 2002 saw a remarkable transformation in Europe as Simon Serfaty describes in "Europe Enlarged, America Detached?" The single currency in 2002, enlargement to 10 new members in 2002, the constitutional convention initiated in 2003, and the intergovernmental Conference (IGC) scheduled for 2004 are marks of the final transformation of Europe. While the EU is achieving a new territorial and political synthesis, NATO has increasingly become an afterthought. Serfaty describes the "new normalcy" and urges more, not less, integration between the EU and NATO. The significance of NATO was evidence by the fact that the Baltic states of Estonia, Latvia, and Lithuania were offered North Atlantic Treaty Organization (NATO) membership and were invited to join the European Union in 2004.

To cope with the upcoming integration, each country is trying to shift the emphasis of its economy to high-tech, "knowledge-based" industrial niches. Only a few years ago the conventional wisdom was that Russia would never let her former Baltic satellite states negotiate formal ties with NATO. However, NATO has been adept at transforming itself into a collective security organization more like the UN than a collective security defense alliance. Most observers assume the full Russian membership in NATO is only a matter of time now that leaders of western countries and Russia share a common interest in the fight against terrorism. Deep fissures continue to exist within the alliance but the recent expansion defies critics who routinely pronounce NATO to be a "dead" organization.

In "America as European Hegemon," Christopher Layne describes the U.S. aim in Europe as one of asserting hegemony, and France and Germany seek to create a European counter balance to U.S. hegemony. Within a widened Europe—France, Germany, with Russia and sometimes China—are developing new habits of diplomatic cooperation to oppose Washington, while the United Kingdom and newer members of a widened Europe work closely with the United States.

As the nature of the threat changes, NATO is creating a special rapid-reaction brigade in response to fears that terrorists could attack military units as well as civilians with nuclear, chemical, or biological weapons. At least 13 member nations, including the United States, have enrolled in the battalion. The unit, known as the Chemical, Biological, Radiological, and Nuclear Defense Battalion (CBRN), will be able to rapidly deploy mobile analysis labs that can work in contaminated areas, operate a specialized infection hospital that would carry stocks of vaccines against biological weapons for deployed forces, do reconnaissance and risk assessments, and perform light and heavy decontamination of people and vehicles. It will enable other NATO troops to carry out missions that otherwise would be threatened by a nuclear, biological, or chemical attack. The battalion will operate both independently and as part of the new NATO Response Force.

European members of NATO continue to maintain major commitments for peacekeeping operations in Kosovo and other states in the former Yugoslavia Federation. At the end of 2001, the European Community assumed the lead role for peacekeeping operations in the former Yugoslavia Federation. However, NATO's military campaign in the region revealed huge military disparity between the United States and European members of NATO. The United States carried 90 percent of the load in the 47-day air campaign. After the Kosovo experience, leaders of several European countries agreed to create a new regional defense structure, the European Security and Defense Identity. The purpose of the new organization is to make it possible for Europeans to operate in situations in which NATO is not engaged. The dilemma is that most European governments lack the military capability to equip or sustain the new force.

Immediately after the military campaign in Afghanistan, Germany declined to serve as the lead nation on the ground for peacekeeping in Afghanistan due to other international commitments. Instead, Turkey served as the first lead nation of allied forces in Afghanistan. Turkey turned over the reigns as the lead in February 2003 to Germany and Dutch nations, who serve as "partners in leadership." This joint leadership is necessary because Germany lacks the capacity to serve as the lead peacekeeping nation-state alone. Neither country has the capabilities to manage the situation if the warlords misbehave or Afghan President Hamid Karzai falls into real trouble. While the lightly armed German Bundeswehr and their Dutch partners can keep the peace, they are ill prepared to fight. Moreover, the international peacekeeping mission controls the capital but little more. The shortcomings of the German and Dutch in Afghanistan illustrate the larger problems of the limited capacity of European militaries for "out-of-area" operations.

As Philip Shiskin notes in "How the Armies of Europe Let Their Guard Down," the 17 European countries in NATO have about 2.3 million active-duty troops, but many of these European national forces are poorly equipped, in part because so much money is spent on pay and benefits for a substantially older military than those maintained by the Untied States. While the U.S. spends 36% of its defense budget on pay and benefits, most NATO members earmark an average of 65% for personnel expenses. This expenditure leaves less for technology, weapons, and other gear needed by a modern force.

Most analysts agree that NATO's future depends less on the military structures and capacities than on the ability of member states to develop a common political purpose. Tensions among members of the western alliance may increase in the future since recent public opinion polls indicate that Americans and Eu-

ropeans do not see eye to eye on many issues. Important differences over such key issues as global leadership, defense spending, and the Middle East are likely to tests the seams of the NATO alliance. The Bush administration's rejection of the Kyoto Protocol on Global Warming, the U.S. refusal to ratify the International Criminal Court, and the detention without charge of several European citizens in the U.S. under the new Patriot Act are three of several issues that are likely to fuel the gap between Americans and Europeans on key political issues. However, U.S. hegemony may not last for long. A month after NATO's historic enlargement, the EU redefined its borders by inviting seven former Communist nations to join: Poland; the Czech Republic; Hungary; Slovakia; the former Soviet republics of Lithuania, Latvia, and Estonia; the ex-Yugoslav Republic of Slovenia; and two Mediterranean islands. Cyprus and Malta will formally join the European Union in May 2004. The historical expansion eastward will put an additional 75 million people under the union's banner and add 23 percent to its territory.

The European Union is preparing a quantum leap eastward by adding 100 to 200 million new citizens from central eastern and southeastern Europe. Most analysts are uncertain whether this unprecedented expansion eastward will endanger the future of this grand experiment in integration. The eastern enlargement will create a "win-win" economic situation while also fueling intense fractions between net payers and net receivers of Brussels's budgetary funds. As the staff writers at *The Economist* noted: "the real issue facing European societies now is not how to keep new foreigners out but how to integrate the minorities they already have."

One immediate loser in the EU enlargement process was Turkey whose demand to set a date to begin EU negotiations was rejected. Instead, the union's leaders agreed to meet in December 2004 to decide whether the largely Muslim country of 70 million people was democratic enough and respectful enough of human rights to begin negotiations. Although the EU and UN failed to reach an agreement on the division of Cyprus, which will become one of the new union members, talks brokered by the United Nations in Cyprus between Greek and Turkish Cypriots finally seemed to reach fruition in 2003. After decades of obstructionism, a breakthrough agreement promised to lead to a reunified Cyprus being admitted to the European Union.

The decision to expand the members of the EU may be the easiest step towards further integration. At present there is no political plan for how the organization will govern itself and no economic plan for how the European Central Bank will create a single monetary policy for dramatically different economies. While the EU currently supports a process to draft a political constitution for the new Europe, no one is suggesting that there will be a "United States of Europe" any time soon. With the expansion, the population of the European club will increase by 20 percent, but the average wealth per person will fall by about 13 percent because most of the newcomers are relatively poor. That means that the new union will have to find ways to balance the interests of a country like Luxembourg, with a per capita GDP of nearly $43,000, with a country like Lithuania with a per capita GDP of $3,200. The new members will also have to adhere to 80,000 pages of European Union laws and regulations. As the staff writers of *The Economist* note, Poland accounts for roughly half the population and half the GDP of all ten incoming countries joining the European Union in 2004. But it must also do the most the get into shape. Poland sees EU membership as playing a big part in its future security but also hopes that the EU keeps rolling eastward.

The expansion also greatly complicates the problems of meeting the monetary and currency requirements of the European Monetary Union. On January 1, 2002 the twelve EU member countries adopted the new single currency, the euro, as their national currency. Three EU members—Britain, Denmark, and Sweden—have not yet agreed to adopt the new currency, but the new currency is widely expected to quickly creep into their economies as well. However, virtually all-European countries are having trouble meeting the EMU criteria, especially those limiting deficit financing in the face of the current international economic slowdown.

Europe Enlarged, America Detached?

"September 11 should be a catalyst for a renewal of the West as a community of action that is shaped by interests that are common even when they are not always equally shared. What the West needs, and must seek in and beyond the EU and NATO—the two central institutions that comprise it—is more, not less, integration."

SIMON SERFATY

Europe, as Americans have known it since 1917, is dead and beyond resurrection. Some, admittedly, remain in a state of denial as they still predict, and even await, the revival of the traditional nation-states whose sovereignty within impermeable boundaries was well worth a war or two.[1] But that prospect ended when the rise of the European Union and the North Atlantic Treaty Organization, institutions created to save the nation-states (from each other, as well as from themselves), progressively eroded their members' sovereignty instead.

"To understand," wrote Isaiah Berlin four decades ago, "is to perceive patterns." Patterns are not shaped by theory but asserted by history. The pattern that has grown out of Europe's history over the past 50 years could not be more evident. With nation-states reinventing themselves as member states of the union they form, or which they hope to join, the EU is achieving a new territorial and even political synthesis that is making much of the continent whole after NATO helped make it free. The single currency (the euro) that was launched in January 2002, the enlargement to 10 new members that was announced in December 2002, the constitutional convention scheduled for spring 2003, and the Intergovernmental Conference (IGC) that will take place in 2004 are the identifiable plays of an endgame known as finality. The agenda is not new—deepen in order to widen, widen in order to deepen, and reform in order to do both—but its scope and urgency are.

The final transformation of Europe should make Americans proud. The deconstructed Old World that twice in just over one generation organized its own collective funeral is now being consolidated *à l'américaine*. In large measure, this transformation is due to inspired United States policies that showed, during 15 glorious weeks in spring 1947, how the peace could be won historically after the war had been won militarily. Indeed, the new Europe is more peaceful, safer, more affluent, and more democratic—in short, more stable and, why not, more likable—than at any previous time. Even the unfinished security business inherited from earlier wars but still feared after the cold war has receded because institutional enlargement has acted as a catalyst for reconciliation and reforms. Yet muting a legitimate United States satisfaction with Europe's current condition, there is increasing exasperation over what is still missing and even some apprehension over what might be about to emerge.

For the growing number of Americans who have at last ceased to view "Europe" as an institutional illusion, causes for concern are varied. Most generally, an ever more united and progressively stronger Europe could conceivably rise as a "counterweight" to United States power. As the cold war was ending, Harvard political scientist Samuel Huntington was quick to forecast Europe's future as the "preeminent power of the twenty-first century." In 2002 the theme became more common: the EU as seen by Harvard's Joseph Nye was the "closest thing that the United States faces" for a world in which, suggests New York Times columnist Thomas Friedman, "two United States are better than one." At its best, such a Europe might explain its interest in assuming a greater role in the world as an obligation to protect Europe from America, but also America from itself.[2] At its worst, the increasingly assertive and even adversarial use of its newly regained power would leave Europe as an alternative to United States leadership at the possible cost of American interests. This latter prospect is all the more troubling since Europe might wish to pull its weight before it is ready, thus leaving Americans once again with the burden of finishing what Europeans started but could not complete. In other words, the new Europe cannot claim to do all it wants so long as it has not become all it can be, especially in the areas of military capabilities and political unity. Pretending otherwise, and viewing this moment as Europe's time, would risk exposing the EU as a mere "counterfeit" of the superpower it claims to be.

In coming years—after the 2003 constitutional convention and the 2004 IGC—Americans will have to be convinced that the new Europe will be the counterpart that successive United States administrations have awaited, rather than the counterweight or counterfeit they might resent or from which they could suffer. To achieve that lofty goal, much will be needed from both sides of the Atlantic. Europeans must take their own commitment to integration seriously— to do what they say they will do. This is especially true for foreign, security, and defense policy, about which "headline goals"—for both capabilities and structure—should also come with robust deadlines if they are to gain the credibility they deserve. In addition, as they move toward their institutional finality, Europeans should acknowledge the United States in the development of their union. More specifically, the United States role and privileges as a nonmember member state of the EU should be made an intrinsic part of Europe's "finality debate" with the invitation of a handful of American observers drawn from Congress to the Convention on the Future of Europe presided over by former French President Valéry Giscard d'Estaing (the convention is expected to produce the first draft of a constitution for Europe).

The EU is achieving a new territorial and even political synthesis that is making much of the continent whole after NATO helped make it free.

Only with a tangible acknowledgment of the United States role can the EU be engaged more directly than it has been—on the same grounds as the bilateral ties maintained by the United States with each of the main EU members. Admittedly, United States-EU relations cannot substitute for bilateral relations if the Europeans themselves do not complete their union. Yet, it is already possible to view the EU as the sixteenth member of the 15-member union—a virtual member that influences its partners no less than they influence each other. The presummit invitation extended to President George W. Bush by the European Council in June 2001 was a first step toward such a privileged United States status relative to, and even within, the EU. Similar consultations should occur not only at the highest political levels but also at lower levels and for each of the various bodies that "represent" the EU. Again, the dynamics of United States-EU ties will remain conditioned by what EU members do with and for their union more than by what the United States seeks from each of them within or outside the union.

ALL THAT NATO IS NOT

Like the EU, the post-cold war North Atlantic Treaty Organization has also evolved. Most fundamentally, NATO, as its members relied on it after 1950, is finished. Although this conclusion had begun to emerge after the 1999 war in Kosovo demonstrated the limits of NATO as a community of action, the events of September 11 clearly reinforced it, and the ensuing United States-led war in Afghanistan, which kept the alliance at a distance, made it explicitly marginal and seemingly irrelevant.

Beginning in 1991, the United States and its 15 NATO partners in Canada and Europe had sought to prepare the alliance for the new security environment caused by the collapse of the Soviet Union. A modest adaptation of NATO, as well as its enlargement, was expected to expand the eastern zone of stability to the rest of Europe. The war in Bosnia and the escalation of violence elsewhere in the Balkans made the process increasingly urgent, leading to NATO's fiftieth anniversary summit in 1999, when the ongoing war in Kosovo further justified a first wave of three new members. After that, the November 2002 summit in Prague was progressively viewed as a point of arrival for the final post-cold war transformation of a NATO that would be opened to a second, larger group of new members.

The importance of such a reconfiguration to its members' security was unclear. A NATO at 19 was not meant to have a reach beyond the European continent, and even a "NATO at 20" that would partner more closely with Russia was not expected to gain the global reach that the Clinton administration had sought in early 1999. Nonetheless, within Europe, enlargement proved decisive in settling some unfinished security business as applicant states resolved long-standing territorial and ethnic disputes in the name of the organizations (including the EU) that they hoped to join. To that extent, post-cold war NATO showed it was still an effective tool of deterrence. But when the Kosovo war also demonstrated the difficulties of waging war by committee, NATO seemed on its way to becoming a cold war relic—an ever-larger, passive, and shallow community lost without the Soviet threat and overwhelmed by the predominance of United States power.

Accordingly, on the way to the Prague summit, the second wave of NATO enlargement seemed to raise little interest. Allies in Europe, openly focused on the EU and its ambitious agenda, were ready to follow the United States the moment that leadership would be exerted. In the United States, however, the attention was elsewhere—in the Middle East but also in Asia, where China was viewed as the most dangerous future challenger. Admittedly, Russia might object to another NATO enlargement, especially when coupled with President Bush's emphatic interest in missile defense and deep cuts in offensive strategic capabilities. But Russian President Vladimir Putin also seemed resigned to these assertive displays of United States preponderance, especially after he was elevated to the status of a "soul brother" during his first meeting with Bush in Ljubljana, Slovenia in June 2001.

Yet what an enlarged NATO would do, and where—and how it would be adapted, and when—were left unclear. That NATO could no longer make decisions by consensus, and that security could no longer depend on increasingly cacophonic "coalitions of the willing," was already true before any enlargement to the east, as was shown during the protracted debates over Bosnia in the early 1990s, when NATO had "only" 16 members. In 1999 a surplus of "willing" partners made the war in Kosovo a challenge to common sense, especially as too many of the willing were not capable, while many of the most capable were also

Article 21. Europe Enlarged, America Detached?

least responsive to the will of the coalition. Even without September 11, and before further enlargement in Prague, unilateral decisions by the United States on behalf of its NATO allies had become more difficult to impose, and a more united Europe with its own security identity was bound to increase that difficulty exponentially.

In addition, because most of the new NATO members were small, weak, or poor, the capabilities gap within Europe would be growing no less dramatically than the capabilities gap between Europe and the United States—between Latvia and France no less than between France and the United States. In any case, the gap was no longer defined by the availability and quality of military capabilities, but also by the will to use them—assuming the ability to agree on the areas, in and out of Europe, where they might be used. Previously, it had been America's will to wield military force everywhere that had been questioned. Now it is Europe's will to use force anywhere that is debated: weakness encourages appeasement, or at least a quest for "solutions" that avoid the use of force even at the cost of additional, occasionally unwanted, and often self-defeating compromises.

> *In Afghanistan and wherever else the wars of September 11 might go, NATO has become an afterthought.*

When the United States undertook military action in Afghanistan in November 2001, its NATO allies expected that their offer of solidarity entitled them to an active role in the war that they were willing to enter but unable to fight because of their own insufficiencies. Yet, reports of American neglect of Europe's offer were no less exaggerated than related accounts of European dissatisfaction with the alleged neglect. The NATO allies were not underused by the United States: their uses—partly military, but mainly nonmilitary—were understated on both sides of the Atlantic. What the United States can do is necessary—indeed "indispensable," as former Secretary of State Madeleine Albright once put it. But it is not, and is unlikely to become, sufficient: the United States alone cannot manage instability around the world, the blowbacks inherited from earlier conflicts. The same is true of NATO. During the cold war, one organization was enough to attend to the common defense of the West: one enemy, one alliance, one theater, and one hegemonic leader. This was United States unilateralism with a NATO prix fixe. Now the security environment is more diffuse, the war more strange, and the enemy more elusive—everywhere and nowhere, about everything and for nothing. Whether at one (the United States), at 20 (with Russia), at 26 (after Prague), or at many more (past 2004), NATO alone will not suffice: however necessary the alliance may be for military purposes, it is not, and never was, a full-service institution. It is Western multilateralism à la carte—a bit of this institution and a bit of that institution, simultaneously or consecutively, and designed to constrain or engage the leading members. Uni-multilateralism is the new imperative—with the United States at the helm, as needed.

Admittedly, other NATO countries must spend more on defense—and after the French and German elections there are indications that they will, at least to a degree. The shared goal is to remain "interoperable" and also maintain appropriate levels of "cooperability," without which the alliance would be too unbalanced to gain the global scope it now needs and was given at the Prague summit. But the criteria of cooperability are not met only by levels of defense spending. They are also satisfied by what European NATO countries can do in relief of their senior partner during the latter innings of a particular contest. Working through the EU, the European allies have the economic capabilities to reward and sanction; the political tools to stabilize and punish; and the know-how to negotiate and isolate. Like sheer military power, these capabilities, tools, and skills are not sufficient to initiate action—but they are undoubtedly necessary to end it.

In short, on the way to Prague, enlargement ceased to be NATO's central issue, and it even ceased to be a contentious issue for Russia. Indeed, it was no longer an issue at all: growth might be better for the organization, but that would be of little significance if NATO itself did not become a more capable and more efficient organization with a global reach. Unless it was decided to let a larger NATO wither away, the NATO agenda had to be broadened. For enlargement to be effective, the organization also needed a new mandate and more capabilities, as well as reform of its structures and governance. In effect, for NATO, as for the EU, the agenda became widen in order to deepen, deepen even as you widen, and reform to do both. Accordingly, in Prague NATO did not merely complete the post-cold war process launched by its 16 members in Rome in June 1991: it also pointed to the rise of a new NATO whose 26 members would rely on new capabilities equipped for global action and placed under the control of new command structures designed to give the organization the means and reach needed to wage a global war on the new enemy.

DEFINING THE "NEW NORMALCY"

The limits of NATO as the security institution of choice, but also its unparalleled potential over that of any ad hoc alliance, were reinforced by the events of September 11, 2001 and the "new normalcy" of "postmodern conflicts" they threatened to inaugurate. Central to the uncertainties that surround NATO and its purpose is a fundamental difference between the allies over the meaning and implications of these events.[3]

The semantic contest that began almost at once between America and its European allies reflected a clash of historic experiences that became increasingly evident in 2002.[4] For Europeans, notwithstanding the spontaneous emotions generated by the extraordinary sight of their bleeding, crippled, and even frightened senior partner, these events were, in a sense, predictable—history as usual. Hegemonic powers cannot live their moment of preponderance without pain. Indeed, judged by standards set by history, the suffering endured by America on September 11 was relatively minor—a few minutes of casualties on a bad day in Europe in 1916 or 1940. Understood as an act of terror, that pain pointed to a way of life that European countries have faced and defeated many times, from Northern Ireland to the Basque region—although the standards of September 11 were especially disconcerting since, in French President Jacques Chirac's words, "next time it might be us." Still, having properly demonstrated their sympathy, Europeans could invoke the *déjà vu* of history to reassert the *déjà dit* of the need for consultation with the allies and for patience in defeating the enemy.

For Americans, such logic hardly applied. Pain may well be the way of history, but it is not the American way. Wars are expected to be waged "over there," where the forward deployment and use of superior American power keeps them by containing foes and even, on occasion, friends. "Over here" acts of terror might be initiated by misguided high school teens or nihilistic misfits. But these acts would be homebred, not exportable to the nation by evil forces abroad. So it had been since the War of 1812, and for 189 years subsequent attempts to violate America's territorial invulnerability had been countered forcefully, whether far away in the Pacific (after Pearl Harbor) or closer in the Caribbean (after the

Cuban missile crisis). Hence, the war that erupted in Afghanistan, where the culprits hid, was more than the "first war of the twenty-first century" (as President Bush called it)—it was the first war in at least half a century that America could truly call its own. The United States goal was not to protect or avenge others but to avenge and protect America's citizens and its institutions. The war would therefore be fought the American way: admittedly brutally, until unconditional surrender or unmitigated annihilation, and somewhat unilaterally, with coalition members used on the basis of need rather than stated availability.[5]

But this transatlantic clash of history is not limited to the experiences enjoyed by the New World relative to, or occasionally at the expense of, the Old. It is also rooted in the differing interpretations that were and continue to consist of the most effective way to contain the threats unveiled through the attacks on the World Trade Center and the Pentagon. As British historian Eric Hobsbawm noted on the eve of the twenty-first century, "single, specific events... are unpredictable," but even post facto the real task for historians and analysts "is to understand how important they are or could be." That certainly is true of the events of September 11: however unpredictable these events may have been, it would be wrong to shy away from assessing their consequences.[6]

Europe's vision of the "new normalcy" does not fit the United States view, let alone its preference. For many in Europe, the perpetrators of this violence, already helped by their enemy's blunders and by chance, aim mostly at the United States and the local conditions it permits or even creates. Accordingly, it is important to influence United States responses whose motivation might be legitimate but whose consequences would be felt in and beyond Islamic countries, including in European countries where the risks of a cultural spillover are especially high. Thus, the Euro-Atlantic community of interests perceived in the immediate aftermath of September 11 has come under threat since President Bush began to emphasize the other dimensions of an "axis of evil" that included, but were not limited to, Iran (and North Korea) as well as Iraq.

Beginning with the president's pronouncement of this axis of evil in his January 2002 State of the Union address, the two sides of the Atlantic appear to have drifted away amid parallel charges of Europe's anti-Americanism and America's anti-Europeanism. In Afghanistan and wherever else the wars of September 11 might go, NATO has become an afterthought—a "spare wheel," suggested its secretary general, George Robertson. In summer 2002, the offensive rhetoric heard during Chancellor Gerhard Schröder's campaign for reelection in Germany provided a catalyst for an American exasperation with, and anger at, countries the United States has saved from themselves and others throughout much of the twentieth century. It also pointed to a potentially dangerous split within the EU between a peripheral ring of Atlanticist states, led by Britain, and a continental core of European countries, led by France and Germany, with both competing for the allegiance of new NATO and EU members in the east.

In January 2003, the rhetoric escalated when United States Secretary of Defense Donald Rumsfeld remarked that "Europe as Germany and France, that's the old Europe." But, added Rumsfeld, "look at vast numbers of other countries in Europe. They're not with France and Germany..., they're with the United States." Rumsfeld's "new" Atlanticist Europe was heard a few days later when eight NATO countries in Europe (Britain, Denmark, Italy, Portugal, and Spain, as well as Poland, the Czech Republic, and Hungary) praised "American bravery, generosity, and farsightedness" and pledged "unwavering determination and firm international cohesion." Compounding this potential split is a growing capabilities gap between the United States and Europe that has prevented common action even when values or interests might otherwise be shared, however unevenly.

With America's traditional margin of security now bridged, there is less room for ambiguity and indecision, at home as well as from the allies. The threats posed by weapons of mass destruction that could be—or have been—acquired by intrinsically hostile groups or "evil states" are real, lethal, and unacceptable. "The depth of the hatred," said Bush in his State of the Union speech, "is equaled by the madness of the destruction they design." The hypothesis is too daunting to be checked for accuracy after the fact. The madness will have to be denied before the hatred can be cured, thereby making it necessary to "be ready for preemptive action when necessary to defend our liberty," as the president put it more cogently in an address in June 2002 at West Point.

This is the defining question that was raised on September 11, 2001, when four hijacked planes and 19 criminals ended America's sense of territorial invulnerability. America and its allies had already perceived the threat (one of the "other risks of a wider nature" first envisioned in NATO's 1991 Strategic Concept, and reasserted in its 1999 Strategic Concept). It is an existential risk written with the invisible ink of an unpredictable future. It carries with it the related danger of an undeclared cultural war that would prove irreversible for the many, although precipitated by the suicidal acts of the few.

The response of Western powers to that risk should not be to go it alone but confront it together. While "Europe must understand we are ready and able to act without them to fight this new war," as Richard Perle, an influential defense analyst, put it, the United States is probably neither ready nor able to end the war on terrorism on its own, even if it continues to win every military engagement it faces or launches. September 11 should be a catalyst for a renewal of the West as a community of action that is shaped by interests that are common even when they are not equally shared. What the West needs, and must seek in and beyond the EU and NATO—the two central institutions that comprise it—is more, not less, integration. Among themselves as a mutually shared right of first refusal, but also with new associates and partners, the NATO countries should be able to agree on immediate priorities and certain key principles on how to define and counter these new threats, as well as on a course of action during the days after these threats are defeated or even preempted. As Harvard's Samuel Huntington has stated, the "idea of integration" is the "successor idea to containment." More specifically, integration is "about locking [the allies] into these policies and then building institutions that lock them in even more."[7] This is the idea that was launched along two parallel paths after World War II, and refined—deepened and enlarged—throughout and since the cold war. It is also the idea that can now be brought to fruition by and between the United States and the states of Europe in the context of the new normalcy envisioned after the cold war for the twentyfirst century. The goal is not merely to do something, let alone everything, together, but to ensure that together, everything, or even something, gets done.

---

[1] In *The Tragedy of Great Power Politics* (New York: W. W. Norton, 2001), John Mearsheimer writes ominously that "[a]lmost every European state, including the United Kingdom and France, still harbors deep-seated albeit muted fears that a Germany unchecked by American power might behave aggressively."

[2] Samuel Huntington, "The U.S.—Decline or Renewal?" *Foreign Affairs*, Winter 1988–1989; Joseph S. Nye, Jr., *The Paradox of American Power: Why the World's Only Superpower Can't Go It Alone* (Oxford: Oxford University Press, 2002), p. 29; Thomas L. Friedman, "I Love the E.U.," *The New York Times*, June 22, 2001.

[3] The reference to a "new normalcy" was first made by Vice President Dick Cheney. Quoted by Bob Woodward, "CIA Told to Do 'Whatever Necessary' to Kill Bin Laden," *Washington Post*, October 21, 2001, p. 22. See also Lawrence Freedman, "Post-Modern Conflict," *Financial Times*, September 12, 2001.

[4] For a lengthier discussion of some of these themes, see Simon Serfaty, "The Wars of 911," *The International Spectator*, October–December 2001.

[5] "Leadership demands a Pagan ethos," writes Robert D. Kaplan, as he acknowledges, or boasts of, the "imperial reality" that "already dominates our foreign policy" and demands that "power politics [be placed] in the service of patriotic values." See *Warrior Politics* (New York: Random House, 2001), pp. 145, 154.

[6] In conversation with Antonio Polito, *The New Century* (London: Abacus Books, 1999), p. 1.

[7] Quoted by Nicholas Lemann, "The Next World Order," *The New Yorker*, April 1, 2002.

SIMON SERFATY is a professor of United States foreign policy and Eminent Scholar at Old Dominion University. He also serves as director of European studies at the Center for Strategic and International Studies. His most recent book is Memories of Europe's Future (Washington, D.C.: CSIS, 2000). EU

From *Current History*, March 2003, pp. 99–105. © 2003 by Current History, Inc. Reprinted by permission.

America as European Hegemon

Christopher Layne

As THE WISEST of all American philosophers, Yogi Berra, has insightfully observed, making predictions is hard, especially about the future. And he might have added—pointing to the predictions of an impending Euro-American rupture that have been a staple of debates about U.S.-European relations at least since the 1956 Suez crisis—prognosticating accurately about the future of Transatlantic relations is extra hard. Through all the ups and downs in U.S.-European relations over the years, those many Chicken Littles who have gone out on a limb to forecast an impending drifting apart of Europe and the United States never have had their predictions validated by events.

Until now, perhaps?

The Iraq War has produced a very different kind of rift. The damage inflicted on Washington's ties to Europe by the Bush Administration's policy is likely to prove real, lasting and, at the end of the day, irreparable. In other words, if the fat lady isn't singing already, she clearly can be heard warming up her voice.

To understand why this crisis is different, we must understand its causes. The rupture between the United States and Europe is not, as some have asserted, mainly about an alleged Transatlantic rift in the realm of culture, values and ideology.[1] It is not about the relative merits of unilateralism versus multilateralism. It is not even about the issues that framed the debate about Iraq during the run-up to war (Should the weapons inspections process have been allowed to play out? Was the United States wrong to go to war without a second resolution from the United Nations Security Council?). For sure, Iraq was a catalyst for Transatlantic dispute, but this crisis has been about American power—specifically about American hegemony.

Of Balance and Hegemony

WHEN FUTURE historians write about how American hegemony ended, they may well point to January 22, 2003 as a watershed. On that day, commemorating the 40th anniversary of the FrancoGerman Treaty negotiated by Charles de Gaulle and Konrad Adenauer as a bulwark against American hegemony, French President Jacques Chirac and German Chancellor Gerhard Schröder jointly declared that Paris and Berlin would work together to oppose the Bush Administration's evident intent to resolve the Iraqi question by force of arms. Later that day, in a Pentagon briefing, Secretary of Defense Donald Rumsfeld responded to the Franco-German declaration by contemptuously dismissing those partners as representing the "Old Europe", thereby triggering a Transatlantic earthquake, the geopolitical after-shocks of which will be felt for a long time. And well they should, for these contretemps reflect what is already a very old issue.

The problem of hegemony has been a major issue in U.S.-European relations since the United States emerged as a great power at the end of the 19th century. The United States fought two big wars in Europe out of fear that if a single power (in those cases, Germany) attained hegemony in Europe, it would be able to mobilize the continent's resources and threaten America in its own backyard, the Western Hemisphere. The conventional wisdom holds that America's post-World War II initiatives—the Marshall Plan, the North Atlantic Treaty—were driven by similar fears of possible Soviet hegemony in Europe. Indeed, many American strategic thinkers define America's traditional European strategy as a text-book example of "offshore balancing."

As an offshore balancer, the United States supposedly remains on the sidelines with respect to European security affairs unless a single great power threatens to dominate the continent. America's European grand strategy, therefore, is said to be counter-hegemonic: the United States intervenes in Europe only when the continental balance of power appears unable to thwart the rise of a would-be hegemon without U.S. assistance. The most notable proponent of this view of America's European grand strategy toward Europe is University of Chicago political scientist John J. Mearsheimer.[2] He argues that the United States is not a global hegemon. Rather, because of what he describes as the "stopping power of water", the United States is a hegemon only in its own region (the Western hemisphere), and acts as an offshore balancer toward Europe. He predicts that the United States soon will end its "continental commitment" because there is no European hegemon looming on the geopolitical horizon. As an offshore balancer, Mearsheimer says, the United States will not remain in Europe merely to play the role of regional stabilizer or pacifier.

There is just one thing wrong with this view: it does not fit the facts.

If American strategy toward Europe is indeed one of counter-hegemonic off-shore balancing, it should have been over, over there, for the United States when the Soviet Union collapsed. By a different but not far-fetched reckoning, it should have been over in the early 1960s, when the Europeans were capable of deterring a Soviet military advance westward without the United States. With no hegemonic threat to contain, American military power should have been retracted from Europe after 1991, and NATO should have contracted into non-existence rather than undergoing two rounds of expansion. Of course, it may be that America will ultimately be ejected from the continent by the Europeans, but there are no signs that the United States will voluntarily pack up and go home any time soon.

It is not a "time lag", or mere inertia, that has kept American military power on the European continent more than a decade after the Soviet Union ceased to exist. There is a better explanation for why U.S. troops are still in Europe and NATO is still in business. It is because the Soviet Union's containment was never the driving force behind America's post-World War II commitment to Europe. There is a well-known quip that NATO was created to "keep the Russians out, the Germans down, and the Americans in." It would be more accurate to say that the Atlantic Alliance's primary *raison d'être*, from Washington's standpoint, was to keep America in—and on top—so that Germans could be kept down, Europe could be kept quiet militarily, and the Europeans would lack any pressing incentive to unite politically. The attainment of America's postwar grand strategic objectives on the continent required that the United States establish its own hegemony over Western Europe, something it would probably have done even in the absence of the Cold War. In other words, NATO is still in business to advance long-standing American objectives that existed independently of the Cold War and hence survived the Soviet Union's collapse.

American Aims

WE USUALLY look to history to help us understand the present and predict the future. But the reverse can be true, as well: sometimes recent events serve to shed light on what happened in the past, and why it happened. Many may react skeptically to the claim that America's postwar European grand strategy was driven at least as much—probably more—by non-Cold War factors as by the Soviet threat. But Washington's post-Cold War behavior provides a good deal of support for this thesis.

For starters, when the Berlin Wall fell and the Soviet Union began to unravel, the first Bush Administration did not feel in the least bit compelled to reconsider the relevance of, or need for, either the U.S. military commitment to Europe or NATO. As Philip Zelikow and Condoleezza Rice, both of whom served that administration as senior foreign policy officials, have observed:

[The] administration believed strongly that, even if the immediate military threat from the Soviet Union diminished, the United States should maintain a significant military presence in Europe for the foreseeable future....The American troop presence thus also served as the ante to ensure a central place for the United States as a player in European politics. The Bush administration placed a high value on retaining such influence, underscored by Bush's flat statement that the United States was and would remain 'a European power.'... *The Bush administration was determined to maintain crucial features of the NATO system for European security even if the Cold War ended.*[3]

The Clinton Administration took a similar view. As one former State Department official avers, NATO had to be revitalized after the Cold War because American interests in Europe "transcended" the Soviet threat.[4] And using phraseology reminiscent of Voltaire's comment about God, then-Secretary of State Madeleine Albright said, "Clearly if an institution such as NATO did not exist today, we would want to create one."[5]

The fact that American policymakers did not miss a beat when the Cold War ended with respect to reaffirming NATO's continuing importance reveals a great deal about the real nature of the interests that shaped America's European grand strategy after World War II, and that continue to do so today. The truth is that, from its inception, America's postwar European grand strategy reflected a complex set of interlocking "Open Door" interests.* These interests are at once economic, strategic and broadly political in nature.

The first of these is that U.S. postwar officials believed that America had crucial economic interests in Europe. Even if there was no communist threat to Western Europe, State Department Policy Planning Staff Director George F. Kennan argued in 1947, the United States had a vital interest in facilitating Western Europe's economic recovery: "The United States people have a very real economic interest in Europe. This stems from Europe's role in the past as a market and as a major source of supply for a variety of products and services."[6] These interests required that Europe's antiquated economic structure of small, national markets be fused into a large, integrated market that would facilitate efficiencies and economies of scale. ** As the U.S. Ambassador to France, Jefferson Caffery, argued in 1947, economic integration would "eliminate the small watertight compartment into which Europe's pre-war and present economy is divided."[7] Paul Hoffman, director of the Economic Cooperation Agency (which administered Marshall Plan aid to Europe), elaborated on the reasons why Washington favored Western Europe's economic integration: "Europe could not be self-supporting until it had made great progress towards unity and until there was a wide, free, competitive market to lower costs, increase efficiency, and raise the standard of living."[8]

To prevent far Left parties (especially the communists) from coming to power on the Continent's western half after World

War II, U.S. aims also required political and social stability there. Washington was not really so concerned that such governments would drift into Moscow's political orbit, but it was very concerned that they would embrace the kinds of nationalist, or autarkic, economic policies that were anathema to America's goal of an open international economy. As Averell Harriman, the U.S. Special Representative in Europe, put it, Washington was committed to multilateral trade and was "opposed to restrictive policies and especially to the creation of an autarkic Europe."[9]

Second, American strategists perceived that U.S. economic interests would be jeopardized if postwar Europe relapsed into its bad habits of nationalism, great power rivalries and realpolitik. To ensure stability in Europe after World War II, the United States sought to create a militarily de-nationalized and economically integrated—but *not* politically unified—Europe. Washington would assume primary responsibility for European security, thereby precluding the re-emergence of the security dilemmas (especially that between France and Germany) that had sparked the two world wars. In turn, Western Europe's economic integration and interdependence—under the umbrella of America's military protectorate—would contribute to building a peaceful and stable Western Europe. In this respect, U.S. economic and security objectives meshed nicely.

Postwar U.S. policymakers viewed Europe's traditional balance of power security architecture as a "fire trap" and, as Undersecretary of State Robert Lovett said following World War II, Washington wanted to make certain that this fire trap was not rebuilt.[10] Starting with those who were "present at the creation", successive generations of U.S. policymakers feared the continent's reversion to its (as Americans see it) dark past—a past defined by war, militarism, nationalism and an unstable multipolar balance of power. For American officials, Europe indeed has been a dark continent whose wars spilled over across the Atlantic, threatened American interests and invariably drew in the United States. Secretary Rumsfeld's disparaging remark about the "Old Europe" thus stands in a long and consistent line of American attitudes toward the Continent and its various historical crimes and misdemeanors.

After World War II, Rumsfeld's cabinet predecessors sought to maintain U.S. interests by breaking the Old Europe of its bad old geopolitical habits. As Secretary of State John Foster Dulles put it in 1953,

> Surely there is an urgent, positive duty on all of us to seek to end that danger which comes from within. It has been the cause of two world wars and it will be disastrous if it persists.[11]

Even during the Cold War, American policymakers acknowledged that, quite apart from the Soviet threat, the United States needed to be present militarily in Western Europe to create a political environment that permitted "a secure and easy relationship among our friends in Western Europe."[12] As Secretary of State Dean Rusk said in 1967, the U.S. military presence on the continent played a pivotal role in assuring stability *within* Western Europe: "Much progress has been made. But without the visible assurance of a sizeable American contingent, old frictions may revive, and Europe could become unstable once more."[13] Former Secretary of State Acheson, too, observed in the mid-1960s that, as the vehicle for America's stabilizer role in Western Europe, "NATO is not merely a military structure to prepare a collective defense against military aggression, but also a political organization to preserve the peace of Europe."[14]

The U.S. goal of embedding a militarily de-nationalized, but economically integrated Western Europe within the structure of an American-dominated "Atlantic Community" dovetailed neatly with another of Washington's key post 1945 grand strategic objectives: preventing the emergence of new poles of power in the international system—in the form either of a resurgent Germany or a united Europe—that could challenge America's geopolitical pre-eminence. Since the 1940s, Washington has had to perform a delicate balancing act with respect to Europe. To be sure, for economic reasons, the United States encouraged Western Europe's integration into a single common market, but the United States sought to prevent that from leading to its political unification.

To prevent the emergence of a politically unified Western Europe, successive U.S. administrations sought to "denationalize" the region by establishing a military protectorate that integrated Western Europe's military forces under, and subordinated them to, American command. The goal was to neuter Western Europe geopolitically and thereby circumscribe its ability to act independently of the United States in the high political realms of foreign and security policy. Embedding West European integration in the American-dominated Atlantic community would prevent the Europeans from veering off in the wrong direction. "An increased measure of Continental European integration", Acheson and Lovett told President Truman,

> can be secured only within the broader framework of the North Atlantic Community. This is entirely consistent with *our own desire to see a power arrangement on the Continent which does not threaten us* and with which we can work in close harmony.[15]

Acheson stated American strategic concerns with crystal clarity when he spoke of the necessity of a "well-knit large grouping of Atlantic states within which a new EUR grouping can develop, thus ensuring unity of purpose within the entire group and precluding [the] possibility of [a] EUR Union becoming [a] third force or opposing force."[16]

Europe's military absorption into the Atlantic Community went hand in hand with its economic integration. By persuading the West Europeans to "pool" their military and economic sovereignty, Washington aimed to strip them of the capacity to take unilateral national action.[17] As Kennan observed, Western Europe should be unified on terms which "would automatically make it impossible or extremely difficult for any member, not only Germany, to embark upon a path of unilateral aggression." But it was the American diplomat Charles Bohlen who cut to the heart of the U.S. de-nationalization strategy when he said,

"Our maximum objective should be the general one of making common European interests more important than individual national interests."[18]

For the United States, therefore, institutions such as NATO, the aborted European Defense Community, the European Coal and Steel Community (ECSC) and the Common Market were the instruments it employed to contain the West Europeans.*† As the State Department said, the United States hoped that "cautious initial steps toward military, political, and economic cooperation will be followed by more radical departures from traditional concepts of sovereignty."[19] The American aim was to create "institutional machinery to ensure that separate national interests are subordinated to the best interests of the community", and achieving this subordination was deemed essential if the United States was to accomplish its grand strategic purposes in Europe.[20]

The Continental Response

JUST AS FEAR of a European hegemon led the United States to intervene in Europe's two great wars of the 20th century, the West Europeans after World War II understood that America had established its own hegemony over them. As realist international relations theory suggests, Western Europe tried to do something about it.

To be sure, West European balancing against the United States was constrained. On the one hand, although the West Europeans feared American power, they feared the Soviet Union even more during the Cold War. In a more positive sense, too, following World War II, Washington was able to use the carrot of economic assistance—notably, the Marshall Plan—to keep Western Europe aligned (albeit very tenuously at times) with the United States. Nevertheless, throughout the post-World War II era, West European inclinations to balance against American power were never far from the surface.

In the five years or so after the end of World War II, it was Britain that hoped to emerge as a "Third Force" in world politics to balance both the United States and the Soviet Union. As the British diplomat Gladwyn Jebb put it, London needed to prevent the geopolitical equilibrium from being undermined "by a 'bipolar' system centering around what Mr. Toynbee calls the two 'semi-barbarian states on the cultural periphery'."[21] The accelerating decline of Britain's relative power, of course, put paid to London's Third Force aspirations, but continental Europe's Third Force aspirations remained. In the late 1940s and 1950s, one of the hopes of the founding fathers of today's European Union was that the European Coal and Steel Community, and then the Common Market, would prove to be the embryo of a united Europe that could act as a geopolitical and economic counterweight to the United States. Commenting on the motives driving the West Europeans to integrate, the diplomatic historian Geir Lundestad observes:

Although they wanted the two sides of the Atlantic to cooperate more closely, in a more general sense it was probably also the desire of most European policymakers to strengthen Western Europe vis-à-vis the United States. This could be done economically by supporting the Common Market and politically by working more closely together on the European side.[22]

Even Jean Monnet, author of the Schuman Plan that led to the ECSC and the "father" of European integration, first toyed with the idea of an Anglo-French federation in the late 1940s because he saw this as the basis of a European bloc that could stand apart from both the United States and the Soviet Union.[23]

The 1956 Suez crisis gave fresh impetus to the arguments that Western Europe needed to counterbalance the United States. Britain's initial reaction to its humiliation by the Eisenhower Administration was to consider reviving the Third Force concept: "We should pool our resources with our European allies so that Western Europe as a whole might become a third nuclear power comparable with the United States and the Soviet Union."[24] Under Harold Macmillan, of course, Britain rejected becoming part of a West European Third Force, opting instead to curry favor—and maintain influence—with Washington through the "special relationship" ("playing Greece to America's Rome"). On the Continent, however, Suez focused French and West German attentions on the need for a West European counterweight to American power. As William I. Hitchcock recounts, Adenauer and French Premier Guy Mollet were meeting in Paris on November 6, 1956, at the height of the Suez crisis (and the simultaneous turmoil in Hungary). Shortly after Adenauer exclaimed that it was time for Europe to unite "against America", Mollet excused himself to take a phone call from the British Prime Minister, Anthony Eden, who informed Mollet that, under U.S. pressure, London had decided to call off the Anglo-French invasion of the Suez Canal Zone. When a crestfallen Mollet returned to the meeting room and conveyed the content of the telephone conversation to his guest, Adenauer consoled him by saying, "Now, it is time to create Europe."[25]

By the early 1960s, French President Charles de Gaulle believed that Western Europe had recovered sufficiently from World War II's dislocations and was poised to re-emerge as an independent pole of power in the international system. De Gaulle, clearly one of the 20th century's towering figures, was well versed in the realities of international politics. Following Washington's successful facing-down of the Soviet Union in the 1962 Cuban missile crisis, he concluded *then* that the world had become "unipolar"—dominated by a hegemonic America. To balance U.S. hegemony, de Gaulle pushed for France to acquire independent nuclear capabilities, and he sought to build a West European pole of power based on a Franco-German axis. That is what the 1963 treaty—the one Chirac and Schröder were commemorating on January 22—was all about, a fact that Washington apprehended clearly. U.S. policymakers were deeply concerned that Paris would lure West Germany out of the "Atlantic" (that is, U.S.) orbit, because such a Euro-centric strategic axis, as a 1966 State Department cable explicitly said, "would fragment Europe and divide the Atlantic world."²⁶ In plainer English, the foundations of America's European hegemony would be undermined.

Washington recognized the Gaullist challenge for what it was—a direct assault on U.S. preponderance in Western Europe—and reacted by re-asserting its own hegemonic prerogatives on the Continent. President Kennedy gave eloquent expression to the fear that Western Europe's emergence as an independent pole of power in the international system would be inimical to U.S. interests, and his doing so shows that U.S. concerns on this score were not limited to the immediate postwar period, as sketched out above. Kennedy voiced concern that U.S. leverage over Europe might be waning because the West Europeans, having staged a vigorous postwar recovery, were no longer dependent on the United States economically. Noting that "the European states are less subject to our influence", Kennedy expressed the fear that "if the French and other European powers acquire a nuclear capability they would be in a position to be entirely independent and we might be on the outside looking in."²⁷ By pushing for a Multilateral Nuclear Force for Western Europe (in reality, one that kept Washington's finger firmly on the trigger), the United States sought—unsuccessfully—to derail France's nuclear ambitions.

With considerably more success, however, the United States did manage to take the teeth out of the Franco-German Treaty. In so doing, Washington played the hardest kind of hegemonic hardball. Threatening to rescind the security guarantee that protected West Germany from the Soviets, the U.S. government insisted that the Bundestag insert a preamble to the treaty reaffirming that Bonn's Atlantic connection to the United States and NATO took supremacy over its ties with Paris.† This intervention by the United States hastened Adenauer's retirement and helped ensure that he would be succeeded by the more pliable Atlanticist, Ludwig Erhard.

What's New?

NOW, FORTY years later, the United States and Europe are still playing the same game. America still asserts its hegemony, and France and Germany still seek (so far without much success) to create a European counterweight. As has been the case in the past, too, Washington is employing a number of strategies to keep Europe apart.

First, the United States is still actively discouraging Europe from either collective, or national, efforts to acquire the full-spectrum of advanced military capabilities. Specifically, the United States has opposed the EU's Rapid Reaction Force (the nucleus of a future EU army), insisting that any European efforts must not duplicate NATO capabilities and must be part of an effort to strengthen the Alliance's "European pillar." The United States is also encouraging European NATO members to concentrate individually on carving-out "niche" capabilities that will complement U.S. power rather than potentially challenge it.

Second, Washington is engaged in a game of divide and rule in a bid to thwart the EU's political unification process. The United States is pushing hard for the enlargement of the EU—and especially the admission of Turkey—in the expectation that a bigger EU will prove unmanageable and hence unable to emerge as a politically unified actor in international politics. The United States also has encouraged NATO expansion in a similar vein, in the hope that the "New Europe" (Poland, Hungary, the Czech Republic and Romania)—which, with the exception of Romania, will join the EU in 2004—will side with Washington against France and Germany on most issues of significance. For the United States, a Europe that speaks with many voices is optimal, which is why the United States is trying to ensure that the EU's "state-building" process fails—thereby heading off the emergence of a united Europe that could become an independent pole of power in the international system.

Finally, the United States has continued to remind the rest of Europe, sometimes delicately, sometimes in a heavy-handed fashion, that they still need an American presence to "keep the Germans down." For example, at his speech in Prague during the November 2002 NATO summit, President George W. Bush—just before invoking the historically freighted memories of Verdun, Munich, Stalingrad and Nuremberg—alluded in a not-so-subtle fashion to the German threat from World War II to make the case for a U.S. role in Europe:

> U-boats could not divide us.... The commitment of my nation to Europe is found in the carefully tended graves of young Americans who died for this continent's freedom. That commitment is shown by the thousands in uniforms still serving here, from the Balkans to Bavaria, still willing to make the ultimate sacrifice for this continent's future.

Washington's aim of keeping Europe apart paid apparent dividends when, at the end of January, the leaders of Britain, Spain, Italy, Portugal, Denmark, Poland, Hungary and the Czech Republic signed a letter urging Europe and the international community to unite behind Washington's Iraq policy. This letter was notable especially because it illustrated that the United States is having some success in using the "New Europe" to balance against the "Old" Franco-German core. Clearly, Washington hopes that states such as Poland, Hungary, the Czech Republic and Romania will not only line up behind the United States within NATO, but will also represent Atlanticist interests over European ones within the EU itself.

In short, U.S. policy seeks to encourage an intra-European counterweight that will block French and German aspirations to create a united Europe counterweight to American hegemony. Indeed, in the wake of the Iraq War, Transatlantic relations are characterized by a kind of "double containment" in Europe: the hard core of Old Europe (centered around France and Germany, and possibly supported by Russia) seeks to brake America's aspirations for global hegemony, while the United States and its "New European" allies in central and eastern Europe seek to contain Franco-German power on the Continent. It is an old game, in a new form.

The Widening Atlantic

IN THE DECADE between the Soviet Union's collapse and 9/11, American hegemony (or as some U.S. policymakers called it during the Clinton Administration, America's "hegemony problem") was the central issue in American grand strategy debates. It still is. Although American policymakers have developed a number of (too) clever rationales to convince themselves that the United States will escape the fate that invariably befalls hegemons, the fallout of the Iraq crisis on the Transatlantic relationship illustrates that concern with America's hegemonic power—and the way it is exercised—is not confined to the Middle East and Persian Gulf.[28]

Why do France, Germany and much of the rest of the world, including other major powers such as Russia and China, worry about American hegemony? The simple answer is that international politics remains fundamentally what it has always been: a competitive arena in which states struggle to survive. States are always worried about their security. Thus when one state becomes overwhelmingly powerful—that is, hegemonic—others fear for their safety.

Doubtless the Bush Administration's fervent hegemonists will scoff at the idea that the United States will become the object of counter-hegemonic balancing. They clearly believe that the United States can do as it pleases because it is so far ahead in terms of hard power that no other state (or coalition of states) can possibly hope to balance against it.[29] They also know, and know that Europeans know, that the United States does not and will never literally threaten Europe with its military power. This confidence is misplaced, however, because it overlooks the effects of what can be called "the hegemon's temptation."

A hegemonic power like the United States today has overwhelming hard power—especially military power—and indeed there is no state or coalition with commensurate power capable of restraining the United States from exercising that power. For hegemons, the formula of overwhelming power and lack of opposition creates powerful incentives to expand the scope of its geopolitical interests. But over time, the cumulative effects of expansion for the United States—wars and subsequent occupations in the Balkans, the Persian Gulf, Afghanistan and the War on Terrorism; possible future wars against North Korea, Iran, Syria, or China over Taiwan—will have an enervating impact on U.S. power.

At the end of the day, hegemonic decline results from the interplay of over-extension abroad and domestic economic weakness.‡ Over time, the costs of America's hegemonic vocation will interact with its economic vulnerabilities— endless budget deficits fueled in part by burgeoning military spending, and the persistent balance of payments deficit—to erode America's relative power advantage over the rest of the world. As the relative power gap between the United States and potential new great powers begins to shrink, the costs and risks of challenging the United States will decrease, and the pay-off for doing so will increase. As the British found out toward the end of the 19th century, a seemingly unassailable international power position can melt away with unexpected rapidity.

There are already today other potential poles of power in the international system waiting in the wings that could quickly emerge as counterweights to the United States. And with the Iraq crisis revealing the stark nature of American hegemony, these new power centers have increasingly greater incentive to do so. Here, by facilitating "soft" balancing against the United States, the Iraq crisis may have paved the way for "hard" balancing as well. Since the end of World War II, policymakers and analysts on both sides of the Atlantic have realized that Europe is a potential pole of power in the international system. Will France and Germany provide the motor to unite Europe in opposition to the United States? Time, of course, will tell.

But for sure, this is not 1963. The Cold War is over, and France and Germany are freer to challenge American hegemony. The EU is in the midst of an important constitutional convention that is laying the foundation for a politically unified Europe. And even as the Iraq War proceeded, there were straws in the wind pointing in the direction of hard balancing against the United States. Most notable are indications that France, Germany, Belgium and Luxemburg may act together to create Europe's own version of a coalition of the willing—by forming a "hard core" of enhanced defense cooperation among themselves.

In the short term, however, Paris and Berlin—supported by Russia—have lead the way in soft balancing to counter American hegemony. By using international organizations like the United Nations to marshal opposition to the United States, France and Germany—and similarly inclined powers such as Russia and China—are beginning to develop new habits of diplomatic cooperation to oppose Washington.

Similarly, it is likely that France and Germany (again, joined by Russia and China) will be more likely to cooperate in propping up key regional powers that might be the next targets in Washington's geopolitical gunsight. Iran is one such potential target. With Washington bidding for hegemony in the Persian Gulf region by establishing a protectorate over postwar Iraq, France and Germany—Russia and China, too—will have strong incentives for collaborating to ensure their own strategic and commercial interests in the region by building up, and supporting, Iran (and perhaps Syria) as a counter-weight to U.S. regional power. It was no coincidence, after all, that Dominique de Villepin showed up in Tehran within days after the fall of Baghdad.

AT THE END of the day, the most telling piece of evidence that the Iraq War marks a turning point in Transatlantic relations, and with respect to American hegemony, is this: Despite widespread predictions that they would fold diplomatically and acquiesce in a second UN resolution authorizing the United States and Great Britain to forcibly disarm Iraq, Paris and Berlin (and Moscow) held firm. Rather than being shocked and awed by America's power and strong-arm diplomacy, they stuck to their guns—just as Britain and France did *not* do at Suez—and refused to fall into line behind Washington. What this shows, at the very least, is that it is easier to be Number One when there is a Number Two that threatens Numbers Three, Four, Five and so on. It also suggests that a hegemon so clearly defied is a hegemon on a downward arc.

Many throughout the world now have the impression that the United States is acting as an aggressive hegemon engaged in the naked aggrandizement of its own power. The notion that the United States is a "benevolent" hegemon has been shredded. America is inviting the same fate as that which has overtaken previous contenders for hegemony. In the sweep of history, the Bush Administration will not be remembered for conquering Baghdad, but for a policy that galvanized both soft and hard balancing against American hegemony. At the end of the day, what the administration trumpets as "victory" in the Persian Gulf may prove, in reality, to have pushed NATO into terminal decline, given the decisive boost to the political unification of Europe (at least the most important parts of it), and marked the beginning of the end of America's era of global preponderance.

Christopher Lane is a visiting fellow in foreign policy studies at the Cato Institute. He is writing a book on America's hegemonic grand strategy for Cornell University Press.

Notes

* The seminal work of the "Open Door" school, of course, is William Appleman Williams' *The Tragedy of American Diplomacy* (New York: Delta, 1962). Williams' work has acted as a powerful stimulus that produced a broad body of historical scholarship that both built upon, and refined, the Open Door interpretation. When read as a whole, it encompasses economics, ideology, national interest and security as key factors in shaping U.S. grand strategy—and underscores their interconnectedness.

** In notes prepared for Secretary of State George Marshall, Kennan argued that the Marshall Plan was necessary for two reasons, the first of which was "so that they can buy from us." The second reason was "so that they will have enough self-confidence to withstand outside pressures." Memorandum Prepared by the Policy Planning Staff, July 21, 1947, FRUS 1947, III, p. 335.

*† Referring to NATO and the EGSC, Secretary of State Dulles observed, "These represent important unifying efforts, but it cannot be confidently affirmed that these organizations are clearly adequate to ensure against a tragic repetition of the past where the Atlantic community, and particularly Western Europe, has been torn apart by internecine struggles." He then underscored the need for even greater unity within the Atlantic Community, not simply to meet the Soviet threat, but "forms of unity and integration which would preserve the West from a continuance of internal struggles which have been characteristic of its past." U.S. Delegation at North Atlantic Council Ministerial Meeting to Dept. of State, May 5, 1956, *FRUS* 1955–57, IV, pp. 68–9.

† As Secretary of State Rusk said, "If Europe were ever to be organized so as to leave us outside, from the point of view of these great issues of policy and defense, it would become most difficult for us to sustain our present guarantee against Soviet aggression. We shall not hesitate to make this point to the Germans if they show signs of accepting any idea of a BonnParis axis." Rusk to the Embassy in France, May 18, 1963, *FRUS* 1961–63, XIII, p. 704.

‡ The two classic elaborations are Robert Gilpin, *War and Change in World Politics* (Cambridge: Cambridge University Press, 1981); and Paul Kennedy, *The Rise and Fall of the Great Powers: Economic Change and Military Conflict from 1500 to 2000* (New York: Random House, 1987).

References

1. Robert Kagan, *Of Paradise and Power: America and Europe in the New World Order* (New York: Alfred A. Knopf, 2003).
2. See Mearsheimer, *The Tragedy of Great Power Politics* (New York: W.W. Norton, 2001).
3. Zelikow and Rice, *Germany Unified and Europe Transformed: A Study in Statecraft* (Cambridge, MA: Harvard University Press, 1995), pp. 169–70 (emphasis added).
4. Ronald D. Asmus, *Opening NATO's Door: How the Alliance Remade Itself for a New Era* (New York: Columbia University Press, 2002), p. 290.
5. Ibid., p. 261.
6. PPS/4, "Certain Aspects of the European Recovery Problem from the United States Standpoint", July 23, 1947, *PPSP*, I, p. 31.
7. Caffery to Marshall, July 10, 1947, *FRUS* 1947, III, p. 317.
8. Memorandum of Conversation, Prepared in the Department of State, September 15, 1949, *FRUS* 1949, IV, p.657
9. Harriman to Hoffman, March 12, 1949, *FRUS* 1949, IV, pp. 375–6.
10. See, Minutes of the First Meeting of the Washington Exploratory Talks on Security, July 6,1948, *FRUS* 1948, III, p. 151; Minutes of the Fourth Meeting of the Washington Exploratory Talks on Security, July 8,1948, *FRUS* 1948, III, pp. 167–8.
11. Statement by Dulles to the North Atlantic Council, December 14, 1953, *FRUS* 1952–54, V, p. 461.
12. Talking Paper prepared in the Department of Defense, undated, *FRUS* 1964–68, XIII, p. 728.
13. Letter from Rusk to Senator Mansfield, April 21, 1967, *FRUS* 1964–68, XIII, p. 562.
14. Memorandum by the Acheson Group, undated, *FRUS* 1964–68, XIII, pp. 406–7.
15. Acheson and Lovett to Truman, July 30, 1951, *FRUS* 1951, III, p. 850 (emphasis added).
16. Acheson to Bruce, September 19, 1952, *FRUS* 1952–54, V, p. 324.
17. Paper Prepared by Kennan, February 7, 1949, *FRUS* 1949, III, p. 92.
18. Minutes of the Seventh Meeting of the Policy Planning Staff, January 24, 1950, *FRUS* 1950, III, p. 622.
19. Policy Statement of Department of the State, September 20, 1948, *FRUS* 1948, III, pp. 652-3.
20. Paper Prepared in the Department of State, n.d., "Economic Benefits of European Integration", *FRUS* 1949, III, p. 133.
21. Jebb quoted in Michael Hogan, *The Marshall Plan: America, Britain, and the Reconstruction of Western Europe, 1947–52* (Cambridge: Cambridge University Press, 1987), p. 113.
22. Lundestad, *"Empire" by Integration* (New York: Oxford University Press, 1998), p. 135.
23. François Duchene, *Jean Monnet: The First Statesman of Interdependence* (New York: Norton, 1994), pp. 186–7.
24. Quoted in David Dimbleby and David Reynolds, *An Ocean Apart: The Relationship Between Britain and America in the Twentieth Century* (London: Hodder and Stoughton, 1988), p. 235.
25. Hitchcock, "Reversal of Fortune: Britain, France, and the Making of Europe, 1945–1956", in Paul Kennedy and Hitchcock, eds., *From War to Peace: Altered Strategic Landscapes in the Twentieth Century* (New Haven: Yale University Press, 2000), pp. 100–1.

26. Department of State to the Embassy in Germany, February 2, 1966, *FRUS 1964–68*, XIII, pp. 308-9.
27. Remarks of President Kennedy to the National Security Council, January 22, 1963, *FRUS 1961–63*, XIII, p.486.
28. See Josef Joffe, "Continental Divides", *The National Interest* (Spring 2003).
29. See Charles Krauthammer, "The Unipolar Moment Revisited", *The National Interest* (Winter 2002/03).

Reprinted with permission of *The National Interest*, Summer 2003, pp. 17-29. © 2003 by The National Interest, Washington, DC.

Forget asylum-seekers: it's the people inside who count

The real issue for European societies is not how to keep new foreigners out but how to integrate the minorities they already have

TWO young men set off with suicide bombs to Tel Aviv. One carries out his deadly mission, the other fails. Embittered Palestinians? No. Both are Muslim Britons, one indeed born in Britain.

Why did they do it? The easy answer is extremism—learned in Britain. Few British Muslims are extreme in their faith; hardly any, however they feel about Palestine, are in favour of terrorism. Yet those two men are not just nasty mavericks. They symbolise a wide-ranging question with no easy answers: can Europe integrate its mainly new, and growing, minorities?

Ask the habitually tolerant Dutch. The most potent phrase in Dutch politics today is *normen en waarden*, norms and values. Traditional Dutch ones, of course; yet few Dutch people two years ago had ever heard the phrase, or thought about the values. Then came September 11th, and then a politician called Pim Fortuyn. Suddenly the Dutch elite noticed what ordinary citizens had long believed, but not dared to say: that many of their immigrant neighbours did not (or so the average Dutchman felt) share these Dutch values.

It was a moment of truth, not only for the Netherlands but for the whole of northern Europe. At last, not just were the long-term effects of immigration openly on the agenda but it was permissible to be open about them; in particular, to admit that they would not go away again if only the plebs would put aside those racial and other prejudices which the better-educated, suburban-dwelling liberal elite wouldn't dream of sharing. Fortuyn was shot dead a year ago; his party was soon in chaos. But the veil that decency and goodwill had cast over discussion of such questions has been decisively torn away.

The main noise since then has been about asylum-seekers and how to keep them out. But the real issue is the immigrants, and their descendants, who are already inside. Integrate these, and European societies could cope well enough with the relatively few asylum-seekers.

That demands changes of attitude in the host societies and among the newcomers. In many European countries it has not been achieved: witness the shaky attempt in France, which has 4.2m Muslims, to set up a council in which they can find a political voice. Yet most of rich Europe is scrambling towards this ideal. Rightly so: social disunity could be a huge long-term threat to Europe, and, as the past two years have shown, harmony does not grow on trees.

Count in their locally born descendants, and there may be 12m-15m poor-country "immigrants" inside the EU: Turks and Kurds, Arabs, Asians (mostly from India, Pakistan and Sri Lanka), all manner of sub-Saharan Africans, Caribbeans, Latin Americans. Some of these communities are long-established, like the West Indians who were first brought into Britain to meet labour shortages in the 1950s, or the Turks who helped to prolong Germany's *Wirtschaftswunder*, the economic miracle that began 40-odd years ago. Some are newer, like the Bangladeshis who have poured into Britain within the past 20 years. But all share two things. First, they are communities, mostly distinct in skin-colour, language and religion from the natives, and not just random collections of individuals. And, second, these communities are not integrated into the society around them.

Just how large they are, from where and settled where, no one knows; in part, because definitions vary. Nordic statistics, for example, tend to lump together new arrivals with the children, even grandchildren, of earlier ones. There are solid reasons for that: the children of, say, brown-skinned, Muslim, poor Pakistanis will certainly be dark, nearly always Muslim, mostly able to speak Urdu, and often, as adults, poor. Yet there are also solid reasons against: nearly all will be vastly more fluent than their parents in the language of their adopted land, and familiar with its ways. And their children, in turn, still more so.

And that is the trap into which most European countries, unwittingly, have fallen. Because natural assimilation has worked in the past, they have sat back to let it do its natural work again.

That was not absurd. Most of Britain's 300,000 Jews are descended from east European immigrants of around 1880-1910. When these arrived, they too were concentrated in poor east London; they too spoke foreign tongues, had their own religion and habits, and were often disliked by the natives, some better-off and long-established Jews included. And officialdom lifted barely a finger to turn them into Britons. That was left to the—often vigorous—efforts of sympathetic, or worried, Jews already in place. Yet by now Britain's Jews (the Hasidim apart) are as assimilated, as British, as any descendants of the Angles or Normans. They did it; why not leave others alone today to do the same?

Because things have changed. Today's newcomers have come fast, and in far greater numbers. They are, literally, more visible to the eyes of native prejudice; and, the spirit of 2000 being far from that of 1900, they—and still more their children—are likelier to resent prejudice than to hunker down, hope not to be noticed and put up with it when they are. Nor have many shown the vigour that saw Britain's Jews spread where they chose and win the acceptance that education, money and a position in the world habitually buy; Britain's Gujaratis, originally from western India, are a parallel case, but a rare one. Maybe all should have assimilated, but the fact is they haven't.

So it is that Oslo has its "little Karachi"; that to Berliners the Kreuzberg district means Turks; that a Parisian calls Montreuil, just to the east, "the second capital of Mali"; that you can count 20 pupils coming out of a Rotterdam primary school before you spot the first obviously Dutch one.

Two halves make a whole

Yet, until recently, few but specialists asked: what is to be done? Britain's Race Relations Board, set up in 1966, has grappled with the question only to reach, in most cases, the usual answer: teach the natives to be less prejudiced. That is a worthy reply, but only half of one. The other half should have been to ask what solid reasons might lie behind the prejudice, and what could be done about them, not least by the minorities at issue. To ask such things was almost like blaming Jews for anti-Semitism. But the answer to both questions is, quite a lot—some of it just the reverse of what good-willed people have done till now.

Go back to the Dutch. Their society was for centuries built on the "mosaic", not "melting-pot", notion of integration: we are Catholics and Protestants (and more), we have each our own churches, schools, even sports clubs. But we live in mutual respect, we're all Dutch. Then in late 2001 they got a shock, symbolised by a magazine poll that asked Muslims their view of the September 11th attacks. A bad thing, said 61%. Fair enough? No: what kind of community is it where 39% do not automatically condemn the murder of 3,000 innocents? Not much of one, said the native Dutch—and "not like us".

So? One response was gut hostility to at least the Arab incomers: the word *Marokkanen*, preceded by an obscenity, was soon in public use. Another was to demand still fiercer immigration controls than the already tough ones brought in earlier in 2001. But the thoughtful answer—it was Fortuyn's, rapidly taken up by other parties, right or left—was to think how to make the country's Muslims more like "us".

In Rotterdam, where the Fortuynists became the largest party on the city council, the resultant coalition made a priority of *inburgering*, the forming of citizens. Get more immigrant children into kindergarten, make sure they master the language, push them to stay longer at school and get better job skills. And act correspondingly for adults, newly arrived or long resident: encourage or shove them into citizenship courses, show them how ordinary Dutch society works and how it thinks.

Such notions can spring from and lead into racism. But it is hardly an act of hostility to make people improve their social or work skills; it happens to all schoolchildren. And to most of the native Dutch, this was simply a reasonable "when in Rome do as the Romans do", and a recognition that this acculturation was not happening fast enough, but needed to be pushed.

This approach is now spreading fast. Indeed the Danes, a people very conscious of their immigrants, would say they pioneered it. They elected a Liberal (ie, free-market) government in late 2001, and a new ministry for "immigrants and integration" began not just fiercely shutting the doors but also pushing the integration that the previous government had merely talked about. The emphasis is on jobs: "Work is the key to integration."

Both sticks and carrots are used. Welfare benefits for all the newly arrived have been cut—for their first seven years!-to well below the rates for most Danes. But they can now work part-time while drawing these. To help the process, the newcomer must sign up to compulsory courses in civics and language and, if need be, compulsory work placements. Fail to comply, and your stingy benefits will become even stingier. But extra money is going into integration, for example, to job counselling for immigrants and to educating foreign women brought in for marriage.

That is the theory, and these are early days. Already problems are plain. The government says what is to be done, but the local authorities have to do it, and don't find that easy. Tighter controls on bringing in a bride are unlikely to drive a young male immigrant into instant marriage with a blue-eyed blonde. Nor will a need for nine years of legal residence, plus other requirements, before he can be naturalised as a Dane help to make him feel like one.

That is the trouble: the clash between the widespread European feeling of "Let's have fewer of these people", and being more welcoming to those already inside. Norway, which is following Denmark down the compulsory "induction" route, has less anti-immigrant feeling. But a new rule won by its most anti-immigrant (and, at this moment, most popular) party bars accepted asylum-seekers from bringing in family members unless they can support them. That will push some people to work, but it will hardly make them feel they belong—and it is not meant to.

Headscarves and cricket

Germany for years exemplified a rather different paradox. It welcomed its "guest-workers" as workers, but no way, least of all by easy naturalisation, did it try to integrate them. The newcomers naturally tended, and cheerfully were left, to stick in their national groups, socialising, shopping and praying with each other, reading their own newspapers—Turkey's *Hurriyet* has a flourishing German-printed edition—and more recently watching their own satellite-television programmes.

An immigration law was passed last year, aimed, among other things, at integration, with publicly financed courses in German language, history and other citizen-like knowledge. Compulsory courses? That was left unclear, as the law itself still is: for procedural reasons, not content, the Constitutional Court last December struck it down. It may yet be revived.

Britain, in contrast, though endlessly alarmed these days about asylum-seekers, has done startlingly little to integrate the millions of immigrants and their offspring, largely from its ex-empire, that it already has. And until recently, and still very largely, the British line has been to accept the resultant mosaic, cross one's fingers and hope: no compulsion here.

Though the state has been slow to finance Muslim schools in Britain (the Netherlands, in contrast, has more than 40 already), multiculturalism is still the rage. Many urban local councils put out documents in several languages. And visiting cricket teams from Pakistan or India win loud support from their ethnic cousins, though most of these are British-born and thereby British citizens; a phenomenon that irritated one of Margaret Thatcher's senior ministers, but worries few people in

Britain and prompts fewer still to suggest any measures that might alter it. Only recently has an authoritarian home minister begun to think of forcing newcomers into British ways, and even he is thinking strictly of newcomers. The case of the Tel Aviv suicide bombers may yet promote fresh thinking; so far it has promoted only fresh security measures.

The French notion of integration, in contrast, is strictly that of the melting-pot, with the heat supplied by "republican values", secularism not least. That noble ideal can produce tortured arguments over the right of Muslim schoolgirls to wear headscarves. But, worse, it led for years to official unreadiness to admit that there was a problem with, specifically, the large Arab and Muslim population, and one requiring active treatment.

Since the September 11th attacks, and amid rising alarm about Muslim terrorists, especially from Algeria, there has been much talk of the need for newcomers to accept French values, but little certainty of how to achieve it. The authorities have long been eager to see Islam "naturalised", with imams trained in France rather than sent and paid for from abroad. France, a secular state by constitution, cannot finance religion. It has, however, been fairly generous in regularising the status of illegal immigrants: of 140,000 who applied when the left came to power in 1997, 80,000 were accepted. This gives some ammunition to the racist right, but it is surely a step towards integration.

This year has brought one more overt effort that way: the setting-up of a single national Islamic council to act as an interlocutor with the authorities. But will this in fact bring more integration—or less? The new body, which met for the first time on May 3rd, was elected in April by delegates from nearly 1,000 mosques. But are these the authentic voice of the Muslim community? The Archbishop of Paris, not alone, doubts it, arguing that most Muslims do not go to the mosque, and that "you can't reduce the issue of North African immigration"—much the largest—"to one of Islam". What's certain is that the election gave a large voice to Muslim traditionalists and fundamentalists, and these were soon challenging the government over headscarves (in identity-card photos, this time). If such clashes occur often, the new council could be a factor against integration; and the interior minister's threat to deport imams who challenge republican values, which is not yet a crime in France, is no great way to teach those values.

Spain and Italy, parts of which centuries ago were actually ruled by Muslims from North Africa, are by northern standards surprisingly relaxed about their immigrant descendants. Spaniards are proud of the Christian *Reconquista*, but also of their Muslim heritage; they hear more about the dozens of Moroccans drowned trying to cross the straits of Gibraltar than about the thousands labouring in Andalusia's horticulture and elsewhere. Italians once had a historic phrase "*Mamma, li Turchi!*"—the Turks (ie, Muslims) are coming! But they are likelier these days to know a not-so-historic north-Italian joke, and it is not immigrants who are its target:

Q: Why did Sicily win the Nobel peace prize?

A: Because it was the only Arab country that didn't make war on Israel.

Italian governments have often acted to legalise illegal workers: measures in 1990, 1995 and 1998 each gave papers to more than 200,000 people. Though one party in the Berlusconi coalition government is openly anti-immigrant, a fresh offer from that government last summer brought almost 700,000 applications. How many will succeed is unclear, given the slow start to the process and the doubts among Mr Berlusconi's governing partners, who see this as a means not of promoting integration but of sorting out who wants to work and who does not, and excluding the latter. The permits in any case will be valid only for a year, though renewable.

More recently, the home minister offered "dialogue" to Italy's moderate Muslims, aiming to isolate the extremists: a move mostly welcomed by Muslims, although the Archbishop of Turin stirred the pot by saying that the church should offer them the Gospel as well. But to Italians the problem with immigration is not so much one of alien values as of the arrival, with the Balkan immigrants, Muslim or not, of Mafia values. And acculturation is still, as ever, left to work largely by itself.

Citizenship and the vote

There is one obvious way of helping it along: citizenship, or at least the vote. Treat people as voteless foreigners, and why would they feel anything else? Let them vote, and maybe they will feel at home. In fact, this remedy may not be much of one. As Commonwealth citizens, most members of Britain's minorities, even those not formally British, can vote already. Anyone born in France (or most European countries) is a citizen automatically. In contrast, Germany until 1998 based citizenship on descent, not birthplace, and required 15 years of residence before an outsider could be naturalised; most of its ethnic Turks are still Turkish citizens. Yet as between France's Arabs, many of whom are French citizens, and Germany's Turks, it is the Arabs who feel, and are seen as, more alien.

Still, political rights must have some integrative value. EU countries already let each other's citizens vote in local elections. Now the EU's economic and social committee is arguing for "civic citizenship", which would give long-term residents from outside equal local-voting rights—and indeed more valuable ones, such as equal access to education and jobs.

Whatever the method, one thing is sure: Europe needs active integration policies. It cannot just sit around and wait for time to sort things out. That has been tried. It has not brought disaster—but it could.

From *The Economist*, May 10, 2003, pp. 22-24. © 2003 by The Economist Newspaper, Ltd. Reprinted with permission. Further reproduction prohibited. www.economist.com

How the Armies of Europe Let Their Guard Down

By PHILIP SHISHKIN

Guaranteed Jobs for Soldiers Leave Little Room to Buy Equipment or Even Train

NEDER-OVER-HEMBEEK, Belgium—Chief Cpl. Rudy Christians, an impeccably coiffed military hairdresser, has been cutting soldiers' hair for 24 years, and he loves his work.

It's a full-time job, guaranteed until retirement, and until then, the 47-year old has enough free time to pursue an amateur singing career featuring Elvis and Tom Jones numbers. When the military does send him on an occasional field exercise, he is amazed by the fellow soldiers lumbering around him. "All the people are so old," he says.

Recruits like this help explain why Europe's military muscle has grown soft, and why the U.S. can't count on substantial military help from many of its European allies.

Even if every member of the North Atlantic Treaty Organization were to back a U.S. strike against Iraq, the military impact might not be huge. The 17 European countries in NATO have about 2.3 million active-duty troops, about a million more than the U.S. does. But many of NATO's forces are poorly equipped, in part because so much money is spent on pay and benefits that there is less left for the technology, weapons and other gear that modern forces need.

Washington has asked NATO for limited contributions to an Iraqi campaign, for both political and military reasons. Its requests to NATO have focused mainly on the defense of Turkey and a reconstruction of Iraq if war occurs. France, Germany and Belgium say it's too early to plan for war, and hope, the Iraq crisis can be resolved peacefully.

While the U.S. spends 36% of its defense budget on pay and benefits, most NATO members in Europe earmark an av-

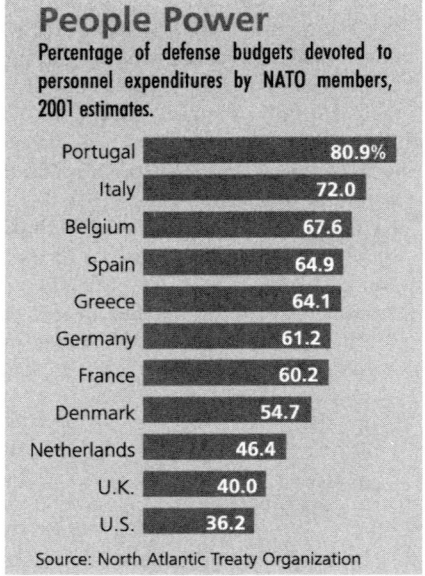

People Power
Percentage of defense budgets devoted to personnel expenditures by NATO members, 2001 estimates.

Country	%
Portugal	80.9%
Italy	72.0
Belgium	67.6
Spain	64.9
Greece	64.1
Germany	61.2
France	60.2
Denmark	54.7
Netherlands	46.4
U.K.	40.0
U.S.	36.2

Source: North Atlantic Treaty Organization

erage of nearly 65%. The U.S. military employs support staff, of course, and also faces rising costs per soldier, especially because of health care. Still, overall, the share of personnel spending in the U.S. defense budget has decreased by six percentage points since the early 1980s. NATO statistics show that, such spending has grown by as much or more in Europe during the same period.

NATO Officials acknowledge Europe needs to upgrade its military capabilities. "We could do with fewer troops, but better troops; better trained, better equipped, more mobile," NATO Secretary General George Robertson said last month at the World Economic Forum. "The problem in Europe is that there are far too many people in uniform, and too few of them able to go into action at the speeds that conflicts presently demand."

Belgium for example, employs hundreds of military barbers, musicians, and other personnel who aren't likely to be called into battle. Yet Belgium doesn't have the money to replace aging helicopters or conduct regular combat-training exercises. Germany drafts 120,000 people every year but can't afford to buy all the vital transport planes it wants; last year, budget crunches forced it to slash an order of planes to 60 from 73. German soldiers who went to Afghanistan as peacekeepers crowded into an aging, leased Ukrainian carrier that had to stop to refuel

In France, one of the few NATO countries to increase its defense budget this year, military-procurement funding fell 14% between 1997 and 2002, leaving its forces wanting in such key areas as refueling aircraft and missiles. The French defense ministry says it will address procurement shortfalls in the new budget. Europe, has 11 troop-transport planes, compared with 250, in the. U. S., and most European members of NATO don't have any modern precision-guided munitions at all.

U.S. Pressure

Since the Sept. 11 terrorist attacks, the U.S. has stepped up calls for Europe to put more emphasis on smart bombs, secure communications, special-forces units and long-hall planes to take them to battle. U.S. officials from President George W. Bush on down have pressed for more investment to offset what U.S. Gen. Joseph Ralston, the former NATO supreme allied commander for Europe, calls European militaries' "outdated and redundant fat." Secretary of Defense Donald Rumsfeld told European defense ministers meeting in Warsaw last summer that unless they start spending more on key defense capabilities, the U.S. won't call on them for backing when it goes to war. "The phone just won't ring," Mr. Rumsfeld said.

European leaders say they want to streamline and modernize their armed forces, and some have started. Outfitting their militaries to be, nimble and high-tech

Article 24. How the Armies of Europe Let Their Guard Down

is vital if Europeans want to influence the U.S. policies with which Europe so frequently disagrees. The U.S. wants Europe to modernize so it can depend on other countries to share the job—and cost—of playing global cop

But swift reform isn't possible in Europe because of labor laws, influential unions and a widespread conviction that defense spending shouldn't be a priority. Beyond that, Europe's economy is weak, and the 12 countries that use the euro are supposed to keep their deficits under 3% of gross domestic product. "It makes it difficult to create a real defense capability, even with all the troops," says Florentino Portero, military specialist for the think tank Grupo de Estudios Estrategicos in Madrid.

One reason Europe has so many soldiers is its strong military labor unions. Unheard of in the U.S. and Britain, these unions trace their history to the end of the 19th century, when disgruntled Dutch soldiers unhappy about living conditions, banded together into a group called Ons Belang (Our Interests). Similar groups soon sprang up around Western Europe. In the 1970s, European military unions gained sweeping collective bargaining rights, though they stay out of war-planning and deployment issues.

In Belgium, military unions are as powerful as anywhere on the Continent. On King Albert's birthday last June, a holiday for the Belgian military, unions deployed thousands of soldiers to Brussels to demand a raise in vacation pay. Soldiers chanted, drank beer and banged their aluminum mess bowls. "Show me the money," one officer shouted to a passing police van. The protest grew so rowdy that police cooled demonstrators off with a water cannon. But it was a success: An emergency session of the Belgian cabinet agreed to give soldiers—already eligible for six weeks' annual vacation—a raise in holiday benefits valued at about $500 each.

For Emmanuel Jacob, an artillery officer and a union leader who was on the front lines of the protest, it was a bittersweet victory. "We must be honest with ourselves," says Warrant Officer Jacob, secretary-general of Centrale Generale du Personnel Miliaire, which represents 6,000 active-duty and 2,500 retired personnel. "Either we have a smaller number of people who are well-trained and equipped or we continue to defend a bigger army and it won't work in the future."

The average age of a Belgian soldier is 40—compared with 28 in the U.S. and 29 in the U.K. Most Belgian military personnel can retire at 56 with full pension benefits. The Defense Ministry, acknowledges too many of its soldiers are too old, and says it is trying to recruit younger people. But Gerard Harveng, a spokesman for Defense Minister Andre Flahaut, says, "I'm not sure that the mission of the Belgian military is to fight." Instead, Belgium sees its military role mostly focused on peacekeeping operations.

During the Cold War, Washington's message to Europe was different than it is today. The U.S encouraged heavy investments in troops to prepare for a Soviet land invasion. People who, were drafted or signed up in the 1970s and 1980s were guaranteed full employment until retirement Though it varies from country to country, some European governments, including Belgium, still have, that policy today.

"Once you enter the military, you are in for life," says Maj. Renaud Theunens, 39, an intelligence officer who took a leave from the Belgian armed forces to work at the International Criminal Tribunal for Yugoslavia in the Hague. "It was quite unusual to ask to leave," he says. In the U.S., by contrast, it is more unusual for a soldier to make the military a career. Under its "up or out" policy, the Pentagon can force officers to leave if they fail to move up the ranks within a certain period of time. The U.S. government also has devised programs encouraging people to take early retirement and get jobs outside the military.

Belgium has cut its military payroll by half since the height of the Cold War, to 44,000. But it still spends some 67% of its annual defense budget of about $2.5 billion on pay and benefits and only about 5.4% on equipment. The U.S., with an annual defense budget of $366 billion; spends 22% on equipment, according to NATO.

Reduction Through Attrition

The Belgian Defense Ministry's goal is to trim its military work force by 10% or so over the next decade, but the reductions can come only through attrition. Early efforts to cut the payroll have already run into opposition from military employees and labor unions. They rebelled in the early '90s, for example, against a proposal to merge Belgium's six military bands into one. "Each band has its own, character and repertoire," says Alain Crepin, director of the Air Force orchestra, as his musicians pack up their instruments after daily practice on a deserted base. The Air Force band's repertoire includes jazz and other modern music, he notes, while the Army is heavy on the classics. Lumping them together to save money would be "stupid," he says. The government compromised, downsizing to three bands, with 260 members.

> The average age of a Belgian soldier is 40—compared with 28 in the U.S.

"We should have a major reform of personnel," says Stef Goris, a member of the Belgian parliament's defense committee and a former tank-battalion officer. The country has more than enough troops, he says, but "it's very hard to send them to a place like the Balkans because they aren't fit enough."

In the meantime, many soldiers are happy with military life. Chief Cpl. Jerome Loos, for instance, is part of an eight-member crew that makes lunch for about 100 people at an army base in Siysele, A typical meal: chicken, french fries and vegetable stew. He says his brother, a private-sector cook, works much harder. "I have lots of free time and good job security," says Cpl. Loos, who was drafted in 1986. An avid runner, he can go jogging between duties and be home by late afternoon to spend time with his kids. He gets 30 days paid vacation and earns about $20,000 a year after taxes. Once or twice a year, Cpl. Loos has to go on shooting exercises but says he feels more comfortable with a knife in the kitchen."

There are cooks in the U.S. armed forces, though few are allowed to make, careers of the job. Most food service on U.S. military bases is handled by private catering contractors. The Belgian military says it tried outsourcing cooking on a limited basis in the early 1990s, but it proved expensive.

For many, Belgium's lopsided spending ratio is frustrating. Belgians in combat positions don't train as regularly as the top brass would like. A lack of funds forced a cutback on training exercises. When, they do practice, troops often use outdated or inadequate equipment. On an army base outside Brussels, Lt. Theo Blomme flies two transport choppers, a 10-year-old Augusta and a 30-year-old Allouette-2, both so small that only two or three people in combat gear can squeeze in at once.

For a safe battlefield rescue, Lt. Blomme says he would need a much bigger helicopter that could land, take in 20 soldiers and leave. With his small helicopters, he would have to evacuate in groups of twos and threes. "The enemy would hear you on the first approach and shoot you down on the second," says Lt. Blomme, 38. He says he feels silly training in an aircraft that will likely never see combat. "It's embarrassing."

The Belgian military says it wants, to update its transport-chopper fleet, but the $500 million price tag is prohibitive right now. The Defense Ministry also has its sights set on a troop transport ship and a fleet of infantry-transport vehicles. Hundreds of millions of dollars will be freed up for such purchases when its work-force-reduction plan is complete in 2013.

Over the next decade, Belgium will eliminate thousands of military jobs, close bases and consolidate operations. Units will be shuttered at the sprawling military hospital in Neder-Over-Hembeek, where many doctors now work four-hour days for full-time pay, allowing some of them to set up private practices. The hair salon where Cpl. Christians works will probably survive, but full-time military hairdressers, jobs that don't exist in the U.S., won't be replaced when they retire.

As a red-haired femme officer sat down for a government-subsidized trim, Cpl. Christians took a break to reflect on his career. He was drafted at 19 following a brief stint as a civilian hairdresser. After a few years doing office work for the military, he landed a job as a barber. His military specialty is defending bases against aerial bombardment, but he has never seen combat. He takes home about $18,000 a year after taxes, and on Saturdays, he is free to work on his pop-singing act, something he didn't have time for in the private sector. "Personally, I think it's important to have people like us in the military," he said. Hairdressers provide part of what Cpl. Christians sees as the three essentials of soldiers' happiness: "Good dress, good food, and feeling good."

From *Wall Street Journal,* February 18, 2003, pp. A1, A7 (Eastern Edition) by Philip Shiskin. © 2003 by Dow Jones & Company, Inc. Reproduced with permission of Dow Jones & Company, Inc. in the format Textbook via Copyright Clearance Center.

Article 25

A nervous new arrival on the European Union's block

Poland is by far the biggest of the countries joining the European Union next year. But it must also do the most to get into shape

"WITH hope and apprehension and anxiety and uncertainty," replies one Polish official when asked how he views his country's entry into the European Union next year. The Polish people appear more enthusiastic: 77% said yes in June to joining the EU in a referendum with a 59% turnout, a strong showing by national standards. But within government the mood is more hesitant, even nervous. Officials worry that Poland has been slow in preparing for the administrative challenges of EU membership, and slow in fixing a clear strategy for advancing the country's interests within the Union.

The same could easily be said of other countries among the ten due to join the EU in May next year, eight of them from central Europe. But Poland is in a category of its own. With almost 40m people, it accounts for roughly half the population and half the GDP of all ten incoming countries. It will have more votes in the Council of Ministers, the EU's main legislature, than any country bar Britain, France, Germany and Italy. How Poland handles itself within the EU will be a matter of vital concern not only to Poland, but to all Europe too.

It may well be in for a rocky start. One foreign diplomat in Warsaw fears that Poland will fail to master all the EU's labyrinthine farming and food-safety laws by the time it joins, allowing other EU countries to block its farm exports, using what the EU calls "safeguard clauses". It might also fail to muster enough well-planned and well-managed projects to claim its full share of EU development funds, leaving it a net payer into the EU budget. The result would be anger among Poland's millions of farm workers and Eurosceptics, playing into the hands of populist politicians such as Andrzej Lepper, a pro-small-farmer nationalist whose Samoobrona (Self-Defence) party vies for second place in the opinion polls behind the ruling Left Democratic Alliance, and who will be hoping to improve on his current 11% of parliamentary seats when Poland next goes to the polls. "If they are net payers, and if there are safeguard clauses, then it will be Lepper in 2005," says the gloomy diplomat.

The European Commission, the EU's executive arm, will freshly assess Poland's preparations for entry in an annual report this autumn. Last year it said Poland needed to make "major efforts" in its farm sector, and expressed "serious concerns" about the country's veterinary standards. It reported progress in preparations to absorb structural funds, but said that administrative capacities still needed improving "substantially".

Poland is likely to get more such warnings and urgings this year, especially on food safety. But civil servants fear that time is short and political leadership is weak. There is strikingly little enthusiasm for the struggling, unpopular, left-wing government of Leszek Miller, the prime minister. The installation of a new agriculture minister last month, at so delicate a stage in preparations for EU entry, can scarcely have helped.

Mr Miller stood back from the EU referendum campaign for fear of turning the vote into one on his own government. But when the happy surprise of a decisive result lifted the national mood, he set about profiting from it. He called an immediate vote of confidence, which his government won despite its minority in parliament, and he seems confident now of holding on to power until the next general election, due in 2005, an outcome far less likely before the referendum.

It is unclear what Mr Miller will do with power while he retains it. His government has been distracted and debilitated by scandals. Economic policy was hostage to a long confrontation between Mr Miller's economics minister, Jerzy Hausner, and his finance minister, Grzegorz Kolodko, until Mr Kolodko resigned on June 11th.

Mr Hausner now needs to find ways of stimulating the economy, hit by slack demand from western Europe and falling investment, without feeding a budget deficit which approached 7% of GDP last year. The business cycle is turning up: the economy may grow by about 3% this year, against a miserable 1.3% achieved last year. But Poland's early entry into the euro zone, which would require a budget deficit of below 3% of GDP, seems less and less likely. Most analysts now expect it in 2009–10 at the earliest, rather than in 2007, as the Polish central bank would prefer.

The weak economy and the continuing scandals have encouraged a mood of national cynicism. Recent polls suggest that less than a quarter of the public supports Mr Miller's government. A perceived rise in corruption has sharpened disdain. More than two-thirds of Poles rank corruption as a big problem, against a half in 2000 and one-third in 1991.

Among Mr Miller's opponents, the have-nots tend to gravitate towards Mr Lepper, the haves to two opposition parties of the centre-right, Law and Justice, and Civic Platform. Further out on the right there is the League of Polish Families, a Catholic nationalist party which alone opposed EU entry.

The jockeying of these diverse opposition parties around a weak and demoralised government means little certainty about the direction or consistency of policy and little scope for the government to inspire or steer public opinion. Poles were in a pro-EU mood when they voted in June, but attachment to the trappings of sovereignty remains strong enough for the national mood to go lurching the other way if membership produces few clear benefits. Poland's ministers and diplomats could wish for better conditions at home when they are trying to signal confidently in Brussels what their country wants from the

EU, and what it can offer the EU in exchange, in the long run.

The lack of vision in domestic politics and the fragile state of the economy mean that Poland will probably enter the EU looking mainly for short-term gains, preferably bankable ones, from any negotiation. "First priority, establish strength; second priority, use that strength," says one Polish official, who makes no secret of his admiration for Spain's self-centred negotiating style when it joined the EU in 1986.

A louder voice, please

Even before it joins, Poland wants to defend the weighting of votes in the Council of Ministers fixed by the Nice treaty of 2001. This will give Poland (and Spain, which has roughly the same population) a generous 27 votes in the enlarged council when issues are decided by majority voting, only slightly less than the 29 votes apiece for Britain, France, Germany and Italy, which have much bigger populations. The EU's draft constitution, presented last month, proposes a new system: a simple majority of countries would be able to carry a vote in the council, so long as those in the majority represent at least three-fifths of the EU's population. Poland would lose clout under this system, so its opposition goes deep—much deeper than its other main professed worry about the draft constitution, the absence of any reference there to God, an argument that comes awkwardly from a country, however Catholic it may be, where the prime minister is an atheist and the president an agnostic.

Once in the EU, Poland will want to maximise the money it gets from the EU budget, mainly through farm subsidies and through the "structural funds" allocated to poorer countries and regions. It will also want to resist any new regimes and rules that impose higher costs or heavier burdens of regulation on business or government in areas such as the environment, labour, taxation, competition and state aids. Poland thinks it has enough on its plate complying with the existing EU rulebook, some parts of which it will need another 12 years to implement.

Poland accepts that Germany, the Union's main paymaster, will not agree to raise the legal limit on total EU budget spending above its current level: 1.27% of the EU's collective GDP. But with planned EU spending this year barely above 1% of GDP, Poland believes that it can and should be raised substantially in future years towards the legal ceiling to meet the development needs of new members.

Be fair to our farmers

Poland is also willing to accept the current rough breakdown whereby almost half of EU spending goes to agriculture, under the common agricultural policy, and one-third to the structural funds. But on agriculture, Poland wants to see its own farmers paid the same subsidies as those in current EU countries—which, under current plans, may not happen until 2013. It also wants the EU to direct more money towards rural development, even if that means giving less to actual farmers. These aims may bring conflict with France, the main beneficiary of the current system of farm subsidies, and with Britain, which would rather see the EU spend less of its money on agricultural subsidies in the first place.

The hopes of Poland, and of the other accession countries, for a big share of the EU's structural funds will also meet tough opposition. The countries getting a lot of cash now, led by Spain and Greece, will not want it cut off just because other, poorer countries are joining the Union. Nor will the countries picking up the bill want to pay out more to anybody. Poland thinks it can win Spain as an ally if it argues for phasing out (rather than cutting off) funds to regions now receiving them, alongside payments to new members. But precedent suggests that Spain will look first to its own interests, if money proves tight.

The getting and spending of quick money risks being the main activity by which Polish governments and Polish public opinion will measure the country's "success" in Europe. That makes for a dangerous course. Reversals are possible, given that Poland will be haggling against equally determined and more experienced countries in Brussels. For some other accession countries, integration into Europe is welcomed almost as an end in itself, but this line goes down much less well in Poland, where even those who favour EU integration tend also to have a strong attachment to national sovereignty. Poland's strong pro-Americanism tempers its European instincts as well. It does not want to go too deep into Europe, if that means leaving America behind.

Poland's attachment to its sovereignty need not mark it out much in Brussels, where all the big EU countries have come to favour co-operation among governments over more surrenders of power to the EU's supranational institutions. But reconciling strongly pro-American and pro-European policies, both of which Poland professes, will be more of a challenge for its diplomats. The Iraq war has helped divide the EU into admirers and resenters of American power, with Britain leading the first camp, France the second. Poland has entered the British camp, but would much rather the EU was not at odds with itself or with America. A choice between America and Europe is one "between mother and father—and we love them both," says an official.

Pals with everyone, if possible

Poland has a useful ally in Britain, but it would be far more comfortable if Germany, too, was squarely in the pro-American camp. Poland had hoped and expected to find itself drawing closer to Germany as enlargement pushed the EU's centre of gravity eastward. Instead, Gerhard Schröder's government has been turning more towards France, opposing America's war in Iraq and even exploring a triangular alliance of sorts along with Russia. The Poles are dismayed, but they say Mr Schröder's anti-war stand has at least helped clarify their own policy for Europe. If even Germany is now considered an unreliable ally by America, the Polish thinking goes, then Poland must be a super-reliable one, so that America will not lose interest in continental Europe entirely.

Poland's tack in the EU will probably be to join in talk about a European security and defence policy, because it wants to be seen as a good European, but to do as little as possible in practice, at least until transatlantic relations improve dramatically. Poland's bottom line is that the EU should do nothing to diminish or challenge the role of NATO, which, for all the doubts about its future, still keeps Europe and America together. "It's the Poles and Hungarians and Czechs who are the 'old Europe' really," snipes a German MP close to Mr Schröder. "They're the ones still wanting nuclear protection from the Russians."

Polish officials twitch at the caricature. They say that NATO is a force for global security, that Russia is no threat to Poland, and that relations between Poland and Russia have never been better—which is probably true, if not saying all that much. But there remains an unspoken part of the Polish assessment. Russia may not be a threat on a one-year or five-year view, but on a 50-year view, who would dare make predictions? And whatever the possible threat, from near or far, Poland still sees America as its best ally.

When Poland sent troops to join America's invasion of Iraq, this was not so much to say thank you for past help but as "a modest investment in reciprocity" for the future, says one official. Public opinion opposed the gesture, but it seems to have worked out well enough for the government. Poland's prestige has risen, and no Polish lives were lost in the war. The experience of Iraq has made it easier, not harder, for Poland to support any future American military actions, says one western diplomat.

Don't forget our eastern friends

For all its Atlanticism, Poland sees the EU, too, playing a big part in its future security, but through foreign relations rather than defence policy. Poland wants to be inside the EU, but it wants the EU to roll on eastwards, stabilising and even embracing Poland's neighbours beyond. This year Poland has been arguing quietly for an "eastern dimension" to the EU's external policy, which would give Ukraine and Moldova—and perhaps even Belarus, if that country can shed or curb its near-dictator, Alexander Lukashenka—deeper ties with the EU. Privately, Poland hopes that they too can join the EU some day, though it hesitates to say so publicly. It wants them as prosperous, stable and accessible neighbours, not as poor and rackety ones cut off by an EU border. With Ukraine especially it shares much history, not all of it happily.

Poland is moving boldly, as an EU newcomer, in reaching for so delicate a dossier. But the European Commission has no very clear plans of its own for an *Ostpolitik*, and most EU governments are too busy worrying about next year's enlargement to think about adding even more countries later. Poland fears that unless it raises its voice now, Ukraine, Moldova and Belarus will be relegated to a place among the many diverse countries with which the EU wants to have good neighbourly relations but which it rules out as future members.

One big problem in all this is that Ukraine, Moldova and Belarus are linked intimately to Russia, through the Soviet Union in past decades and through the Russian-dominated Commonwealth of Independent States now. Any policy for drawing them much closer to the EU would have to include some very good ideas for keeping Russia happy at the same time. It is hard to imagine what those ideas could be. Russia has no serious ambitions to join the EU, and economic recovery is helping its re-emergence as an effective regional power. It may resent talk that its protégés in the region would be better off under the EU's influence instead.

Poland's view seems to be that, even so, it has nothing to lose by prodding the EU to look east. One modest result could be more EU money for cross-border projects benefiting not only Ukraine, Belarus and Moldova but also the eastern provinces of Poland, which next year will become the poorest regions in the whole EU. And from Poland's point of view, almost any policy which encourages the EU to engage collectively with Russia will be better than the lack of an EU policy which leaves national capitals vying for Russia's favours.

If the EU can only export the "soft" security of democratic values and trade ties to its eastern hinterland, while America offers the "hard" security of smart bombs and nuclear shields through NATO, the fit for Poland could hardly look better. The two mechanisms might not always mesh perfectly, but that would be a secondary worry. "If you have the historical experience of Poland," says Janusz Reiter, a Warsaw think-tank boss and former ambassador to Germany, "then you have a strong need of security."

From *The Economist*, August 30, 2003, pp. 15, 17-18. © 2003 by The Economist Newspaper, Ltd. Reprinted with permission. Further reproduction prohibited. www.economist.com

UNIT 7
Former Soviet Union

Unit Selections

26. **US-Russian Relations: Between Realism and Reality**, Celeste A. Wallander
27. **The Terrorist Notebooks**, Martha Brill Olcott and Bakhtiyar Babajanov

Key Points to Consider

- What do you think about the call for the United States and Russia to form a formal alliance dedicated to fighting terrorism and stopping the proliferation of weapons of mass destruction worldwide?
- What will Vladimir Putin's main foreign policy priorities be over the next few years if he wins the 2004 election?
- What does it mean to say that Uzbekistan is the regional hegemon of Central Asia?
- If you grew up in one of the Central Asian republics today, do you think you might become a terrorist? Why or why not?

 **Links: www.dushkin.com/online/**
These sites are annotated in the World Wide Web pages.

Russia Today
http://www.russiatoday.com

Russian and East European Network Information Center, University of Texas at Austin
http://reenic.utexas.edu/reenic/index.html

The former USSR is a region composed of 15 independent nation-states, with each state trying to define separate national interests as it experiences severe economic problems. Many ex-Soviet citizens share a sense of disorientation and "pocketbook shock" as their standard of living is lower today than it was under communism. About half of the states are experiencing political instability and growing discontent.

While uncertainty remains as a principal characteristic of the current Russian political system, the government increasingly looks like it has made a permanent transition from a communist state while retaining many features of the old authoritarian regime. The transition was fueled by economic imperatives. After the Russian government failed in attempts to impose austerity measures and collect taxes in the late 1990s despite massive aid from IMF-secured loans, it devalued the ruble, defaulted on domestic bonds, and placed a moratorium on paying overseas creditors. The actions worsened the economic crisis in Russia.

Russia's parliamentary election at the end of 1999 was widely viewed as a step toward democracy. Optimistic assessments were made despite widespread mudslinging, dirty tricks at the polls, and corruption. The vote was also an endorsement of then, Prime Minister Vladimir Putin, the sixth prime minister to be appointed to Yeltsin's government during a two-year period. Boris Yeltsin abruptly stepped down as president in January 2000 after naming Prime Minister Vladimir Putin as his successor.

One of Putin's first acts as President was to grant Yeltsin blanket amnesty from any future investigation of his family finances. After winning the 2000 presidential elections, Putin displayed a willingness to use military force in Chechnya, to change the reliance on nuclear weapons in the military doctrine, and to implement reductions in conventional armed forces and the size of the defense budget. Putin also built bridges to rival politicians and implemented measures designed to reign in the power of the country's economic oligarchs. After a bungled rescue attempt of the Kursk submarine cost 118 sailors their lives and became a national scandal, Putin formulated a plan to transform Russia's internal balance of power in his favor by reorganizing the country into seven "super-regions" with Kremlin loyalists in charge. Putin argued that only with a strong state can Russia establish the rule of law and the foundations of a civil society.

Provincial elections at the end of 2000 suggested that Putin's approach had wide support with Russian voters. The overwhelming majority of Russians in recent public opinion polls indicated a willingness to trade away some democratic freedoms if it was necessary to achieve order. While progress was made on the political front, funds for conventional military forces continued to decline. The dismal state of the Russian army was indicated by reports that desertions and suicides were on the rise again during 2002. In 2004 Putin ran for re-election virtually unopposed. Many observers expect him to increase his already tight rule on the Russia state after the election. Several recent incidents involving political opponents or critics of Putin lead some analysts to speculate that Putin is re-instituting a slow moving security state in Russia.

Despite these recent suspicious incidents, most Russians also support Putin's hard-line approach to dealing with Chechen rebels. In recent years, the long-simmering conflict received little news coverage until it burst into the national consciousness during October 2002 after Chehen rebels seized 700 patrons in a Moscow theatre. President Putin refused to negotiate with the terrorists and authorized Russian special forces to pump gas into the Moscow theatre before storming the building in a dramatic pre-dawn raid. The raid left 50 militants and 90 hostages dead. The Russian government refused to identify the type of gas used by security forces to secure the building. The gas is widely thought to be a BZ incapacitating agent and it remains unclear whether the use of the gas violated international conventions. By supporting the U.S. war on terrorism, Putin has also succeeded in placing several Chechen rebels on the U.S.-generated list of wanted terrorists.

Even before the September 11th attacks in the U.S., Putin was moving to implement far-reaching changes in Russia's foreign policy by proposing greater cooperation between Europe and the U.S. on security matters. The Bush administration also dropped its campaign rhetoric about Russia after the attacks and moved quickly as well to re-engage Russia. As Celeste A. Wallander notes in "US-Russian Relations: Between Realism and Reality," the United States and Russia have tried to form a strategic partnership but "competing interests, divergent domestic views, and mismatched political and economic system keep getting in the way." As the relationship between the countries' two leaders grew stronger after September 11, it was the War on Terrorism, rather than the pace of Russia's internal reform or continued fighting in Chechen, that came to dominate the new Russian-American relationship. Putin expressed his interest in working with NATO and the U.S. in the War on Terrorism. During 2002, representatives of NATO and Russia met to form the NATO-Russia Council that gives Russia a formal but limited role in the alliance. Russia and NATO continue to explore other ways for their armed forces to cooperate, including an agreement to sign a Military Cooperation Pact in 2003 to cooperate on rescue missions at sea. Recently, NATO Secretary General George Robertson said the alliance is also prepared to assist Russia in modernizing, downsizing, and professionalizing Russia's armed forces.

On the eve of Putin's first summit meeting with President Bush in the United States, the Russia leader in a television interview on a U.S. news show emphasized that the most serious proliferation threat that Russia and the West shared was that terrorists might get a hold of and use tactical nuclear weapons. At his meetings with President Bush in Crawford, Texas, Putin has allied himself with the West in the War on Terrorism. Both countries agreed to make deep cuts in their arsenals of nuclear weapons over the next decade. Putin acknowledged publicly that the Anti-Ballistic Missile Treaty was probably an outdated relic of the Cold War. In response, President Bush announced support for Russia gaining most-favored nation trading status and admission into the World Trade Organization.

Despite dramatic signs of an emerging strategic realignment in geo-political relations, it is still too soon to know the longer term domestic and international consequences of the current period of instability on Russia's future leaders or foreign policy. Since 1991 Russia has been trying to fashion a national-security

policy to fit its changed status in a new era. Today, Russian strategists are deeply concerned by the failure to secure Russia's links to CIS states and by the paucity of promising options. Most Russian foreign policy elite consider themselves *derzhavniki*-believers in strong central government and Russia as a nuclear superpower and great power in Eurasia and East Asia. Whether this elite group can accept the tenets of a more modest Russian foreign policy remains uncertain.

Russia, the U.S., and China are all interested in countering the rising influence of Islamic fundamentalism in Central Asian States. The U.S. military had negotiated military pacts with several states in the region prior to the September 11th attacks. The agreements paid off for the U.S. in the war to oust the Taliban in Afghanistan. The existence of these agreements permitted U.S. military personnel to operate from several states in the Caucuses during the war including: Tajikistan, Uzbekistan, and Kyrgyzstan. Today, Uzbekistan has become a regional hegemon at least in part because of Uzbekistan's support for the United States in the War on terrorism. The United States chose to ignore charges levied against Heydar Aliyev's regime in Azerbaijan of using torture against political opponents in 2003 as this former Soviet republic is also supporting the U.S. war on terrorism in the region.

Western states, along with Russia and China are also anxious to maintain good relations with the Central Asian states bordering the Caspian Sea in part because all of these states are poised to undergo dramatic changes because of their oil and gas reserves. When developed these reserves are estimated to be worth between $2.5 and $3 trillion dollars. The untapped resources stimulated a growing web of recent deals in Central Asia and increasingly make the area appear as a new kind of post–cold war zone of competition where the interests of three former military rivals—China, Russia, and the United States—and a variety of multinational corporations intersect.

Uncertainty remains about how long pro-western governments will remain in charge of the governments of former "ikistan" states. Martha Brill Olcott presents excerpts from a young man recruited for jihad as one of a group of Central Asians, mostly Uzbek by nationality in "The Terrorist Notebooks." Many of these recruits, who were trained at local terrorist schools in the mid-1990, were killed during U.S. bombings in Afghanistan. However, there remain many young people in the region with limited education and diminishing economic prospects throughout Central Asia that are likely to be future recruits for radical forms of Islam.

Article 26

US-Russian Relations: Between Realism and Reality

"If only realism could prevail, one is tempted to hope, the United States and Russia could work together to meet their common interests in security, stability, and prosperity. Reality, however, just keeps getting in the way."

CELESTE A. WALLANDER

Late summer and early autumn have set a series of striking markers in the evolution of US–Russian relations: the failed coup against Mikhail Gorbachev in August 12 years ago that put in motion the final steps leading to the disintegration of the Soviet Union and the end of the cold war; the October 1993 attacks that Boris Yeltsin ordered against reactionary opponents in the Duma, which made it clear the Russian transition to liberal democracy and a market economy would be far more sporadic and problematic than US policymakers had assumed; and the meltdown of the Russian economy in August 1998, which led to debt default because of irresponsible macroeconomic policy, a weak currency, and Potemkin reforms.

What do the events of the autumn of 2003 portend for US–Russian relations? Will they lay the foundation for more counterterrorist and counter-proliferation cooperation, as was exercised this August in a joint Russian-American sting to thwart the smuggling of a Russian missile launcher into the United States? Will they establish great power cooperation through the United Nations Security Council for a robust, internationally based force to secure and build a new Iraq following the attack in late August on the UN headquarters in Baghdad? Or will US–Russian relations founder on the Bush administration's failure to deliver Russia's release from cold war–era trade restrictions because of pressure from American domestic poultry exporters?

The answer is that cooperation between the two countries will continue in core areas related to common security interests. Cooperation not only supports a constructive and cordial relationship between the United States and Russia; it also strengthens broader global stability. The US–Russian relationship over the next decade likely will prove a model of realist foreign policy, serving basic security interests. The relationship will fall far short of a strategic partnership, however, because realism is not enough to support it. A global strategic partnership can be built only on a strong foundation of common purpose and stable domestic support, which is lacking in both countries.

Russia and the United States can be allies in the best traditions of far-sighted traditional great power diplomacy, but the realities of domestic constraints and the imbalance of their national power will prevent their alliance from meeting the requirements of deep security and economic integration in the first decade of the twenty-first century. If only realism could prevail, one is tempted to hope, the United States and Russia could work together to meet their common interests in security, stability, and prosperity. Reality, however, just keeps getting in the way.

RETURN TO REALISM

Realism has a general appeal—who would want to be unrealistic in foreign policy?—but as a concept in international relations it has a specific meaning: in the conduct of foreign relations, national interests defined in terms of power and security guide national leaders. According to realist thought, a country's leaders should not be misled by moral imperatives, driven by cooperation for cooperation's sake, or unduly constrained by international institutions if such policies would cause the leaders to neglect balance-of-power calculations or the rational pursuit of national interests. Leaders should not be misled by the belief that the political or economic composition of other countries—whether they are liberal democracies or market economies, for example—will or should significantly affect foreign policy choices. By implication, a responsible leader should not base foreign policy on whether a potential ally or partner state is democratic; rather, cooperation is possible when states have common interests and when policies are shaped to take into account the realities of their capabilities.

The problem with a US–Russian strategic partnership is not at the strategic level, but within the competing domestic interests, divergent domestic views, and mismatched political and economic systems.

The Bush administration came into office articulating a clear realist premise for its foreign policy, particularly toward Russia. It criticized the Clinton administration's emphasis on engagement and reform of Russia's domestic political and economic order and declared that it would seek cooperation where interests coincided, but would not shrink from confronting Russia in areas where interests diverged, such as nuclear technology sales to Iran. Bush administration officials stated early on that the United States would not seek or adhere to arms control agreements merely for the sake of the habit of cooperation. The new administration said it would withdraw from the Anti-Ballistic Missile Treaty if it could not achieve modifications that would allow the United States to develop and deploy new systems to provide defensive coverage of the American homeland. Most important, Russia was to be downgraded from its preeminent role in US foreign policy in accordance with its decline in power: Russia was not viewed as irrelevant, but simply one among the ranks of other great powers, meriting neither constant high-level attention nor special status.

Similarly on the Russian side, President Vladimir Putin appeared to base his new foreign policy pragmatism on a cold assessment of his nation's strategic interests. Russia's foreign policy concepts and national security doctrine were reformulated to identify Russian weakness as the greatest threat the country faced, and to support development of a vibrant and successful economy as the main foreign policy task. Terrorism supplanted the United States and NATO as the main external threat to Russian security, reflecting Russian preoccupation not only with the war in Chechnya, but also with transnational criminal and terrorist networks—often although not exclusively with Islamic links—extending from Central Asia through the Caucasus. Even before the terror attacks of September 11, Putin's foreign policy rhetoric was characterized by a startling degree of self-critical realism: looking at the country's weaknesses unflinchingly and finding Russia wanting and vulnerable.

It certainly was possible that a more forceful Putin regime, waging a brutal war in Chechnya and bent upon asserting its own national interests, would set back US–Russian relations when joined with a confident Bush administration dismissive of sentiment in its Russia policy. Yet, by mid-2001, the relationship looked better than it had in years, seemingly bolstering realist prescriptions. On the two security issues that dominated the agenda and on which the two countries seemed headed toward confrontation—US withdrawal from the ABM Treaty and NATO's plans for a second round of enlargement—Putin declared that Russia did not agree with the American position, but would not become "hysterical" or sacrifice its relations with the West in a vain attempt to block US policy.

Instead, Putin made a priority of economic reform and integration, seeking US support for foreign investment and Russian membership in the World Trade Organization (WTO). He found a receptive partner in the Republican, business-oriented American president. Faced with growing European criticism and concerns that it preferred unilateral action to cooperation, the Bush administration shifted its focus as well. The administration sought Russian acceptance of its preferred policies on missile defense. It negotiated a strategic arms agreement based on deep cuts in deployed weapons. And it attempted to engage Russia in a special relationship with NATO.

PILLARS OF PARTNERSHIP

The realist groundwork for focusing on common strategic national interests was thus already laid when Al Qaeda attacked the United States on September 11, 2001. The attack made transnational terrorism—a terrorism rooted partly in Russia's Eurasian borderlands—the core threat to American security. This common strategic interest with Russia was not abstract: Al Qaeda was based in Afghanistan, and the Taliban regime that harbored it was viewed as a threat to Russia and the newly independent Central Asian states. Many within Putin's government opposed the Russian president's decision to accept US military bases in Central Asia and to support the US military mission in Afghanistan with intelligence and aid to the anti-Taliban Northern Alliance. But Putin's motives were far from altruistic. Russians had viewed the Taliban as a major threat throughout the 1990s, but had not been successful in eliminating it. By supporting the United States, Putin achieved a significant security objective that Russia had been unable to achieve alone. And by working with Russian intelligence and the military, the United States was able to adapt a creative and unexpectedly effective military strategy that resulted in the relatively swift collapse of the Taliban regime.

Likewise, the early common interest of the United States and Russia in the securing, storing, and disposing of Russia's inherited arsenal of weapons of mass destruction (WMD) took on new dimensions and urgency after 9-11. Russia and the United States suddenly had a very large and very serious common security agenda in the combination of terrorism and WMD that former Democratic Senator Sam Nunn has labeled "catastrophic terrorism."

Along with terrorism and WMD, common strategic interests include a third pillar: economic concerns. Russia clearly sees economic growth and international integration as necessary for national power and security, but what is America's strategic interest in an economically successful and integrated Russia? One consideration is that an enfeebled Russia invites the spread of terrorist bases and networks in Eurasia and increases the chances of WMD proliferation. As national security adviser Condoleezza Rice has suggested, Russia's weakness, not its strength, is the greater threat to America. It is in America's strategic interest to support Russia's economic development so that the government can improve conditions for Russian citizens—including those who work in WMD-related industries and who might be led by a lack of alternatives to sell their knowledge or access. It is also in America's interest to foster an economic environment in which the Russian state can build competent institutions that will support security and stability in the region. Yet another strategic economic interest stems from Russia's position as the world's second-largest producer of oil. Russia's future productive capacity could support a diversification in energy resources that might reduce the dependence of the United States and its allies on Middle Eastern oil.

At a summit meeting this June in St. Petersburg, Russia, both Putin and President George W. Bush portrayed the basis of their countries' strategic partnership in terms of common interests in these areas. They signed the Treaty of Moscow at the meeting, limiting each country's strategic nuclear arsenal to between 1,700 and 2,200 deployed warheads. They announced plans for cooperation in research on missile defense, agreed that North Korea must dismantle its nuclear program, and said that Iran must comply with its obligations under the nuclear Non-Proliferation Treaty. Both highlighted the potential of Russian energy supplies to support a far-reaching strategic relationship. President Bush pledged again to work for Russian membership in the WTO, and to remove Russia from the provisions of the Jackson-Vanik Amendment, the cold war legislation that denies Russia most-favored-nation trading status. They agreed that the US–Russian relationship had emerged intact after Russia's opposition to America's use of force against Iraq and the US decision to act without a UN Security Council resolution. Unlike German Chancellor Gerhard Schröder, who was perceived as personalizing his opposition to President Bush, and unlike French President Jacques Chirac, who was perceived as actively leading the opposition to US policy in Iraq as a pretext for resisting American hegemony, Putin was seen as advancing legitimate Russian interests in a professional manner and was thus "forgiven."

REALITY AND THE RELATIONSHIP

But appearances can deceive. While the two countries' presidents enunciate the strategic rationale for a partnership rooted in common interests, four aspects of reality undermine a US–Russian strategic partnership forged solely in realism.

First, despite acknowledgement that stopping terrorism and WMD proliferation are the two core strategic interests held in common, there is little agreement on the concrete nature of each problem and how to prioritize the threats. Although the Bush administration recognizes Al Qaeda's involvement in the war in Chechnya and does not strongly or publicly criticize the Putin leadership for ongoing human rights violations, it continues to draw a distinction between international terrorists and Chechen separatists, angering Russian officials and limiting the degree to which the countries can cooperate. Domestic critics in the United States limit the extent to which the administration can ignore Chechnya even if the White House sought such leeway. Similarly, Russian critics point out that the US focus on state sponsors of terrorism conveniently neglects American allies Pakistan and Saudi Arabia and centers attention on countries such as Iran and Iraq, which had developed lucrative commercial relationships with Russia.

The Bush administration came into office articulating a clear realist premise for its foreign policy, particularly toward Russia.

The United States and Russia also have different priorities when it comes to the nations that each believes must be a focus of nonproliferation efforts. Russian policy on North Korea has shifted during 2003 to more closely support the US view that a firm and united line must be taken against North Korean nuclear programs. Yet, while highlighting Russian and US agreement that Iran must comply with its international commitments for inspections of nuclear facilities, Putin has not entirely conceded that the key issue with Iran is the proliferation danger. At his St. Petersburg summit appearance with President Bush, Putin said Russia seeks to cooperate on Iran, but also expects the United States not to use nonproliferation efforts to unfairly compete in international markets for nuclear reactor technology.

Without agreement on the primary terrorist or WMD threats, it is difficult to see where the United States and Russia can turn their strategic partnership to operate as effectively as it did against the Taliban. That achievement may have been the high point rather than the model for future cooperation. This brings us to the second reality that checks the US–Russian partnership: the imbalance of power between the two countries and the mistrust this nurtures. Russian officials welcomed the US military presence in Central Asia and the Caucasus as temporary and tied to specific counterterrorist missions. But they suspect that the United States may intend a long-term presence to limit Russia's own political and military influence in both regions, and they will consider their suspicions justified if Americans settle in for a long stay. One scenario they fear is US pressure on Chechnya. Another is active support of the present or a successor government in Georgia. A third is US military protection of the Baku-Tbilisi-Ceyhan pipeline under construction that will enable Azerbaijan to ship oil without relying on Russian pipelines. Russians focus on the potential of US military pressure in the region to undermine Russia's strategic and commercial interests through political pressure backed by superior military force.

The Russian government has also made clear that it views the potential relocation of US bases in Europe from Germany to Poland as contrary to Moscow's interests. It sees the potential as a violation of commitments made not to expand NATO's military eastward under both the 1990 agreements on German unification and the NATO-Russia Founding Act of 1997, as well as a likely violation of limits on national deployments allowed under the Conventional Forces in Europe Treaty. If the United States and Russia deeply held common values and understandings of one another's motivations, the imbalance of power between them would not be a source of mistrust. But this is where reality confronts realism: many Russians continue to fear and resent US power, and Americans continue to view that resentment as evidence of ill will.

The reality effects of Russian weakness and American preponderance were largely behind the near confrontation on Iraq. Russia had clear financial interests in preventing the US attack in March 2003, but these had been discussed and could have been addressed to Russia's satisfaction by post-conflict contracts and energy deals. In joining with France and Germany to try to force the United States to work through the UN, Putin was responding to domestic pressure to resist US power and wield one of Russia's few remaining great power instruments: its permanent seat on the Security Council and the veto power it carries. This was clear in Putin's repeated appeals to the United States to abide by international law and his call for "multipolarity" rather than American hegemony as the basis for dealing with terrorism and WMD threats. His appeal was answered by national security adviser Rice in a speech in London, in which she reminded her audience that multipolarity led more often to conflict, and had in fact led to World War I; multipolarity was "a necessary evil," no longer required among partners with common interests and common values. The United States as a preponderant power that can choose to cooperate with like-minded partners when it wants to—but will rely on itself when it must—has been unwilling to be constrained by international law when such constraints prove inconvenient in securing priority objectives. It is impossible to overstate the importance of this reality in limiting Russia's whole-hearted commitment to strategic partnership with the United States.

ENTER THE BUREAUCRATS

Even more important in limiting realization of a strategic partnership is a third reality: the weight of domestic politics and bureaucratic inertia. Both presidents appear to have a genuine commitment to a deeper bilateral relationship, but this does not implement or fund programs, and it does not resolve competing viewpoints on the relationship within both governments. Presidents Bush and Putin can get along famously at their summits and agree to an ambitious menu of initiatives, but the most rational and strategically self-interested national goals can be undermined by the realities of bureaucracies or domestic interests that do not care, or worse, do not agree.

It is no secret that the Russian military continues to view the United States as a threat to Russia, either out of genuine belief or to protect perks and budgets. As long as senior Russian officers continue to see America and NATO as a threat, they go through the motions of military-to-military programs mandated by the civilian leadership to create "trust" and the bases for joint operations without achieving the trust and cooperative capacities. It is well known that officers who participate in international exchanges or tours of duty, such as in Kosovo, return to Russia to find their careers effectively over. Others tasked with leading cooperation in the strategic relationship can hardly be thinking seriously about common terrorist threats when they are planning to counter the perceived threat posed by the United States.

The civilian leadership of the US Defense Department is also widely perceived to have a more skeptical view of the field of common strategic interests than does Bush, national security adviser Rice, and Secretary of State Colin Powell. Since Russian military and defense officials are more likely to deal with representatives from the Defense Department who themselves see Russian presence in the Caucasus and Central Asia as evidence of Russia's pressure on its weaker neighbors, mutually reinforcing assumptions of conflicting rather than common strategic interests have prevailed.

It is not only wary military officials who are responsible for dragging the potential partnership through the mud of non-cooperation; the US Congress also has played a role. President Bush has called repeatedly on Congress to lift Jackson-Vanik Amendment restrictions on granting Russia permanent most favored nation status. The amendment was enacted in 1974 to sanction countries that did not permit free emigration. Russia has been certified by the US government as permitting free emigration for years, and has been granted waivers of the sanction, but Congress has used the sanction as a way to punish and pressure Russia on unrelated issues. The situation is a source of embarrassment and annoyance in a relationship that is supposedly leading to a strategic energy partnership and that is meant to facilitate Russian membership in the WTO. Congress currently seems unlikely to approve Russia's "graduation" to most favored nation status because of a combination of domestic lobbying efforts to punish Russia for restrictions on importing US meat and for Russia's opposition to the United States on Iraq, and as leverage in negotiations with Russia on its terms of membership in the WTO.

The signature case of internal obstacles to realizing vital strategic interests is implementation and funding of the Cooperative Threat Reduction (CTR) program. Battles within the Congress have limited funding for programs to secure and dismantle Russia's WMD arsenal. In addition, implementation of this core program in the strategic relationship has been delayed and constrained by battles among US players in the legislative and executive branches on whether Russia can be certified as living up to its arms control obligations. In some cases there are legitimate concerns about Russia's commitment, and about activities of Russian recipients of CTR support that may actually contribute to proliferation. In other cases, however, the process has been used by those who oppose the CTR program's premise and objectives to stall its implementation. On the Russian side, officials have consistently thrown up obstacles by preventing access to US personnel required to monitor use of the funds and implementation of the programs on grounds that the facilities have security restrictions.

The fourth reality impeding a strategic relationship is that common interests between countries are rooted not only in strategic security issues, but also in commercial business ties, and these ties are minuscule between the United States and Russia. Russia accounts for less than 1 percent of total US trade, and the United States accounts for less than 5 percent of Russian trade. Cumulative US foreign direct investment in Russia is only $6 billion.

If the Russian economy continues to perform well, if more Russian firms move to international accounting standards and transparency in their corporate governance, and if the Russian legal system enforces contracts in accordance with the rule of law, these numbers will surely increase. Although it is unlikely that Russian oil exports will create solid commercial or economic interests for US private industry or the government (since oil is sold on international markets and is rarely customer-specific, unlike natural gas), it is likely that American energy firms will increase their investment and presence in Russia. As trade and investment ties grow, US interests and stake in the relationship will grow as well. America's relations with strategic partners such as Germany and Japan are rooted in a mix of security and commercial interests, which helps to base support for relations with those countries in wider constituencies within American society.

In short, a robust strategic partnership between the United States and Russia would need more engagement of national interests beyond the narrow if important areas of vital security. Realism might recommend a strategic partnership, but the reality of limited common interests matters more in day-to-day policy.

REALITY CHECK

A realist conception of US-Russian relations in the twenty-first century provides a clear understanding of just where the strategic security and economic interests of the two countries overlap. Without this basis, there are no joint objectives for a partnership to target. Similarly, a realist approach focuses on getting the structure of the relationship in place so that it is more than a house of cards to be blown over in the first crisis to test the relationship's commitment and capabilities.

But a foreign policy that stops with realism will not be very productive, nor is it likely to address the broad array of issues that confront states in the modern world. It was enough in the nineteenth century to conduct a spare foreign policy based on national power and interests because states interacted primarily in the military and political spheres, and their societies and economies were not very integrated or interactive. Foreign policy bureaucracies were small and managed a limited set of requirements that focused on diplomacy.

In the modern world, countries interact much more intensively in official and private contexts. Mobility, technology, and integration have supported economic growth through trade, more efficient global production, and global investment. They have also created the capacity for the transnational terrorism and global military reach that globalize vulnerability and the potential to defend against it. Realism is spare and elegant, but the US–Russian relationship has to embrace the realities of the twenty-first century. The sense that many observers express that US–Russian relations are cordial but hollow arises from a failure to seize the challenge of the realities.

Among the most important of these realities for US–Russian relations in the next decade is the failure of domestic constituencies and institutions in both countries to support the overall structure of strategic objectives. The problem with a US–Russian strategic partnership is not at the strategic level, but within the competing domestic interests, divergent domestic views, and mismatched political and economic systems. The onus in this respect is on Russia to create a functioning democratic state that is accountable politically and subject to societal oversight, including a free media. It is also a parallel obligation to continue economic reforms and to establish market institutions to encourage American business interest and investment.

The United States for its part also could improve the coherence and accountability of executive agencies responsible for implementing the president's policy, but the focus of US efforts to build a partnership should be different. American leaders should reread the classics of realist theory to remind themselves that one of the consistent conclusions of theorists from Thucydides to Kissinger is that hegemonic powers ultimately fall when they fail to build reliable alliances (or in modern terms, partnerships) based on common strategic interests. Self-interested cooperation is more likely than reflexive self-reliance to succeed. It is less likely to create fear and hostility among potential competitors, which historically have fueled their efforts to build capabilities to protect themselves.

The United States can do little about the imbalance of power that creates uncertainty and concern about US intentions, but it can do more to establish self-interested partnerships for the pursuit of national interests. While Russia builds its democratic and market capacity for a real partnership, America can build a strategic partnership with Russia as a showcase of a responsible realist foreign policy deeply rooted in reality.

CELESTE A. WALLANDER *is the director of the Russia and Eurasia program and a senior fellow at the Center for Strategic and International Studies.*

THE TERRORIST NOTEBOOKS

During the mid-1990s, a group of young Uzbeks went to school to learn how to kill you. Here is what they were taught.

By Martha Brill Olcott and
Bakhtiyar Babajanov

> "Jews, Russians, and Americans are always against Muslims and kill Muslims. And the Muslims are sound asleep."
>
> —Notebook #3 page 16

The world of a young man recruited for jihad or holy war is a frightening one. His training teaches hatred in the name of religious purification. He learns to divide people into those who embrace the true faith and properly follow its precepts and those who do not. His former colleagues and neighbors become enemies he must destroy with deadly weapons he learns to fashion out of everyday objects.

That reality describes the world of a group of Central Asians, mostly Uzbek by nationality, who went through local terrorist schools in the mid-1990s. Their course of study is laid out in 10 remarkable notebooks we acquired in 2001–2002. Covering topics such as the use of weapons, the making of poison, and the ideology of jihad, the notebooks offer a unique window into a frightening mind-set that predates the expansion of Osama bin Laden's network in the region and still holds sway in much of Central Asia.

References in the notebooks suggest that much of this training took place in Uzbekistan's Fergana Valley. Long a center of Islamic revival in the region, the Fergana Valley is a mix of scrub desert, low hills, and lush oases. It is the most densely populated area of Central Asia and one of the most densely populated regions in the world. Throughout Soviet rule, the valley was home to a host of underground mosques and religious "schools" that thrived even as Islamic teachings were banned or re-

This page is from the notebook belonging to "Ayub," probably a Tajik from the city of Namangan, whom we nicknamed "the gunner" because of his single-minded focus on weapons and targeting. Ayub notes in his text that the AKM-59, a version of the AK-47, is "pretty and well put together and is comfortable to handle."

stricted. When the Soviet Union began to collapse, graduates of these schools played an important role in the revival of Islam in Central Asia, as thousands of new mosques and religious schools opened. Clerics who preached radical Islam gained new contacts and sources of financing when the mujahideen started fighting the Soviets in the Afghan war and when Saudi groups began what became a global crusade.

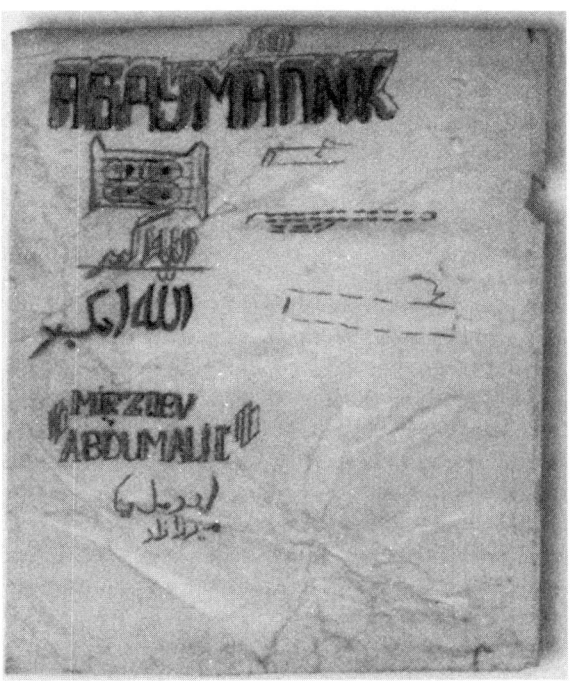

Covering the period from 1993 to 1994, this notebook belonged to "Abdumalik," also from Namangan, whose name is written on the cover in Russian, Uzbek, and Arabic. Among other subjects, his notes cover the characteristics of various pistols and detailed diagrams for making bombs and fuses.

The late 1980s and early 1990s were difficult and confusing years for young people living in Central Asia. A seemingly invincible state had virtually disintegrated and was replaced by fragile new ones. Conditions were almost apocalyptic: The economy was in disarray, an expansive social safety net had shredded, and the powerful Red Army was in tatters, with those who served it selling off their weaponry to survive. Muslim activists who claimed that moral turpitude brought down the Soviet regime found it easy to muster arguments to bolster their cause, and they organized the Islamic Renaissance Party (IRP). Although the Uzbek government refused to register the IRP, a number of charismatic clerics who preached rejection of the secular state continued to gain supporters, especially in the Fergana Valley. And these men in turn developed armed supporters, who in the first months of Uzbekistan's independence briefly took control of key government buildings in the city of Namangan. Fearing the outbreak of civil war, Uzbek President Islam Karimov authorized a purge of the official Islamic establishment and the arrest or disappearance of prominent unlicensed clerics and leaders of "extremist" Islamic groups.

Several prominent figures escaped the official dragnet, fleeing with followers into neighboring Tajikistan and the Tajik- and Uzbek-dominated parts of northern Afghanistan, long a host site for jihadi training camps. Thus was born the Islamic Movement of Uzbekistan (IMU), led by Soviet Army veteran Juma Namangani.

ABDUMALIK'S WORLD

During the mid- to late 1990s, hundreds, and, some claim, even thousands, of young Uzbeks belonging to the IMU passed through terrorist camps in Afghanistan, Pakistan, and elsewhere in the region. Some of the Uzbek mujahideen went home to train their countrymen, and they created clandestine terrorist schools for this purpose. The notebooks we acquired belonged to students who attended such courses during the period of 1994 to 1996. [For more information on the origins of the notebooks, consult the section Want to Know More at the end of this article.] We purchased or otherwise acquired these books through various intermediaries, each unaware that we were collecting material from others as part of an effort to document the Islamic revival in Uzbekistan.

Taken collectively, the notebooks allow us to reconstruct the training of the young mujahideen. Students seem to have spent the bulk of their time on military subjects. Once they mastered these subjects, the students focused on when and how to make jihad—and some of the students may have heard lectures on jihad by Namangani himself, or one of his close associates.

We don't know much for certain about the students themselves. Some of the notebooks have the names (or pseudonyms) of the fighters in training who wrote them—for example, Abdumalik or Ayub. We have reason to think some of them studied in Namangan, possibly in the basement of the Juma mosque; reopened during the 1990s under pressure from the community, the mosque had been used as a storehouse for alcoholic beverages during the Soviet era. Our sources told us that all of the students were eventually arrested—in one case, for smuggling consumer goods (and "trade" was, in fact, their livelihood). Uzbek security forces picked up most of the others as suspected terrorists. Their parents, who gave us or our intermediaries the notebooks, were reluctant to talk about them, save to disassociate themselves from their children's "mistakes."

We do not know whether the young men who studied in these schools were devout Muslims, but their notes suggest they were not very knowledgeable about Islam. The same may also be said about their teachers: In the lessons on jihad, for example, references to the Koran, offered by chapter and verse, sometimes cite passages unrelated to the subjects under discussion. These errors are clearly those of the teachers; most students at this early stage of religious education would not have possessed their own copies of the Koran, and they also lacked

Dangerous neighborhood: The Fergana Valley where most of the notebooks were acquired cuts across Tajikistan, Uzbekistan, and Kirgizstan and is near Afghanistan.

the necessary Arabic-language skills to read the Holy Book in the original.

We can also say with certainty that the students were not very educated. They made many grammatical mistakes when writing in Arabic, Russian, and even their native Uzbek. Some of the students seem to have had poor attention spans, and they were careless in taking notes and studying.

THE ABCs OF TERROR: HOW TO KILL

One thing is clear, though: These students learned how to make deadly weapons. As their notes show, these pupils "learned by doing" in every field of terrorism from instructors proficient in their respective subject matter. The teachers who used Russian terminology clearly had experience with the Red Army and Soviet system of military instruction, and those who used Arabic likely passed through terrorist camps in Afghanistan and maybe even those of the Middle East. In many cases, several different instructors taught the various military subjects.

Cartography Students first learned to orient themselves to their surroundings. When we showed some of the notebooks to a professor of cartography in Tashkent—without revealing the source of the material—he was able to identify them as terrorist manuals and was certain the instructor was a cartographer. All high-school students in the Soviet Union were required to receive paramilitary

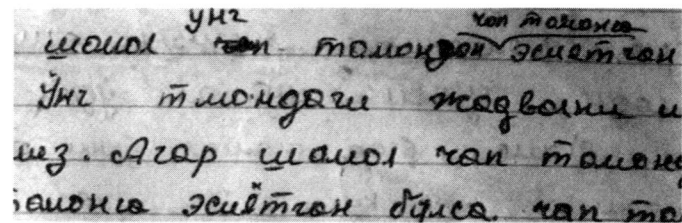

The basic language of instruction in the notebooks was Uzbek; those who taught the technical subjects knew Russian or Arabic, and in some cases both languages.

training, so there was no shortage of people capable of teaching cartography or most other military subjects, even in the remotest areas of Uzbekistan. Moreover, with some modification, textbooks from the Soviet courses would have been a good starting point of instruction.

Small Arms Students then went on to study how to handle small firearms—a fixture of life in a region where military service was compulsory and where anyone familiar with the black market could buy an AK-47. During the years of the Soviet occupation of Afghanistan (1979–1989), more local youth acquired combat experience than at any time since the end of World War II. Much of our knowledge about this field of study comes from the notebook of Ayub, a Tajik from Namangan (writing in Uzbek), who spent so much time mastering this material that we dubbed him "the gunner."

Article 27. The Terrorist Notebooks

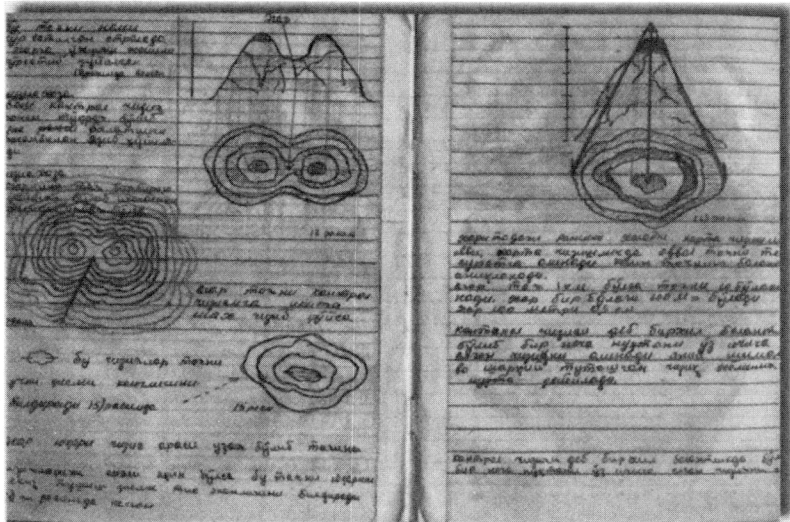

Detail from the eight lessons on maps contained in the notebook acquired in Tashkent oblast, in which students learned to use compasses and natural measurements.

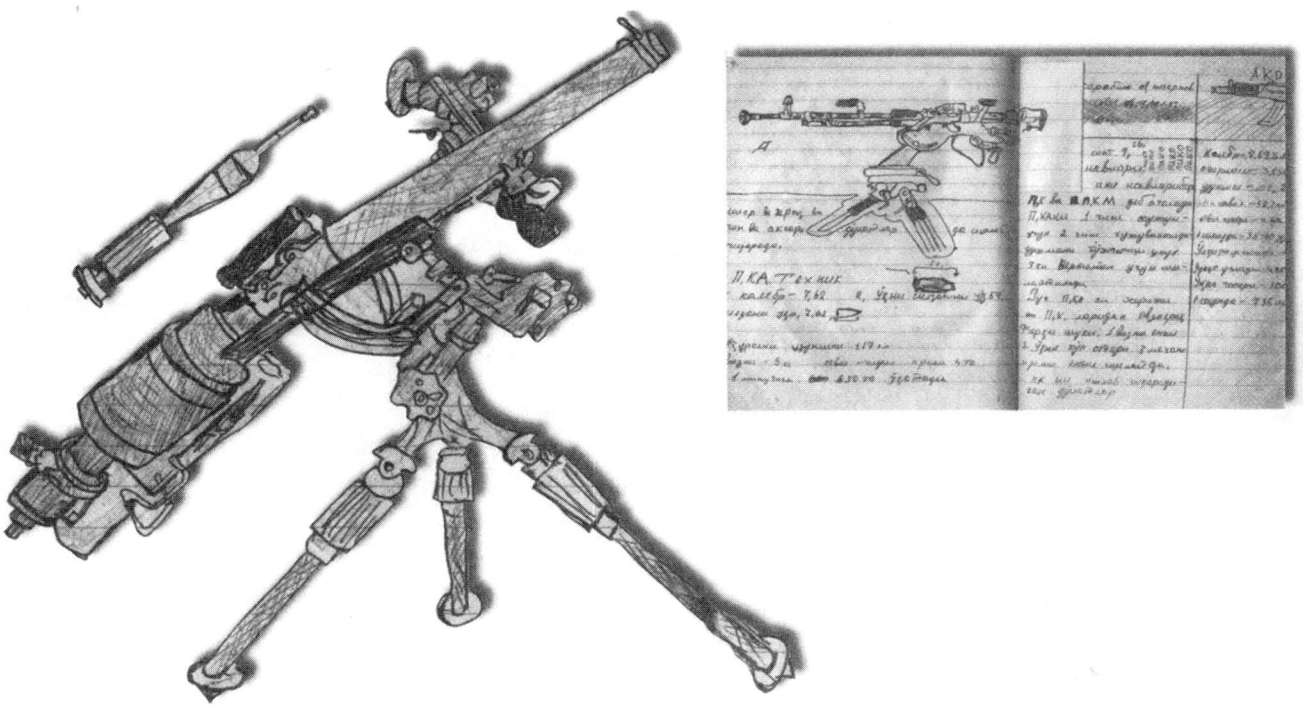

This drawing is Ayub's depiction of a rocket-propelled grenade launcher (left). "This will be demonstrated later in the field," he writes. Description of a machine gun, including discussion of speed, caliber, and other technical specifications (right).

Many of the weapons the students learned to use were common Soviet-era ones, including various forms of the Kalashnikov rifle (AK-47, AK-56, and AKM). Ayub, though, also learned to handle several weapons of choice from Afghanistan, including the Egyptian rocket-propelled grenade launcher. This 82-millimeter weapon is based on a Russian or Chinese modification of an earlier U.S. weapon, writes Ayub in broken Arabic. All of these weapons appear in detailed illustrations, with accompanying notes on their functions and maintenance.

Targeting Ayub was also diligent in learning how to target the enemy, on the ground and in the air. His notebook includes tables with elaborate calculations on how to target planes and helicopters in varying wind and weather conditions. His teacher used both Arabic and Russian military terminology. In the course of his lessons, Ayub handled various forms of sighting instruments, writing in one case that "the front glass reflects many colors" and "the plus sign that regulates distance is easily obscured by finger marks."

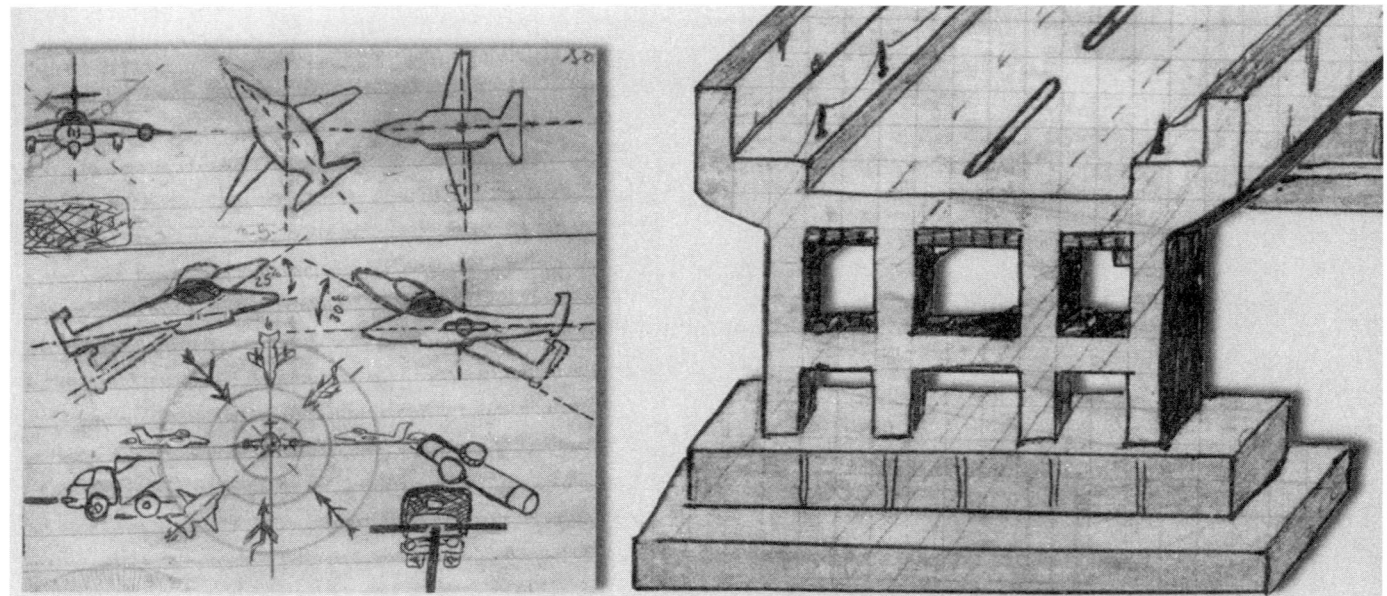

Left: Part of Ayub's notes on calculating target data and using sighting mechanisms. Right: One of several drawings by Abdumalik on the correct placement of explosive charges for blowing up a bridge.

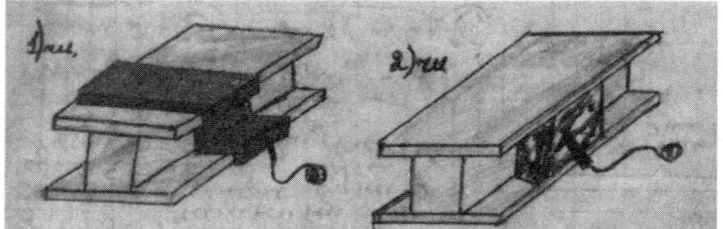

Left: Part of a series of diagrams on how to set mines, from a notebook shared by two students. Right: details from another notebook on the best explosive placement in wells, bridges, and support structures.

Mines and Demolitions The notebooks suggest this subject was a standard part of the instruction that all young mujahideen in Uzbekistan received. Many of the mines the mujahideen learned to make had been commonly used in Afghanistan and other guerrilla war settings, including the MI8AI antipersonnel mine—a plastic-bodied directed fragmentation mine that has ball bearings embedded in the facing of the target. Variations of this mine were produced in the Soviet Union, Pakistan, South Africa, South Korea, and Chile. The students also received instruction in the POMZ-2 antipersonnel mine, activated through the use of a trip wire, with a lethal radius of 4 meters. Variations of this mine were manufactured in the Soviet Union, China, North Korea, and throughout Europe's Eastern bloc.

One notebook includes information on making 16 different explosive devices. The two students who prepared this notebook learned reaction times and temperatures for blowing up buildings, bridges, railroad ties, and electricity relay stations. They were taught everything necessary to become competent arsonists, including how to escape unharmed, a subject emphasized in some of the lessons. These young men were not trained to be suicide bombers but guerrilla fighters who would endure long periods of battle.

Poisons Students also learned how to make poisons with readily accessible substances, such as tobacco or toxic mushrooms. Precise information on the amounts of each ingredient, how to mix them properly, and reaction times are carefully documented. The students penned lengthy instructions on safety techniques, when to wear gloves and masks, and how to conceal noxious odors so potential targets would not be alerted. Alongside instructions for making and using cyanide, one student writes, "And the power of Allah is mightier"—a phrase commonly used at deathbeds—as if to see off his future victims.

WHO DESERVES TO DIE?

The section on jihad—the final course of study—is also the most terrifying. At one level, the lectures on jihad were designed to mobilize students for battle with the enemy. But stripped of their pedagogical intent, these lectures make clear that the explicit goal of the students' military studies was to kill people, preferably as many as possible.

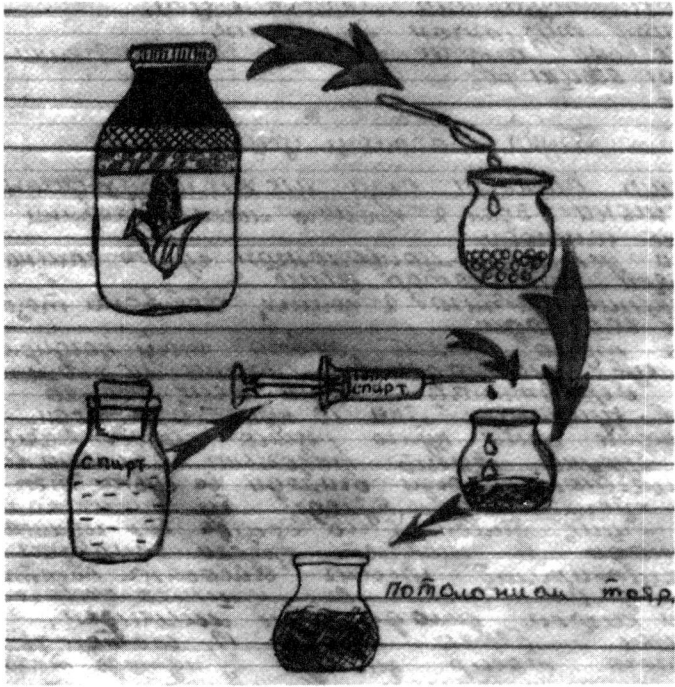

Entries in the notebook from Tashkent on how to make poison using corn flour, beef, yak dung, alcohol, and water.

Since Islam was spread by the sword, holy war is an important theme in the Koran. But since the time of Mohammed, theologians have fought over when jihad is required and when it is forbidden. The view of jihad presented in the notebooks is both simplistic and uniform—so much so that the same person may have taught students studying in different cities. The teacher likely lacked even a middling religious education (8–10 years of study) and was instead a fighter with a religious background, perhaps someone like Namangani or Tohir Yuldashev, the leaders of the IMU.

Jihad is depicted as a cleansing act, as "Jafar" (owner of one of the notebooks) writes, "so the old ideology makes way for the new"—by which he supposedly meant that Uzbekistan's dominant Hannafi school of religious law would make way for Salafi (or fundamentalist) Islam. Central Asian theologians from the Hannafi school preached accommodation with secular rulers; most, in fact, argued that Islamic law demanded such accommodation, for to do otherwise was to put the community of believers at risk.

By contrast, these students learned that Uzbekistan's secular rulers were betrayers of the faith, and, as Jafar writes, holy war is imperative: "for our faith of Islam, to make Allah pleased with us, to eradicate oppression against Muslims, to establish Islamic rule in perpetuity."

Jafar and his fellow mujahideen were taught that jihad has multiple goals—that economic, political, ideological, and military goals have to be mutually reinforcing. The propaganda that precedes military action, they learned, is critical.

As another student writes, the goal of this propaganda is to raise popular awareness of the enemies among them:

> To make a declaration of the fact that unbelievers and the government are oppressors; that they are connected with Russians, Americans, and Jews, to whose music they are dancing; and that they don't think about their people. We spread true knowledge about Islam in our country [Uzbekistan]. We speak of the fate of faith betrayers, according to Islamic law, and about how people should distance themselves from those who breach the faith and should side with the mujahideen. At the same time, it has to be announced that jihad is a necessary religious requirement, for all social groups of people. And in life, everyone must either be a Muslim or a non-Muslim, that is, no one can remain in the middle. After this, the declaration will be done, the mujahideen will inform the people of the beginning of jihad.

Targeted enemies are depicted in political cartoons, which the students appear to have been asked to draw outside of class. In a perverse manifestation of continuity with the Communist years, many of these cartoons are variants of the anti-imperialist and anti-Zionist pictures that Soviet students sometimes drew during their studies. The drawings in these notebooks, however, include caricatures of Russians alongside those of Americans and Jews.

The anti-Semitism taught to these Uzbek students in their classes was primitive, based on perverse distortions of history, but effective:

> All the countries of the world are today ruled by Jews. This people who is cursed by God began to rule everyone 120 years ago, at the time of Napoleon. It was so. At the time of the fighting between the armies of Napoleon and the British, the Jews spread rumors among the people of England that Napoleon won the battle. Upon hearing these rumors, the British fell into panic and began to sell their stores, factories, plants, and

"Now the first task is to kill Jews, and then the rest," reads the caption above this drawing of a three-headed dragon. The knife of jihad has already cut off the head of the Jews, but the heads of the United States and Russia remain.

Left: "This is how America destroys Muslims. Muslims, rise up in holy war." The drawing on the right is of an insignia to be worn by "the most faithful troops, those especially prepared for purging enemies." Note the faint swastika at the bottom.

other kinds of enterprises. They thought as following: "After the victory of Napoleon, he will arrive in England, and we will lose everything." And so lots of enterprises were sold, and very cheaply. The Jews took advantage of this opportunity and started buying everything very cheaply. A week later, it became known that the British won the battle against Napoleon. Upon hearing this news, all the people began to buy back their things. The Jews sold all this, but for 5 to 10 times more than they paid, and received enormous profit. That the Jews are cursed by God is demonstrated in Ayat 14 of sura al-Khashr. (59:14)

The first Jews came to the region long before the Arab conquests in the mid-seventh century. Traditionally, anti-Semitism was much worse in the Slavic parts of the Russian Empire than in Central Asia. And historically, Uzbeks have had more resentment for the Russians, who conquered Uzbek lands in the late 19th century and restricted the practice of their faith. Russians remain a target in the notebooks, despite their withdrawal from Uzbekistan after the country gained independence in 1991.

Now, the mujahideen are determined to rout out these enemies and kill them, as part of larger economic, political, and ideological goals. Such economic goals mentioned in one notebook include:

1. To attack the joint ventures that have been organized by the officials of our city [perhaps Namangan—M.B.O. and B.B.]. That is, in the first instance, those enterprises with Russians, Jews, and American [partners] at the head.

Article 27. The Terrorist Notebooks

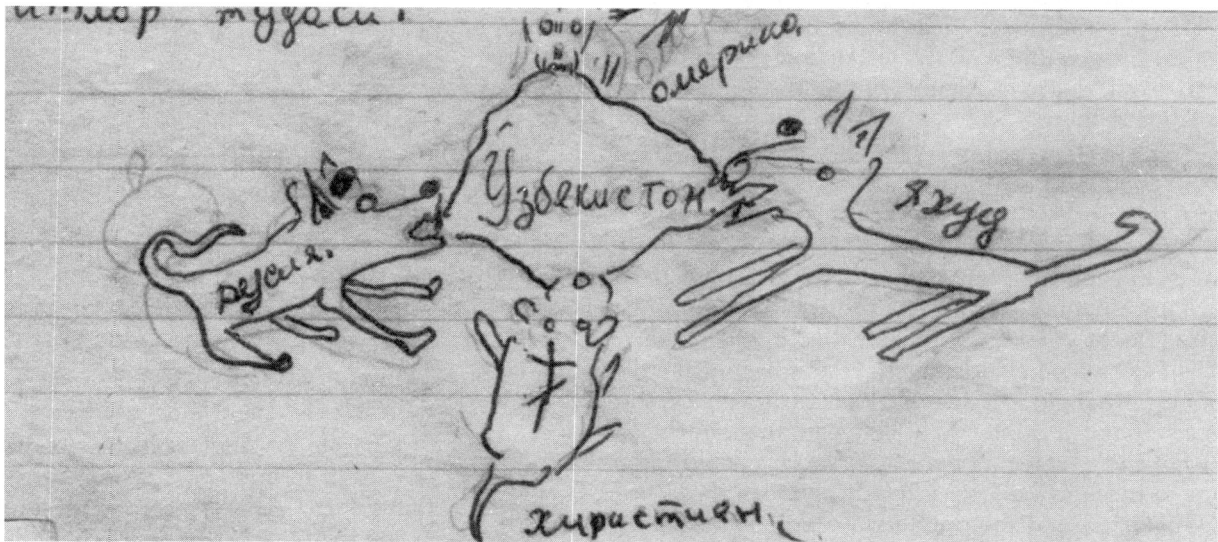

Four dogs (Jews, the United States, Russians, and Christians) attack a map of Uzbekistan.

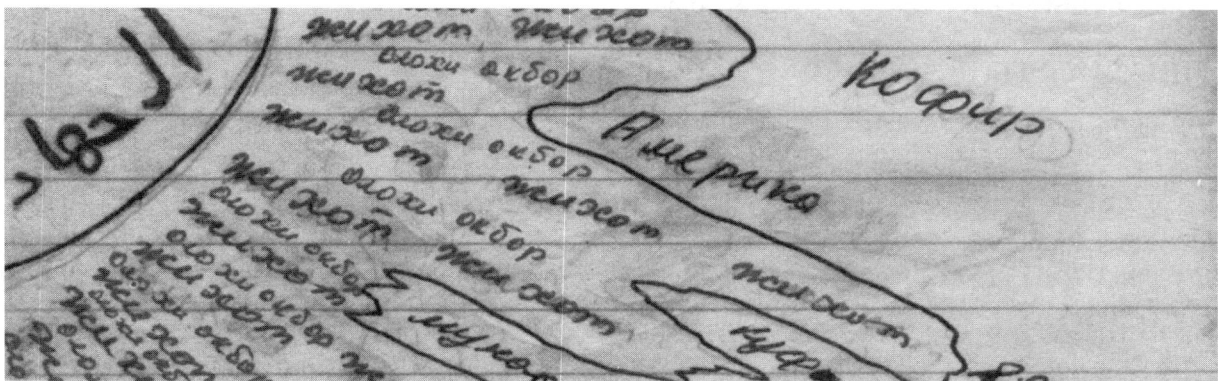

An incantation to jihad and a list of enemies: The drawing is of an explosion, with jihad at its epicenter. The fire of the explosion is an incantation of Allah Akbar (God is great) and jihad. Those who have been blown up include unbelievers, Americans, Jews, two-faced people who put themselves forward as Muslims, Russians, and Christians.

2. To destroy all that is imported from the countries of the enemy, whatever it may be, food, clothes, etc. This, too, is an economic and political blow.
3. To destroy all raw materials exported from the country by unbelievers. This includes fruit… one or two cases of fruit should be poisoned, and when this is discovered, it should be announced that all the fruit that was sent (for example) to Russia, is poisoned.… [This threat is very serious, since Uzbekistan is such an important source of fruit and vegetables for Russia—M.B.O. and B.B.] Those who transport things for personal use will be warned once or twice, and then everything will be confiscated from them.
4. Specialists from Russia, Jews, and Americans working in the economy will be destroyed.

The same groups were targeted under political goals: "At the time of the political strike against the state, we should also kill Russians, Americans, and Israeli citizens. That is, ambassadors, or others of them, who live here, they all must be beaten."

Clerics and missionaries of other faiths are slated for extermination as part of the ideological program:

From among religious people we will kill:

1. Those who try to gain converts to Christianity on Muslim soil
2. Spies who work as Christian clerics [During Soviet times, there were many KGB employees among Christian clerics—M.B.O. and B.B.]
3. We will kill those Christians and Jews who speak against the mujahideen and those who propagate against Islam
4. Those Christians who collect money for the struggle against Muslims, and those who speak against Muslims. They will be stabbed or shot or hung or beaten to death.

The Christian missionaries targeted here were fairly recent arrivals in Central Asia. Many belonged to U.S. evangelical groups that saw the fall of communism as a signal to expand proselytizing efforts throughout the former Soviet Union.

But it's important to remember that the mujahideen who were trained in Uzbekistan at this time were mobilized to fight a local war, for local causes. Their goal was to prevent enemies of Islam from using new economic structures, like joint ventures, to keep down true believers. Such arguments echo those of radical Muslim thinkers such as Egypt's Sayyid Qutb, who died in 1966. Qutb's works had circulated clandestinely among Islamic activists in Central Asia for decades. Only occasionally do the notebooks make a connection between the efforts in Uzbekistan and a larger, global cause. Those teaching and studying in these schools were keenly aware of the situation in Tajikistan and the ongoing struggle in Afghanistan. But the notebooks make no mention of or link between their efforts and the ongoing Chechen war nor to conflicts in more distant places such as Bosnia or Somalia.

THE FIRE NEXT TIME

The good news is that the owners of these notebooks were never able to execute the number or kind of operations planned with the deadly knowledge they acquired. True, in February 1999, the IMU was credited with masterminding the simultaneous bombings of key government offices in Uzbekistan's capital, Tashkent, killing 13 people. But these attacks did not set off the panic or chain reaction of other violent acts predicted in the notebooks. After the bombings, the Uzbek government successfully pressured the United States to list the IMU as an international terrorist group. And faced with heightened Uzbek security, the IMU made do with taking hostages and raiding parts of nearby Kirgizstan. The group did become part of the al Qaeda network, with camps in Afghanistan and safe havens over the border in Tajikistan. But its founder, Namangani, and many of his fighters were reportedly killed during the U.S. bombings in Afghanistan; the whereabouts of another prominent leader, Yuldashev, are still unknown.

The bad news is that the threat posed by such terrorist groups is infinitely renewable in states such as those in Central Asia, where large numbers of young people with limited education and diminishing economic prospects live in densely populated communities. Moreover, popular resentment toward these countries secular leaders remains high: Many of these leaders were local masters of the openly atheistic Soviet regime, and most of them have profited mightily from the unprecedented increase in corruption since independence.

Each of the budding Central Asian states has attempted to carve out an identity in the past decade. But conditions have not favored the development of authentic moderate Islamic clerics. State authorities view leaders who are credible to religious believers as too threatening, and religious believers are suspicious of those championed by state authorities.

These conditions are made-to-order for those preaching more radical forms of Islam. The best known of these groups is Hizb-ut-Tahrir (Party of Liberation), which attracts young people despite the extraordinary efforts of the Uzbek government to harshly punish those associated with the group. Its numbers are increasing in Kirgizstan, Kazakhstan, and Tajikistan. This movement is committed to the reestablishment of the Caliphate—the rule of Islam as it was practiced by the Prophet Mohammed. For now, the group maintains, this goal can be advanced only through persuasion, not force.

Whatever the fate of Hizb-ut-Tahrir, other radical groups seem certain to emerge from the turmoil of the transitions that Central Asian states are still undergoing. Notwithstanding the presence of new U.S. military bases in Uzbekistan and Kirgizstan and expanded assistance in the war on terrorism, no amount of force alone will defeat such groups. Any security agency capable of routing out all potential terrorists would inevitably become a source of terrorism. Not only would such an organization tread on the basic civil rights of peaceful citizens, but, by targeting "radical" Islam, it would invariably cause those who consider themselves devout Muslims to see the government as an enemy of Islam.

In every part of the world, there are heroes who have died fighting for their faith and who make ready role models. In Central Asia, it is Namangani or Ahmed Shah Massoud, the Lion of Panjshir in Afghanistan. As the disturbing contents of these notebooks attest, purveyors of jihad supply their own credentials and design their own curricula. They require no licenses for their undertakings. The proof of their success is whether they can gain recruits and successfully teach them how to kill.

Want to Know More?

The authors acquired the 10 notebooks between 2001 and 2002. Six were obtained in the Fergana Valley, three in the Tashkent region, and one from an Uzbek village just over the Uzbek border in Kazakhstan. Those interested in more information about the notebooks should contact Martha Brill Olcott at the Carnegie Endowment for International Peace.

For more guidance on Islam, readers can refer to Cyril Glassé's *The Concise Encyclopedia of Islam* (San Francisco: Harper & Row, Publishers, Inc., 1989) and *The Oxford History of Islam* (New York: Oxford University Press, 1999), edited by John L. Esposito. Ira Lapidus's *A History of Islamic Societies* (New York: Cambridge University Press, 2002) is also recommended as an all-purpose guiding tool. The authors used J.M. Rodwell's translation of the *Koran* (London: J.M. Dent, 1994).

Readers interested in learning more about political Islam should consider Graham Fuller's **"The Future of Political Islam"** (*Foreign Affairs*, March/April 2002), Gilles Kepel's *Jihad: The Trial of Political Islam* (Cambridge: Harvard University Press, 2002), Ahmed Rashid's *Jihad: The Rise of Militant Islam in Central Asia* (New Haven: Yale University Press, 2002), and Olivier Roy's *The Failure of Political Islam* (Cambridge: Harvard University Press, 1994). The Carnegie Endownment's Husain Haqqani provides a unique firsthand view of life in a *madrasa* in **"Islam's Medieval Outposts"** (FOREIGN POLICY, November/December 2002). In addition, astute accounts of fundamentalism are provided by Daniel Pipes in *Militant Islam Reaches America* (New York: W.W. Norton, 2002) and Daniel Benjamin and Steven Simon's *The Age of Sacred Terror* (New York: Random House, 2002).

Insightful books on Central Asia include Roy's *The New Central Asia: The Creation of Nations* (New York: New York University Press, 2000) and Rashid's *The Resurgence of Central Asia: Islam or Nationalism?* (Karachi: Oxford University Press, 1994). Another comprehensive source is a recent collection of essays edited by Boris Rumer, *Central Asia: A Gathering Storm?* (New York: M.E. Sharpe, 2002).

The **Institute for War & Peace Reporting** and **EurasiaNet** provide information and analysis about political, economic, environmental, and social developments in the countries of Central Asia and the Caucasus in both Russian and English. Some of the regional sources of reliable information are the Web sites of the information agency **"AkiPress,"** the **Central Asia Information Center**, and the **Central Asian Information Agency**.

Martha Brill Olcott is a senior associate at the Carnegie Endowment for International Peace (CEIP) and author of Kazakhstan: Unfulfilled Promise *(Washington: CEIP, 2002). Bakhtiyar Babajanov is a senior research fellow at the Institute of Oriental Studies of the Academy of Sciences of Uzbekistan in Tashkent.*

From *Foreign Policy*, March/April 2003, pp. 31-40. © 2003 by the Carnegie Endowment for International Peace (www.foreignpolicy.com). Permission conveyed through Copyright Clearance Center, Inc.

UNIT 8
The Pacific Basin

Unit Selections
28. **Changing Course on China**, Elizabeth Economy
29. **How to Deal with North Korea**, James T. Laney and Jason T. Shaplen
30. **Can India Overtake China?**, Yasheng Huang and Tarun Khanna
31. **Dangerous Neighbours**, Ahmed Rashid

Key Points to Consider
- Why has the United States changed it's policies towards China in recent years?
- Explain why you agree or disagree with the conclusion that China is an over-rated emerging power.
- How significant for future regional relationships is the fact that Japan has now permitted its military self-defense force to serve overseas?
- What should the U.S., South Korea, China, Japan, and Russia do now that North Korea has unilaterally left the NPT Treaty and violated its agreement under the 1994 Framework Agreement by withdrawing nuclear fuel from a reactor in order to produce nuclear weapons?
- China or India? Which do you believe will become the dominant economic power in Asia?
- Explain why you believe the U.S. will or will not remain engaged in Afghanistan over the next decade.

 **Links: www.dushkin.com/online/**
These sites are annotated in the World Wide Web pages.

Crisisweb: The International Crisis Group (ICG)
http://www.crisisweb.org/home/index.cfm

ASEAN Web
http://www.asean.or.id

Inside China Today
http://www.insidechina.com

Japan Ministry of Foreign Affairs
http://www.mofa.go.jp

Conventional wisdom about economic development was shaken by the economic instability experienced by the Asian "tigers" and by the prolonged recession in Japan. Analysts now worry about political fallout from the economic downturn on domestic political stability. The economic slowdown occurred at the same time as a relatively unstable Pacific Basin security environment experienced a series of shocks, including the continued development of missiles by North Korea and the mobilization of Indian and Pakistani troops on their shared borders all at the same time that the United States was waging a War on Terrorism in Afghanistan and preparing for a military intervention in Iraq.

Some analysts now view the Southeast Asian currency crises of 1997 as part of a pattern of financial instability that often accompanies rapid economic growth. These analysts predict that Asian economies will continue to grow and account for over half the world's income by 2025. Economists and financial analysts debate whether recent economic recovery is a meaningful upswing fueled by consumer demand in the United States or a temporary recovery. Most analysts agree that sustained economic recovery will require growth in the economies of both Japan and China.

While its economy also slowed, China managed to avoid a currency devaluation. The transfer of control in Hong Kong by Great Britain to China went smoothly, despite a dramatic fall in tourism and a sell-off of property. Instead, China's most serious economic problems stem from dislocations caused by government efforts to reform its "iron rice bowl" economy based on socialistic principles. Economic reforms are eliminating many of the subsidies for food, housing, and jobs that had been promised to all citizens under socialism. China is also eliminating nonperforming state entities. The result is massive unemployment, growing resentment over increased disparities among urban and rural areas and across regions, and a fundamental change in the relationship between the state, the Communist Party, and workers. The Chinese Premier Wen Jiabao acknowledged these problems at the opening of China's annual legislative session in March 2004 by emphasizing the importance of ensuring that the fruits of China's economic growth are more equitably distributed through policies based on "putting people first." This unusual public statement gave voice to a growing governance crisis in China that has created fundamental contradictions in the economic reforms implanted by a communist party over the past two-and-a-half decades. The rising tensions between the state and society may eventually threaten China's economic development.

While China's drive for continued rapid economic growth dominates much of its foreign policies, Elizabeth Economy describes in "Changing Course on China," how relations between China and the United States were dramatically altered from strategic competitors to partners against terror after September 11, 2001. The two countries are currently acting like status quo powers in their dealings with each other despite a number of issues, including Taiwan, human rights, and the U.S. growing trade deficit with China.

At the same time China is undertaking bold but quiet new foreign policies towards neighboring states. China is using quiet diplomacy and modest pressure to cement relations with its 14 territorial neighbors.

China's admission to the WTO in 2001 was greeted with a great deal of trepidation in Japan. Many Japanese fear that the continuing recession means that Japan will be unable to compete with China in the economic or political realm. Sino-Japanese relations is a vital national interest for Japan because half of Japan's largest companies are planning to increase production overseas-and 71 percent of those planned to do it in China. Japanese fears are also fueled by the lingering recession. While the current morass is widely viewed as being a temporary condition, economic problems are fueling growing insecurities as the country is facing increasingly powerful and assertive neighbors on the Asian continent. At the end of 2003, Japan took the unprecedented step of deploying thousands of its self-defense forces to Iraq.

Growing instability on the Korean Peninsula is a major factor influencing Japan's changing foreign policies. The North Korean regime confirmed Western suspicions in 2003 by announcing that the regime had withdrawn all of the spent enriched uranium from fuel rods in a nuclear reactor frozen by the 1994 Framework agreement in order to build nuclear warheads. China, Japan, and Russia joined with the United States in sponsoring six-nation talks in an effort to ease a growing crisis. Six-nation talks held at the end of February 2004 broke up without an agreement despite the fact that North Korea had said they were willing to consider nuclear disarmament. The United States refused to negotiate until North Korea met additional conditions for direct negotiations. While everyone expects talks to resume, the crisis continues to be one of the most serious unresolved conflicts in international relations today. In "How to Deal With North Korea," James T. Laney and Jason T. Shaplen explain why "Pyongyang's belligerent behavior should not obscure other dramatic conciliatory steps North Korea has taken in recent years-steps suggesting that, even now, a solution is within reach."

During the campaigns among democrats in the United States' 2004 primary season, all of the democratic candidates criticized the Bush administration for policies that facilitated the contracting out of high paying service jobs abroad. The country that has received a large number of contracts from U.S. businesses over the past three years is India. Yasheng Huang and Tarun Khanna explain in, "Can India Overtake China?," that India now has a number of domestic entrepreneurs who can compete internationally with the best that Europe and the United States has to offer. India's stronger infrastructure, and more efficient and transparent capital markets and legal system, in comparison to China's export-led manufacturing boom, are additional reasons why India's entrepreneurs may have a long-term advantage over China. Whether India's future growth will be aided by additional service jobs contracted out from the United States and other western countries promises to be one of the most hotly debated issues of the 2004 Presidential election in the United States. At the same time, international analysts will continue to debate whether India or China are in a position to be the most successful "rising power" in Asia.

While much of the public's attention has focused on Iraq, the battle for the rural areas of Afghanistan and the search for bin Laden in the border areas between Afghanistan and Pakistan promise to be major goals of the future war on terrorism. As Ahmed Rashid describes in "Dangerous Neighbours," several of Afghanistan's neighbors are keen to sponsor rival warlords much like they did in the early 1990s. The United States fears that the country will split along ethnic lines if U.S. troops leave. Meanwhile, 90 percent of the attacks they face in the country are coming from groups based in Pakistan. Peace will require a stronger central government and greater economic development. Future events in Afghanistan, India, and Pakistan are likely to determine future regional relationships and stability throughout South and West Asia even if the threat posed by al Qaeda is eliminated. Many analysts, including Rashid, remain concerned about the future since the United States does not seem to have a long-term strategy for implementing political or economic change in Afghanistan. Ahmed Rashid shares the fear of many that without a long-term strategy, ethnic divisions and Islamic fundamentalism will reassert themselves and divide Afghanistan, and possibly other countries in the region.

CHANGING COURSE ON CHINA

Elizabeth Economy

"Relations between China and the United States are perhaps the best they have been since 1989...What accounts for this seemingly dramatic transformation?"

In the immediate aftermath of his election in November 2000, President George W. Bush proclaimed China a strategic competitor and asserted that US policy in Asia should be reoriented toward American allies in the region, including Taiwan. Consensus within the administration on how to implement this new policy, however, remained elusive.

Indeed, for much of the early tenture of the Bush White House, the administration seemed divided on how best to approach China. The US trade representative and members of the State Department preached the virtues of engagement with the mainland while the Pentagon formulated its own policy to enhance US relations with Taiwan. Some within the administration went so far as to place the mantle of the former Soviet Union on China, calling the People's Republic the next great threat to US security. Within months after Bush took office, an ugly altercation over a US spy plane, a sizeable arms sale to Taiwan, and aggressive talk of American missile defense increased tensions markedly.

Nearly three years later, relations between China and the United States are perhaps the best they have been since 1989. President Bush appears to be following in the footsteps of his predecessors in recognizing both the importance of China to US foreign policy interests and the benefits of a more proactive approach to the mainland. Apparent divisions within the Bush administration over how to approach China have resolved themselves, at least temporarily, in favor of a more engagement-oriented policy.

What accounts for this seemingly dramatic transformation in US policy toward China? Above all else, the evolution in policy reflects a new set of strategic realities that have confronted America since September 11, 2001. The events of September 11 caused the Bush adminstration to reorient its international priorities, both diminishing the centrality of China in US foreign policy as a potential long-term threat and offering a new and important opportunity for the two countries to cooperate.

Soon after September 11, when the Bush administration began its campaign to enforce Iraqi compliance with United Nations weapons inspections, US officials actively courted the Chinese leadership. The administration first sought China's support for stepped-up inspections of Iraq's weapons arsenal, then for a UN resolution authorizing military action against Iraq. In the process, the administration turned its agenda with China on its head. More recently, the nuclear crisis on the Korean peninsula has reinforced the importance of China to US strategic interests while opening a new avenue for cooperation. For the United States, China has become an essential partner in meeting its new geostrategic challenges.

China's leaders, in turn, have recognized an opportunity to use these new strategic realities to meet their own fundamental objectives: stability in the Sino-American relationship to ensure continued economic modernization, and the maintenance of China's own domestic security. As Chinese President Hu Jintao has argued, the importance of the United States to China's economic development requires a flexible and accommodating posture that keeps US–China relations on an even keel.

This does not mean that significant policy differences between the two countries have disappeared, or that this partnership will continue indefinitely at its current level of mutual accommodation. The Sino-American relationship remains fragile and continues to require high-level intervention to ensure its stability. Critical differences still mark the manner in which the two countries approach international relations generally, as well as specific issues such as Taiwan, missile defense, and human rights. In addition, President Bush does not make China policy in a vacuum. Differences within the administration over relations with China remain. An active congressional lobby on China and Taiwan, while quiescent for most of 2002 and 2003, may again become energized in an effort to redefine US-China policy.

THE FIRST SIX MONTHS

During the presidential campaign, candidate Bush set the tone for a distinctly new China policy. He promised to refocus US attention in the region away from the mainland and toward Japan, South Korea, and Taiwan. The Bush team stressed that these East Asian powers were democratic and capitalist and thus natural allies of the United

States. Candidate Bush also referred to China as a strategic competitor, emphasizing its human rights abuses and role in the proliferation of missile technology.

> For the United States, China has become an essential partner in meeting its new geostrategic challenges.

While this rhetoric received significant media attention, subtle hints also signaled a degree of continuity with the previous administration's policies. Candidate Bush argued that the United States had to remain deeply engaged on the trade front with China, noting that the development of an entrepreneurial class and the advent of the Internet were cornerstones in a process of long-term political liberalization in China and that American farmers would benefit from China's entry into the World Trade Organization. Bush even reiterated the classic engagement line: "I think if we make China an enemy, they'll wind up being an enemy."

The first testing ground for the new administration and its approach to China came less than three months after Bush assumed the office. In late March 2001, a US Navy EP-3 surveillance aircraft flying near China's Hainan Island collided with a Chinese fighter jet. The Chinese pilot was lost at sea and the EP-3's flight crew was forced to make an emergency landing on Hainan. It was not until April 4, five days after the accident, that diplomatic exchanges occurred.

This was a tense and difficult time, with many members of Congress calling for tough action and some policy analysts predicting military conflict between China and the United States. Once China opened the door to US diplomats, however, President Bush moved quickly to effect a resolution. On April 5, Secretary of State Colin Powell issues a statement of "regret" that progressed to "sorry" and "very sorry" by April 7. On April 11, the US flight crew departed Hainan to return to the United States. In the aftermath of the incident, the consensus in the United States was that President Bush had handled a difficult situation well by preventing the conflict from escalating. Still, some conservative policy analysts critized the administration for appeasing Chinese misbehavior. They argued too that the strategy President Bush adopted represented a victory for Secretary of State Powell and National Security Adviser Condoleezza Rice over Vice President Dick Chaney and Secretary of Defense Donald Rumsfeld.

Still, the EP-3 incident, as well as the initial difficulty in reaching and negotiating with senior Chinese military and party officials, provided the rationale that China hawks within the administration needed to take a series of significant steps. The Defense Department broke off all informal and regular military-to-military contacts, noting that such contacts would have to be approved on a case-by-case basis. The administration announced an arms sale package to Taiwan worth as much as $4 billion that included up to eight diesel submarines and four guided-missile destryoers. It cited as justification China's continued missile buildup opposite Taiwan and the mainland's refusal to renounce the right to use force to reunify with Taiwan. For the first time, the Defense Department allowed Taiwanese military officials to participate in courses at its Hawaii-based think tank, the Asia-Pacific Center for Security Studies.

Meanwhile, discussions of the Bush administration's plans for missile defense, which the Chinese regarded as extremely threatening, filled the news media. And when asked whether he would use military force to defend Taiwan in case of a Chinese attack, the president said, "Whatever it takes to help Taiwan defend itself." This assertion caused great consternation in mainland China, where it was perceived by some as dropping the American commitment to "strategic ambiguity" (whereby the United States had refused to discuss various scenarios in which conflict between the mainland and Taiwan might emerge in order to preserve the full range of options for US action).

Yet even as the security relationship clearly deteriorated, Bush did not permit the EP-3 incident to become entangled with other areas of the Sino-American relationship, such as China's impending entry into the World Trade Organization. Morever, the manner in which the EP-3 incident itself was handled demonstrated that, whatever the administration's rhetoric, the White House was committed to keeping its relationship with China on track and to preventing tensions in the security realm from spilling over into other arenas. In fact, during the summer of 2001 Secretary of State Powell pursued a noticeably more engagement-oriented approach than others within the administration, visiting the People's Republic and elucidating a new foundation for US dialogue with China dubbed the "Three C's": candid, cooperative, and constructive. Thus, for much of the early part of the Bush administration's tenure, China policy proceeded along two distinct tracks: one directed by the Department of Defense and vice president's office and a second navigated primarily by the State Department.

AFTER SEPTEMBER 11

The devasting terrorist attcks on the World Trade Center in New York and on the Pentagon in Virginia caused the United States to reorient radically its foreign policy priorities. The central focus shifted from promotion of free trade, democracy, and stability to a global war against terrorism. No longer was the Bush administration preoccurpied with defining the next Soviet-like menace; it needed to identify terrorist cells throughout the world.

For Sino-American relations, this transformation is US foreign policy had two important implications. First, China policy became simply one of many issues rather than a top preoccupation for the administration, Congress, and the media. In a world in which America's physical integrity had been so violently breached, concern over the potential economic and security threat posed by China greatly diminished. Second, September 11 provided a clear opportunity for China to establish common interest with the United States on the latter's number one priority: combating terrorism.

China initially hedged its bets, calling for the United States to support China in battling terrorists in Xinjiang, Tibet, and Taiwan. Just one week after the attacks, the government spokesman Zhu Bangzao remarked that "the United States has asked us to help it fight terrorism. Equally, we have reasons for asking the United States to lend us its support and understanding in our struggle against terrorism and separatism. There can be no double standards. We are not suggesting any horse trading; but China and the United States has a common interest in opposing the Taiwanese independence movement which constitutes the main threat to stability in the [Taiwan] Strait."

China soon amended its request to ask for assistance only in fighting the terrorist threat in Xinjiang Autonomous Region, perhaps realizing that equating Al Qaeda with separatist movements in Taiwan and Tibet would be poorly received by many in the United States. Still, China reiterated its concerns that America establish concrete proof of Osama Bin Laden's guilt, that any military strike it might carry out accord with UN rules, and that any action taken be in the long-term interest of world peace and development. Since September 11, the Bush administration has generally given China high marks for its help in tracking down terrorist financing, cooperating on law enforcement, and providing humanitarian aid to Afghanistan.

But, even as China and the United States began to forge new bonds on the issue of global terrorism, Taiwan remained a sticking point. In March 2002, for example, the US Taiwan Business Council hosted in the United States Taiwanese Defense Minister Tang Yiau-ming and other military officials—including US Deputy Secretary of Defense Paul Wolfowitz—thereby violating an undeclared prohibition against meetings between senior Taiwanese and senior US officials. During the meeting, moreover, Wolfowitz repeated President Bush's earlier controversial statement that the United States would do "whatever it takes" to defend Taiwan from military strikes by China. Wolfowitz's affirmation that "China is not an enemy" was lost in the firestorm that his other rhetoric provoked.

At the same time, the Defense Department's Nuclear Posture Review, a summary of strategic planning for America's nuclear weapons over the next decade, was leaked to the media. The document noted that a conflict between China and Taiwan could lead to the use of nuclear weapons by the United States. This added to the growing tensions in Sino-American relations. Congress, too, always an active player on Taiwan issues, became energized by the Bush administration's more proactive Taiwan policy. On April 9, 2002—the twenty-third anniversary of the Taiwan Relations Act—the House of Representatives established the Taiwan Caucus. One of the forum's primary goals was to promote US military ties with Taiwan. (The 1979 act guarantees continued trade and cultural relations with the island and provides assurances for Taiwan's security.)

Beijing was quick to respond to the Defense Department's increased attention to Taiwan. In reply to Wolfowitz's remarks, Foreign Ministry spokeswoman Zhang Qiyue stated, "The comments of the senior US defense official seriouly violated the clear-cut promises laid out in the three joint communiques [that have framed Sino-American relations] and moreover rudely interfered in China's internal political affairs." In addition, Beijing denied the destroyer the USS *Curtis Wilbur* a port call in Hong Kong for early April, and threatened to cancel the impending visit of then Vice President Hu Juntao later in the month. Thus, only weeks before Hu's visit to the United States, the Pentagon's priority on enhancing ties with Taiwan at the expense of relations with the mainland threatened yet again to derail the Sino-American relationship.

SETTING A NEW AGENDA

While the war against terrorism opened the door to the creation of a new foundation for Sino-American relations, the most significant factor in the evolving relationship was the Bush administration's almost singular focus on Iraq. In working to secure backing for a US-led attack on Iraq, the Bush administration arrived at an entirely new agenda with China. Beginning in April 2002 with the visit of then Vice President Hu Jintao, the White House moved quietly and effectively to set the stage for a newfound unity in US-China relations.

The first sign of this emerged during Hu's visit. Both before and after the visit, the White House downplayed the contentious issues of human rights, Taiwan, and weapons proliferation. Although these concerns came up during a session between Vice President Chaney and Vice President Hu, the thrust of the meetings was positive. One high-ranking but unidentified White House official said: "Mr. Hu is bright, amenable, and very pleasant ...she is a pleasure to do business with." Another senior administration official offered even higher praise: "He can come across as warm and even flexible, yet gives nothing away. I can see the day when Mr. Bush feels he can pick up the phone and call Mr. Hu. I don't think he has ever quite felt that way with [President] Jiang Zemin. They are of different eras."

Just one month after Hu's visit, Pentagon hawks, who had defined China as the next Soviet Union, were flying to the People's Republic to discuss conditions for restoring military-to-military contacts. Senior administration officials no longer identified China as a strategic competitor but rather as Secretary of State Powell had—as a participant in a "candid, cooperative, and constructive" dialogue. Deputy Secretary of Defense Wolfowitz began to speak of the need to engage China.

In August 2002 the Bush administration labeled the East Turkestan Islamic Movement—a small group committed to the independence of China's Xinjiang region—a terrorist organization. Beijing, which had long claimed connections between Al Qaeda and ethnic Uighur separatists in Xinjiang, welcomed the move. Perhaps most telling, the Bush administration also indicated in August that it would grant President Jiang Zemin his long-sought invitation to President Bush's ranch in Crawford, Texas, an invitation previously extended only to President Bush's closest perceived allies, such as British Prime Minister Tony Blair and Russian President Vladimir Putin. Moreover, the administration virtually ignored the release of a long-anticipated report by the congressional US-China Security Review Commission in fall 2002, which widely condemned China for its indirect sponsorship of rogue states through the sale of missle technology and called for a range of new sanctions. The report was an embarrassment for an administration intent on making a new friend in China.

China responded positively to th US steps to improve relations. In late August 2002, the Chinese government issued a new set of regulations governing the export of missile technology and reacted only moderately to strong separist rhetoric by Taiwan's leader, Chen Shui-bian, and a visit to the Pentagon by a senior Taiwanese defense official. In October Beijing also announced regulations to control material and technology for biological weapons and took steps to improve the situation of a few political dissidents, including AIDS activist Wan Yanhai, who was permitted to establish his own non-governmental organization to combat AIDS.

On the issue of greatest importance to the United States—reining in any Iraqi program to develop weapons of mass destruction—China supported the United States in its initial effort to press for more aggressive sanctions. It voted in favor of UN Security Council Resolution 1441, which warned Iraq that it would face "serious consequences" if it did not comply with UN weapons inspections. Prime minister Zhu Rongji insisted that "Iraq must cooperate unconditionally with the United Nations" on weapons inspections "At the same time," he continued, "we must respect the sovereignty and territorial integrity of Iraq. If arms inspections do not take place, if there is not clear proof and if there is no authorization from the Security Council, there cannot be a military attack on Iraq."

As the United States pressed for an additional resolution authorizing the automatic use of force in the face of Iraqi noncompliance, China stood with France and Russia in opposition to the proposal. But it did so much less forcefully and publicly than its Security Council partners; there was nothing to be gained from unnecessary antagonizing the United States, particularly in the midst of such a positive overall bilateral relationship. In return, China escaped the far more vocal criticism the United States directed at France and Russia.

Iraq engendered a new US approach toward China, but it has been the crisis on the Korean peninsula that has served to reinforce the importance of a US-China partnership. In October 2002, North Korea admitted that it had not forsaken its nuclear program, triggering a crisis in US-North Korean relations and setting off shock waves throughout the rest of East Asia. The United States and China had long professed "peace and stability on the Korean peninsula" as a common policy goal. Yet rarely had the policy been put to such a test, and the two countries soon found themselves articulating signficiantly different approaches to the resolution of the crisis.

China, like Japan and South Korea, favored a strategy that engaged North Korea, offering economic incentives for compliance. The United States argued for a much tougher policy, possibly involving sanctions and not ruling out the use of force. China also supported North Korea's desire to resolve the issue through bilateral US-North Korea negotiations; the United States insisted on multilaterial talks. As North Korea pushed the envelope, withdrawing from the nuclear Non-Proliferation Treaty, the United States sought a resolution within the UN security Council condemning North Korea for its actions. But China resisted resorting to the United Nations, believing that doing so would only isolate North Korea further and put China in the awkward position of having to ally itself openly with either North Korea or the United States.

Despite their different policy prescriptions, the United States and China continued to work together behind the scenes at the behest of the United States. Some Chinese scholars within elite policy circles also began to suggest that China needed to reevaluate its position, arguing that a nuclear North Korea posed a threat to China. In February 2003, China shut down an oil supply line to North Korea for three days—ostensibly due to "technical problems" but more likely in an effort to signal to North Korea that its belligerent approach was costing it China's support. Soon thereafter, China brokered an agreement between the United States adn North Korea to hold three-way talks in which China would be the third participant. While the negotiations produced no tangible results, the United States and China had moved their relationship to a new level of partnership in the process and agreed to continue to work together to resolve the crisis.

MARGINALIZED: HUMAN RIGHTS, TAIWAN, TRADE

After 1989, the issues that defined the Sino-American relationship were human rights, Taiwan and trade. These concerns no longer dominate the bilateral agenda. Securing stability on the Korean peninsula, jointly fighting terrorism, and meeting US objectives in Iraq, as well as China's desire for continued stability to ensure economic growth, are now the dominant factors in the relationship.

This transformation in the bilateral agenda has been marked by substantive changes in policy. For the first time since 1989, the United States in April 2003 did not pursue a resolution at the UN Human Rights Commission in Geneva condemning China for its human rights practices. The United States has been noticeably quiet about human rights violations that otherwise would have given rise to comment from the White House, including China's decision to prevent labor activists in Liaoning province from meeting with US diplomats and foreign reporters, and the continued imprisonment without councel of long-time American resident Yang Jianli, a prominent democracy activist who illegally entered China to observe mass labor protests in northeast China. Whereas previously the Bush administration might have exploited Beijing's initial mismanagement of the SARS crisis as an opportunity to criticize the regime, at least publicly, the Bush team instead congratulated the Chinese leadership on its handling of what was feared might become a global epidemic.

With regard to Taiwan, the Bush administration continues to act as the island's primary interlocutor in international forums over the objections of the People's Republic: supporting Taiwan's accession to the World Health Organization (blocked by China) in the midst of the SARS crisis, for instance, and backing Taiwan as it wrangles with Beijing over its formal name within the World Trade Organization. The United States continues to promote military relations with Taiwan through exchanges and the sale of advanced weaponry.

Yet the administration has taken steps to reassure Beijing of its commitment to a one China policy, moving beyond previous US administration to state definitively that America does not support Taiwan independence. Beijing, in turn, has begun to develop policy proposals that might persuade the Bush administration to reduce its commitment to Taiwan. In October 2002, for example, President Jiang proposed withdrawing some of the several hundred missiles targeted at Taiwan in exchange for US cutbacks on arms sales. Although the administration has yet to reply, Bush has reportedly asked his staff to develop a formal response. At the same time, within some quarters of China's elite, growing economic integration and personal links between the mainland and Taiwan have begun to foster a new sense of confidence concerning the eventuality of reunification.

In meetings between US and Chinese officials, Taiwan remains an important matter, but there is little evidence of the rancor that marked bilateral exchanges on the issue prior to September 11. In June 2003 at the Group of Eight summit in Evian, France, Hu Jintao—who replaced Jiang as president this March—stated that he appreciated President Bush's declaration not to support Taiwanese independence.

Unlike human rights and Taiwan, trade issues typically have had an ameliorative effective on Sino-American relations. China's accession to the World Trade Organization in December 2001 was celebrated both in China and the United States as signaling signficant future trade and economic opportunities. America is China's largest export market, with overall exports from China to the United States last year totaling more than $100 billion.

Still, many in the American business community and within the US trade representative's office have begun to express concern with a growing trade deficit that, according to US Trade Representative Robert Zoellick, will exceed $100 billion in 2003. In several critical areas such as agriculture, telecommunications, and finance, Zoellick and various China watchers have noted an increasing number of bureaucratic impediments to foreign access to China's markets.

In addition, while successive US administrations had felt comfortable with Prime Minister Zhu Rongji and his committment to effective implementation of China's World Trade Organization obligations, his retirement in April 2003, as well as the elimination of the Ministry of Foreign Trade and Economic Cooperation, has left some uncertainly in the trade relationship. In May 2003, the United States sanctioned a major Chinese conglomerate, NORINCO, for transferring dual-use technology to an Iranian company known to produce missiles. (The Chinese government's muted response lent weight to the US claim.) Thus, although trade remains one of the pillars of the Sino-American relationship, it is possible that friction will increase over time, particularly if China's WTO implementation proves problematic. Indeed, the United States is the third-largest source of foreign direct investment (FDI) in China after Hong Kong and the Virgin Islands, through which FDI from a number of countries flows. (In 2002, total US FDI contracted totaled more than $7 billion.)

WHAT NEXT?

Even as China and the United States continue to strengthen their bilateral relationship by working closely on issues of global security and downplaying traditional rifs, their recent accommodation may well prove ephemeral. Fundamental divisions persist. The United States coninued to desire, and work toward, evolution in China's political system. Many in China still perceive America as a signficant obstacle to China's growing status as a regional, if not global power.

The uncertainty generated by the recent leadership transition in China and the potential for a new policy agenda in Beijing could also open the door to significant change in China's approach to the United States. Former President Jiang Zemin continues to exercise power beind the scenes, often in conflict with Hu Jintao. While unlikely, it is not impossible that policy toward the United States could fall victim to elite power politics as one side or the other attempts to play an anti-US/nationalism card in hopes of currying popular support. President Hu's new emphassi on slowing down the pace of economic reform to redress the vast social inequities that emerged over the course of Jiang's tenure might also contribute to a slowdown in the implementation of China's trade commitments and a consequent downturn in Sino-American relations.

Important opportunities remain, however, for strengthening the foundation of the relationship and helping it endure beyond the current, potentially transitory accommodation that has resulted from new geostrategic realities. Cooperative ventures in areas where common interest exists naturally, such as public health and the environment, should be fostered. Even more critically, the administration should seek opportunities to advance Sino-American relations in the area most difficult to negotiate: security, human rights, and Taiwan.

In the security realm, the Department of Defense should move quickly to reestablish military-to-military relations. Although the EP-3 incident reinforced the importance of maintaining open channels of communication and fostering military exchange, the Pentagon has yet to follow through on President Bush's directive to restart these exchanges. American military officers must still obtain approval for every meeting or exchange with their Chinese counterparts. This is inefficient and does little to help develop the open dialogue and personal relationships that can prove effective for crisis management as well as longer-term understanding of each country's security priorities and appraches. Military contracts are particularly critical as the United States and China continue to work closely on a highly sensitivie security issue such as North Korea.

In working to secure backing of a US-led attack on Iraq, the Bush administration arrived at an entirely new agenda with China.

With regard to human rights, the new accommodation in Sino-American relations offers the Bush administration the opportunity to call directly on President Hu to embrace a more aggressive program of rights protection. The United States is already cooperating, both directly and indirectly, with the Chinese government on law enforcement and the development of the rule of law. Recent high-profile cooperation in dismantling a major drug-trafficking network in Asia and North America, for example, may mark a breakthrough for future cooperation on law enforcement. Various US and NGO efforts focus on the training of Chinese judges and lawyers, cooperation on intellectual property enforcement, and support for more open media in the hopes of assisting enforcement. The State Department's Bureau of Human Rights, Democracy and Labor in particular has mounted a variety of ambitious initiatives designed to advance the cause of human rights, rule of law, development of NGOS, and freedom of expression.

The administration should also renew the dialogue on human rights at the highest level. While the long-stalled dialogue on human rights between the State Department and the Chinese Foreign Ministry resumed formally in December 2002, it has failed to produce follow-up visits or any tangible results, ostensibly because of personnel shuffling within China. The United States should make it clear to Beijing that dialogue and action on human rights are as important a part of the bilateral relationship as discussions on restoring direct military-to-military relations or providing additional assistance to combat terrorism.

Finally, the administration appears to be successfully navigating the often treacherous waters of cross-strait politics. As the White House presses forward with various security initiatives in Asia, including the restationing of troops in South Korea and Japan and the development of missile defense, administration officials should take the opportunity to consult and even cooperate with the People's Republic. If the United States pursues a missile defense architecture that permits targeting of Chinese missiles, for example, early consultation with China to establish mutually acceptable limits on the number of Chinese missiles deployed in response would be extremely useful in avoiding a potential regional arms race and a downward spiral in US-China relations. In the spirit of the new Sino-American relationship, China as indicated that, while it is concerned about the development of theater missile defense, it is willing to conduct "constructive dialogue" on the issue. The only aspect of a potential US missile defense system that China considers non-negotiable is the inclusion of Taiwan behind the shield.

For the foreseeable future, a strong and stable bilateral relationship serves both Chinese and American interests. China considers US support central to its own continued economic development and security. The United States considers China's support necessary for addressing challenges in North Korea and Iraq, and to combat terrorism. As long as these strategic realities continue to dominate each country's domestic and foreign policy agendas, positive relations between China and the United States are likely to remain a high priority for the leaders of both countries. By developing long-term strategies on key contentious issues, the leaders could improve chances that this stability will continue.

ELIZABETH ECONOMY is a senior fellow and director of Asia studies at the Council on Foreign Relations.

How to Deal With North Korea

James T. Laney and Jason T. Shaplen

MIXED MESSAGES

PROGRESS in reducing tensions on the Korean peninsula, never easy, has reached a dangerous impasse. The last six months have witnessed an extraordinary series of events in the region that have profound implications for security and stability throughout Northeast Asia, a region that is home to 100,000 U.S. troops and three of the world's 12 largest economies.

Perhaps the most dramatic of these events was North Korea's December decision to restart its frozen plutonium-based nuclear program at Yongbyon—including a reprocessing facility that separates plutonium for nuclear weapons from spent reactor fuel. Just as disturbing was the North's stunning public admission two months earlier that it had begun building a new, highly-enriched-uranium (HEU) nuclear program. And then came yet another unsettling development: a growing, sharp division emerged between the United States and the new South Korean government over how to respond.

But recent events have not been entirely negative. In the two months prior to the October HEU revelation, North Korea had, with remarkable speed, undertaken an important series of positive initiatives that seemed the polar opposite of its posturing on the nuclear issue. These included initiating an unscheduled meeting between its foreign minister, Paek Nam Sun, and Secretary of State Colin Powell in July—the highest-level contact between the two nations since the Bush administration took office; inviting a U.S. delegation for talks in Pyongyang; proposing the highest-level talks with South Korea in a year; agreeing to re-establish road and rail links with the South and starting work on the project almost immediately; demining portions of the demilitarized zone (DMZ) and wide corridors on the east and west coasts surrounding the rail links; sending more than 600 athletes and representatives to join the Asian Games in Pusan, South Korea (marking the North's first-ever participation in an international sporting event in the South); enacting a series of economic and market reforms (including increasing wages, allowing the price of staples to float freely, and inaugurating a special economic zone similar to those in China); restarting the highest-level talks with Japan in two years; holding a subsequent summit with Japanese Prime Minister Junichiro Koizumi, during which Pyongyang admitted abducting Japanese citizens in the 1970s and 1980s; and finally, allowing the surviving abductees to visit Japan.

Viewed individually, let alone together, North Korea's initiatives represented the most promising signs of change on the peninsula in decades. Whether by desire or by necessity, the North finally appeared to be responding to the long-standing concerns of the United States, South Korea, and Japan. Equally important, Pyongyang seemed to have abandoned its policy of playing Washington, Seoul, and Tokyo off one another by addressing the concerns of one while ignoring those of the other two. For the first time, the North was actively (even aggressively) engaging all three capitals simultaneously.

Until October, that is, when North Korea acknowledged the existence of its clandestine HEU program—ending the diplomatic progress instantly. Once the news broke, Pyongyang quickly offered to halt the HEU program in exchange for a nonaggression pact with the United States. But Washington, unwilling to reward bad behavior, initially refused to open a dialogue unless the North first abandoned its HEU effort. In November, the United States went a step further: saying that Pyongyang had violated the 1994 Agreed Framework and several other nuclear nonproliferation pacts, Washington engineered the suspension of deliveries of the 500,000 tons of heavy fuel oil sent to the North each year under the 1994 accord. The Agreed Framework had frozen the North's plutonium program—a program that had included a five-megawatt experimental reactor, two larger reactors under construction, and the reprocessing facility—narrowly averting a catastrophic war on the Korean Peninsula.

In the weeks following the suspension of fuel shipments, the United States hardened its stance against dialogue with the North—despite the fact that most U.S. allies were encouraging a diplomatic solution to the situation. North Korea responded by announcing plans to reopen its Yongbyon facilities. It immediately removed the seals and monitoring cameras from its frozen nuclear labs and reactors and, a few days later, began to move its dangerous spent fuel rods out of storage. Pyongyang subsequently announced its intention to reopen the critical reprocessing plant in February 2003. On December 31, it expelled the inspectors of the International Atomic Energy Agency (IAEA). And on January 9, it announced its withdrawal from the nuclear Nonproliferation Treaty.

Although Washington, strongly urged by Seoul and Tokyo, ultimately agreed to talks, the situation appeared to be worsening almost daily. Depending on how it is resolved, the standoff could still prove a positive turning point in resolving one the world's most dangerous flash points. But it could also lead to an even worse crisis than in 1994. The proper approach, therefore, is to now re-engage with North Korea without rewarding it for bad behavior. Working together, the major external interested parties (China, Japan, Russia, and the United States) should jointly and officially guarantee the security of the entire Korean Peninsula. But the outside powers should also insist that Pyongyang abandon its nuclear weapons program before offering it any enticements. Only when security has been established (and verified by intrusive, regular inspections) should a prearranged comprehensive deal be implemented—one that involves extensive reforms in the North, an increase in aid and investment, and, eventually, a Korean federation.

THE NORTH GOES NUCLEAR

TO UNDERSTAND how the most promising signs of progress in decades quickly deteriorated into nuclear brinkmanship, it is necessary to first understand the origins and motivation behind the North's HEU program and Pyongyang's subsequent decision to restart its plutonium program. Even before North Korea admitted that it was building a new HEU program, the United States had long suspected the country of violating its relevant international commitments. Three years ago, such concerns had led to U.S. inspections of suspicious underground facilities in Kumchang-ni. Although those inspections did not reveal any actual treaty violations—in part because Pyongyang had ample time to remove evidence before the inspectors arrived—suspicions lingered. These doubts proved justified in July 2002, when the United States conclusively confirmed the existence of the North's HEU program.

It now seems likely that Pyongyang actually started its HEU program in 1997 or 1998. Although Kim Jong Il's motives for doing so will probably never be clear (his regime has a record of confounding observers), there are two plausible explanations. The first focuses on fear: namely, North Korea's fear that, having frozen its plutonium-based nuclear program in 1994, it would receive nothing in return. Such a suspicion seems unreasonable on its face, since, under the 1994 Agreed Framework negotiated with Washington, Pyongyang was to be compensated in various ways for abandoning its nuclear ambitions. But from the perspective of a paranoid, isolated regime such as North Korea's, this concern was not without justification. Almost from its inception, the provisions of the 1994 accord fell substantially behind schedule—most notably in the construction of proliferation-resistant light-water reactors in the North and improved relations with the United States.[1] North Korea may thus have started its HEU program as a hedge against the possibility that it had been duped, or, more likely, that new U.S., South Korean, or Japanese administrations would be less willing to proceed with the politically controversial program than were their predecessors.

A second, darker, and more likely explanation for Pyongyang's decision to start the HEU program holds that the North never really intended to give up its nuclear ambitions. Whether motivated by fear, honor, or aggression (the determination to stage a preemptive strike if threatened), Pyongyang views a nuclear program as its sovereign right—and a necessity.

Whichever of these theories is true, the North seems to have undertaken its HEU program slowly at first, ramping it up only in late 2000 or 2001. And it was able to hide the program until July 2002, when U.S. intelligence proved its existence. Although Bush administration officials insist otherwise, it is possible, as North Korean officials have suggested, that Pyongyang decided to step up its nuclear program in response to what it perceived as Washington's increasingly hostile attitude—a hostility demonstrated to North Koreans by President Bush's decision to include them in the "axis of evil" and to set the bar for talks impossibly high. This perceived hostility was further encouraged when the administration announced its new doctrine of preemptive defense. Notwithstanding the president's remarks to the contrary, Pyongyang views the new defense doctrine as a direct threat. After all, if Washington is willing to attack Iraq, another isolated nation with a suspected nuclear program, might it not also be willing, even likely, to do the same to North Korea?

This fear helps explain why the North decided to restart its plutonium program. Many within the senior ranks of the North Korean military believe that if the United States attacks, Pyongyang's position will be strengthened immeasurably by the possession of several nuclear weapons. North Korean planners thus reason that they should develop such weapons as quickly as possible, prior to the American attack that may come once Washington has concluded its war with Iraq.

HIGH-STAKES POKER

THERE ARE AGAIN two plausible explanations for why the North revealed its HEU program in October 2002. Since its earliest days in office, the Bush administration has made clear that it favors a more hard-line approach to North Korea than did the Clinton team. Even prior to the North's HEU admission, Bush's support for the 1994 Agreed Framework was lukewarm at best. His administration considered the accord a form of blackmail signed by his predecessor—even though, after a long review of North Korea policy in 2001, the Bush administration found it could not justify abandoning the pact without having something better with which to replace it. In short, Washington grudgingly considered itself bound by a diplomatic process it viewed as distasteful—if not an outright scam.

When U.S. Assistant Secretary of State James Kelly visited North Korea in early October, he took with him undeniable evidence of the North's HEU program. He also took with him very narrowly defined briefing papers, hard-line marching orders that reflected the influence of the Defense Department and the National Security Council.

Anticipating isolation and a worsening of already strained relations in the face of Washington's evidence, Pyongyang opted to play one of its few remaining trump cards: open admission of its nuclear program. This openness, Kim may have hoped, would keep the Bush administration from disengaging entirely. By acknowledging its HEU effort, Pyongyang essentially sent Washington the following message: "We understand that despite everything we've done over the past several months you want to isolate or disengage from us. Well, we admit we have a uranium-based nuclear program. You say you don't want to deal with us. Too bad—you can't ignore a potential nuclear power. Deal with us."

Another hypothesis to explain the timing is that Pyongyang simply miscalculated. North Korea watchers learned long ago to expect the unexpected, but even the most jaded observers were surprised in September 2002 when Kim admitted to Koizumi that the North had abducted 13 Japanese in the 1970s and 1980s to train its spies. Kim apologized for the abductions and, with remarkable speed, subsequently authorized a visit of five of the surviving abductees to Japan. In doing so, he removed a decades-old barrier to normalization of relations between the two nations (and to the payment of billions of dollars in hoped-for war reparations from Tokyo).

Kim's gamble on coming clean about the abductions appeared at the time to have paid off. Notwithstanding the predicted public backlash in Japan, further talks between Tokyo and Pyongyang took place in October (after the HEU admission).[2] Having experienced better-than-expected results in admitting to the abductions, Kim may have hoped for the same by confessing to his HEU program. His thinking may have been that, in view of Washington's evidence, Pyongyang would eventually have had to come clean anyway. That being the case, it was better to do so sooner rather than later, thereby removing one of the primary obstacles to improved U.S.–North Korea relations. Kim may further have surmised that the timing of such a revelation in October was advantageous, given recent progress in talks with Japan and South Korea. He probably hoped that Tokyo and Seoul would pressure Washington to mitigate its response.

In the weeks immediately following Kelly's visit, Washington made it clear that it did not see a military solution to the crisis on the Korean Peninsula. This left isolation, containment, and negotiation as the only viable alternatives. A policy of isolation would seek the North's collapse but would not address the HEU problem and would likely result in the North's restarting its plutonium-based nuclear program. Containment, or economic pressure designed to squeeze the North, would seek to punish Pyongyang while leaving the door open to future negotiation. It too would not address the HEU problem but, it was hoped, might maintain the freeze on the plutonium program. Negotiations, meanwhile, would seek to address the nuclear problem but could be viewed by some as a reward for bad behavior.

If a successful isolation or containment policy wins the day, the North will have miscalculated in coming clean. If, however, a policy of dialogue and subsequent negotiation ultimately emerges—or if isolation or containment fails (in part because Washington is unable to persuade China, South Korea, and Russia to endorse it over a sustained period)—Kim will have played his cards exceedingly well.

BEST OF A BAD SITUATION

MANY PUNDITS and policymakers in Washington, on both sides of the aisle, argue that the revelations about Pyongyang's clandestine HEU program prove that President Clinton's policy of engaging the North was a mistake. This argument maintains that giving in to blackmail leads only to more blackmail.

Although it is inherently valid, such analysis is too simple. In 1994, the United States was on the edge of war with North Korea. Washington had beefed up its forces in the theater, installed Patriot missile batteries in the South, and was reviewing detailed war plans. The White House had even begun to consider the evacuation of American citizens. The 1994 Agreed Framework, although deeply flawed, represented the best deal available at a far from ideal time. It remained so for several years. And although it has been disappointing on many levels, the agreement has not been useless.

Indeed, it averted a potentially catastrophic situation. Instead of a war (which the U.S. military commander in South Korea, General Gary Luck, estimated would have killed a million people, including 80,000 to 100,000 Americans), Northeast Asia has experienced eight years of stability. This has had vast implications beyond security. In 1994, South Korea's GDP was 323 trillion won; today, even after the 1997 financial meltdown, its GDP is approximately 544 trillion won.[3] This transformation would have been unlikely in the face of imminent armed conflict. China has similarly experienced explosive growth, much of which might also have slowed had there been a major confrontation on its porous border with North Korea.

The Agreed Framework also provided the parties with critical breathing room, which has allowed new realities to emerge both within North Korea and among the United States and its allies—developments that improve the chances for a better, more comprehensive deal today. To cite one example, in 1994, Kim Jong Il had only recently succeeded his father, North Korea's founder Kim Il Sung. Viewed as weak, mentally unstable, and without a power base of his own, Kim was expected to last a mere two weeks to several months. Today, however, he is acknowledged as the only power in North Korea and has established diplomatic relations with scores of nations, including many of Washington's closest allies in NATO and the European Union. This puts him in a vastly better position to strike a deal.

For its part, the United States in 1994 could not have counted on Russia or China to support its position toward North Korea. Today, however, Washington is likely to receive baseline support—albeit not carte blanche—from both. Indeed, although there has hardly been unanimity among the outside powers, there has already been evidence of such cooperation, in the form of a joint Chinese-Russian declaration issued in early December stating that the two powers "consider it important ... to preserve the non-nuclear status of the Korean Peninsula and the regime of non-proliferation of weapons of mass destruction."

Another benefit of the breathing room created by the 1994 accord is the North's economic dependence on the South. South Korea today is North Korea's largest publicly acknowledged supplier of aid and its second-largest trading partner. Although not as successful as he would have liked, former South Korean President Kim Dae Jung's "Sunshine Policy" of engaging the North has, in conjunction with the North's economic collapse, given Pyongyang a strong economic interest in avoiding a crisis. (Although the numbers are much smaller, the situation is not wholly unlike that between Taiwan and China.) Should the North exacerbate current tensions, the economic fallout would be traumatic, and the loss of South Korean investment could destabilize the North.

THE WAY OUT

THE TIMING of the steps now taken to resolve the current crisis will be crucial to their success. Indeed, timing is important to understand because the North's HEU program does not pose an immediate threat. Although it has the potential to eventually produce enough uranium for one nuclear weapon per year, it has not yet reached this stage and is not expected to do so for at least two to three more years, according to administration officials and the Central Intelligence Agency.

The North's decision to reopen its plutonium-based nuclear program at Yongbyon poses a more critical and immediate threat, however. Prior to its suspension in 1994, most experts believe this program had already produced enough plutonium for one or two nuclear weapons. The 8,000 spent fuel rods from the five-megawatt reactor contained enough plutonium for an additional four to five nuclear weapons.[4] The IAEA monitored the freeze via seals, cameras, and on-site inspectors. It also canned the 8,000 existing spent fuel rods, placed them in a safe-storage cooling pond, and monitored them until its inspectors were expelled from North Korea on December 31.

The five-megawatt reactor, when operational, will produce enough plutonium for one or two additional nuclear weapons per year. But the 8,000 rods represent an even more immediate challenge. If the North follows through on its threat to reopen the reprocessing facility in February, it would take just six months to reprocess all of its spent fuel and extract enough plutonium to make four or five additional weapons. This would bring Pyongyang's nuclear arsenal to between five and seven weapons by the end of July. It could have enough plutonium for one to three weapons even sooner.

Thus there exists only a short window of opportunity before the North's recent action translates into additional nuclear-weapons material on the ground. The trick to unraveling the current impasse is to avoid rewarding the North for its violations of past treaties with a new, more comprehensive agreement. Blackmail cannot and should not be condoned. The starting point for future discussions should therefore be that the North must completely and immediately abandon its HEU and plutonium-based programs. This pledge must be accompanied by intrusive, immediate, and continuous inspections by the IAEA.

It is a tenet of all international negotiations, however—particularly those that involve the Korean Peninsula—that all crises create opportunity, and this one is no different. At its core—politics stripped aside—the current standoff will allow Washington to scrap the flawed Agreed Framework and replace it with a new mechanism that better addresses the concerns of the United States and its allies. In many ways, the North's HEU admission and its subsequent decision to reopen its plutonium program might therefore be viewed as a blessing in disguise. The Bush administration can finally rid itself of a deal it never liked and never truly endorsed and replace it with one that addresses all of Washington's central concerns, including the North's missile program and its conventional forces. Washington must, however, be willing to make such a deal attractive to the North as well.

Yet timing poses an immediate barrier to negotiating a new mechanism. Pyongyang has insisted it will give up its HEU and plutonium programs only after Washington signs a nonaggression pact with it. But the Bush administration, while publicly reassuring the North that it has no intention of invading, has justifiably insisted that Pyongyang give up these programs before there is any discussion of a new mechanism. The North seems unwilling to lose face by giving up this trump card without a security guarantee, and Washington is unwilling to take any action that appears to reward Pyongyang before it has fully dismantled its nuclear programs.

Those who think they can outwait Pyongyang by isolating it or pressuring it economically, as the Bush administration proposed in late December, are likely to be proved wrong. North Koreans are a fiercely proud people and have endured hardships over the last decade that would have led most other countries to implode. It would therefore be a mistake to underestimate their loyalty to the state or to Kim Jong Il. When insulted, provoked, or threatened, North Koreans will not hesitate to engage in their equivalent of a holy war. Their ideology is not only political, it is quasi-religious. Pyongyang also enjoys an inherent advantage in any waiting game: Beijing. Although China might initially support a policy of economic pressure, Beijing is afraid that it will face a massive influx of unwanted refugees across the Yalu River should the North collapse. To guard against this event, it will ultimately allow fuel and food (sanctioned or unsanctioned) to move across its border with the North. Similarly, South Korea, which also wants to avoid a massive influx of refugees, is unlikely to support a sustained, indefinite policy of squeezing the North. In mid-December, it elected by a larger margin than predicted a new president who ran specifically on a platform of expanding engagement with Pyongyang.

The way to cut the Gordian knot of who goes first is through a two-stage approach. The first stage would provide the North with the security it craves while also ensuring that Pyongyang is not rewarded for its bad behavior. To achieve this end, the four outside interested powers (the United States, Japan, China, and Russia—each of which has supported one side or the other in the past) would jointly and officially guarantee the security and stability of the entire Korean Peninsula. Washington may not be able or willing to convene a meeting of the four powers to this end. If not, back channels or unofficial initiatives should be used to encourage Moscow or Beijing to take the lead. Both Russia and China have sought to increase their influence on the Korean Peninsula in recent years. This plan would solidify their places at the table.

Once the security of the peninsula has been guaranteed by the outside powers, it will be time for stage two: a comprehensive accord, again broken into two parts. The North must completely give up its HEU and plutonium programs and allow immediate, intrusive, and continuous inspections by the IAEA; end its development, production, and testing of long-range missiles in exchange for some financial compensation; draw down its conventional troops along the DMZ (although there will be no reduction of U.S. troops at this time, and only a very limited reduction of U.S. troops in five years, should the situation permit); and, finally, continue to implement economic and market reforms.

In exchange for the above, Japan would normalize its relations with the North within 18 months of the agreement's coming into effect. This normalization would include the payment of war reparations in the form of aid, delivered on a timetable extending five to seven years. Both halves of the peninsula would also enter a Korean federation within two years of the agreement's coming into effect. And as soon as the IAEA had verified that the North has dismantled its nuclear weapons programs, Washington would sign a nonaggression pact with Pyongyang. This pact, which by prior agreement would automatically be nullified by subsequent signs that the North was not cooperating or was initiating a new nuclear program, would include the gradual lifting of economic sanctions over three years.

The United States, South Korea, Japan, and the European Union—the primary members of the Korean Peninsula Energy

Development Organization (or KEDO, which was set up to administer the Agreed Framework)—would further maintain the organization and provide the two new light-water reactors stipulated in the original deal. KEDO would also resume delivery of heavy fuel oil until the first reactor was completed.

In addition to the above measures, China and Russia would agree to support the North economically via investment. All outside parties to the deal—the United States, South Korea, Japan, China, and Russia—would also contribute to the compensation the North would receive in return for ending its long-range missile program.

Finally, five years after the above accord is signed, a Northeast Asia Security Forum, consisting of the four major powers plus South and North Korea, would be created to ensure long-term peace and stability throughout the region.

The timing of the various parts of stage two will be critical to its success. To this end, the leaders of all the countries involved (or their high-ranking representatives) should meet in person to negotiate the deal. North and South Korea, Japan, China, Russia, and the United States must all sign on if the plan is to work.

Certain components of the comprehensive deal (such as the U.S.-North Korea nonaggression pact and the missile accord) should exist as separate agreements, referenced in but not attached as appendices to the main text. They should be fully agreed and initialed prior to signing the comprehensive deal. Immediately after signing the comprehensive agreement, the North would have to take the first step by fully dismantling both its HEU and its plutonium programs and allowing IAEA inspections to verify these steps. Only after the IAEA had certified the dismantling would the nonaggression and missile pacts be signed: in the case of the nonaggression pact, by Pyongyang and Washington alone, and in the case of the missile pact, by Beijing, Moscow, Pyongyang, Seoul, Tokyo, and Washington.

THE SUM OF TWO PARTS

INITIALLY, Washington's response to North Korea's HEU and plutonium programs consisted mostly of condemning Pyongyang. Then, in early January, President Bush and Secretary of State Powell took steps to ease the tension. Following a trilateral meeting with South Korea and Japan (during which Seoul and Tokyo pressed for a diplomatic approach), Washington finally agreed to open a dialogue with Pyongyang. The Bush administration, however, limited the scope of the meetings to discussion of how North Korea could abide by its international commitments. It is now time to move beyond this narrow agenda to a policy of resolution—one that addresses all concerns on the Korean Peninsula.

Such a shift is particularly important given the very serious rupture that has opened between Washington and Seoul. At precisely the time that the situation in North Korea has reached a crisis stage, U.S.–South Korean relations have hit their lowest level ever. Korean anti-Americanism—far more than just a difference of opinion on how to deal with the North—was responsible for the election of Roh Moo Hyun as president in December. Roh beat a more hard-line rival specifically by distancing himself from Washington's position on the North and by promising to continue Kim Dae Jung's Sunshine Policy. More critically, he promised a new, more prominent role for South Korea in its relationship with the United States. America will therefore no longer be able to force its position on the more assertive and restless South Korean population.

The process above, fortunately, will address the major concerns of all the parties involved. It will assure North Korea of the underlying security it seeks, without requiring Washington to sign a nonaggression pact until after Pyongyang has dismantled its HEU and plutonium programs. If the North balks despite a security guarantee by all major outside powers and the prospect of a comprehensive accord, isolation or economic pressure by Washington and its allies will not only remain a viable alternative, it will be stronger and more fully justified than it would be otherwise, and will more easily win the unified, sustained support of major players in the region. The upside to exploring the path presented above is therefore massive, and the downside very limited. Doing nothing, meanwhile, could become the most dangerous option of all.

JAMES T. LANEY is President Emeritus of Emory University and Co-Chairman of an independent task force on "Managing Change on the Korean Peninsula," sponsored by the Council on Foreign Relations. He served as U.S. Ambassador to South Korea from 1993 to 1997. JASON T. SHAPLEN was Policy Adviser at the Korean Peninsula Energy Development Organization (KEDO) from 1995 to 1999 and is a member of the task force.

NOTES

1. The 1994 Agreed Framework called for best efforts to be made to deliver two light-water reactors to North Korea: one in 2003 and one in 2004. Even before the North admitted to having an HEU program—which has cast the future of the Agreed Framework into doubt—it had become unlikely that the first light-water reactor would be completed before 2008 or the second before 2009. To be fair, however, responsibility for the delay is borne by both sides, and primarily by the North, since it has frequently been intransigent on practical issues related to the agreement's implementation.

2. The visit of the surviving abductees to Japan in October was originally scheduled to last two weeks. After the North acknowledged its HEU program, Japan refused to allow their return and pressed for their North Korean relatives to be allowed to join them. Interestingly, although the dispute remains unresolved, Tokyo and Pyongyang have opted to handle it quietly—even as Japan has made clear that future progress with the North is tied to this issue.

3. In U.S. dollar terms these figures equal a 1994 GDP of $404 billion (using the 1994 exchange rate of 800 won to the dollar) and a 2001 GDP of $422 billion (using today's much lower exchange rate of 1,290 won to the dollar).

4. Had the Agreed Framework not been signed in 1994, the North's plutonium-based program would by today have produced enough plutonium for up to 30 nuclear weapons. Critics of the accord should not ignore this fact.

Can India overtake China?

By Yasheng Huang and Tarun Khanna

What's the fastest route to economic development? Welcome foreign direct investment (FDI), says China, and most policy experts agree. But a comparison with long-time laggard India suggests that FDI is not the only path to prosperity. Indeed, India's homegrown entrepreneurs may give it a long-term advantage over a China hamstrung by inefficient banks and capital markets.

Walk into any Wal-Mart and you won't be surprised to see the shelves sagging with Chinese-made goods—everything from shoes and garments to toys and electronics. But the ubiquitous "Made in China" label obscures an important point: Few of these products are made by indigenous Chinese companies. In fact, you would be hard-pressed to find a single homegrown Chinese firm that operates on a global scale and markets its own products abroad.

That is because China's export-led manufacturing boom is largely a creation of foreign direct investment (FDI), which effectively serves as a substitute for domestic entrepreneurship. During the last 20 years, the Chinese economy has taken off, but few local firms have followed, leaving the country's private sector with no world-class companies to rival the big multinationals.

India has not attracted anywhere near the amount of FDI that China has. In part, this disparity reflects the confidence international investors have in China's prospects and their skepticism about India's commitment to free-market reforms. But the FDI gap is also a tale of two diasporas. China has a large and wealthy diaspora that has long been eager to help the motherland, and its money has been warmly received. By contrast, the Indian diaspora was, at least until recently, resented for its success and much less willing to invest back home. New Delhi took a dim view of Indians who had gone abroad, and of foreign investment generally, and instead provided a more nurturing environment for domestic entrepreneurs.

In the process, India has managed to spawn a number of companies that now compete internationally with the best that Europe and the United States have to offer. Moreover, many of these firms are in the most cutting-edge, knowledge-based industries—software giants Infosys and Wipro and pharmaceutical and biotechnology powerhouses Ranbaxy and Dr. Reddy's Labs, to name just a few. Last year, the *Forbes* 200, an annual ranking of the world's best small companies, included 13 Indian firms but just four from mainland China.

India has also developed much stronger infrastructure to support private enterprise. Its capital markets operate with greater efficiency and transparency than do China's. Its legal system, while not without substantial flaws, is considerably more advanced.

China and India are the world's next major powers. They also offer competing models of development. It has long been an article of faith that China is on the faster track, and the economic data bear this out. The "Hindu rate of growth"—a pejorative phrase referring to India's inability to match its economic growth with its population growth—may be a thing of the past, but when it comes to gross domestic product (GDP) figures and other headline numbers, India is still no match for China.

However, the statistics tell only part of the story—the macroeconomic story. At the micro level, things look quite different. There, India displays every bit as much dynamism as China. Indeed, by relying primarily on organic growth, India is making fuller use of its resources and has chosen a path that may well deliver more sustainable progress than China's FDI-driven approach. "Can India surpass China?" is no longer a silly question, and, if it turns out that India has indeed made the wiser bet, the implications—for China's future growth and for how policy experts think about economic development generally—could be enormous.

THE STIFLING STATE

The fact that India is increasingly building from the ground up while China is still pursuing a top-down approach reflects their contrasting political systems: India is a democracy, and China is not. But the different strategies are also a function of history. China's Communist Party came to power in 1949 intent on eradicating private ownership, which it quickly did. Although the country is now in its third decade of free-market reforms, it continues to struggle with the legacy of that period—witness the controversy surrounding the recent decision to officially allow capitalists to join the Communist Party.

India, on the other hand, developed a softer brand of socialism, Fabian socialism, which aimed not to destroy capitalism but merely to mitigate the social ills it caused. It was considered essential that the public sector occupy the economy's "commanding heights," to use a phrase coined by Russian evolutionary Vladimir Lenin but popularized by India's first prime minister, Jawaharlal Nehru. However, that did not prevent entrepreneurship from flourishing where the long arm of the state could not reach.

Developments at the microeconomic level in China reflect these historical and ideological differences. China has been far bolder with external reforms but has imposed substantial legal and regulatory constraints on indigenous, private firms. In fact, only four years ago, domestic companies were finally granted the same constitutional protections that foreign businesses have enjoyed since the early 1980s. As of the late 1990s, according to the International Finance Corporation, more than two dozen industries, including some of the most important and lucrative sectors of the economy—banking, telecommunications, highways, and railroads—were still off-limits to private local companies.

These restrictions were designed not to keep Chinese entrepreneurs from competing with foreigners but to prevent private domestic businesses from challenging China's state-owned enterprises (SOEs). Some progress has been made in reforming the bloated, inefficient SOEs during the last 20 years, but Beijing is still not willing to relinquish its control over the largest ones, such as China Telecom.

Instead, the government has ferociously protected them from competition. In the 1990s, numerous Chinese entrepreneurs tried, and failed, to circumvent the restrictions placed on their activities. Some registered their firms as nominal SOEs (all the capital came from private sources, and the companies were privately managed), only to find themselves ensnared in title disputes when financially strapped government agencies sought to seize their assets. More than a few promising businesses have been destroyed this way.

This bias against homegrown firms is widely acknowledged. A report issued in 2000 by the Chinese Academy of Social Sciences concluded that, "Because of long-standing prejudices and mistaken beliefs, private and individual enterprises have a lower political status and are discriminated against in numerous policies and regulations. The legal, policy, and market environment is unfair and inconsistent."

Foreign investors have been among the biggest beneficiaries of the constraints placed on local private businesses. One indication of the large payoff they have reaped on the back of China's phenomenal growth: In 1992, the income accruing to foreign investors with equity stakes in Chinese firms was only $5.3 billion; today it totals more than $22 billion. (This money does not necessarily leave the country; it is often reinvested in China.)

THE MOGUL AS HERO

For democratic, postcolonial India, allowing foreign investors huge profits at the expense of indigenous firms is simply unfeasible. Recall, for instance, the controversy that erupted a decade ago when the Enron Corporation made a deal with the state of Maharashtra to build a $2.9 billion power plant there. The project proceeded, but only after several years of acrimonious debate over foreign investment and its role in India's development.

While China has created obstacles for its entrepreneurs, India has been making life easier for local businesses. During the last decade, New Delhi has backed away from micromanaging the economy. True, privatization is proceeding at a glacial pace, but the government has ceded its monopoly over long-distance phone service; some tariffs have been cut; bureaucracy has been trimmed a bit; and a number of industries have been opened to private investment, including investment from abroad.

As a consequence, entrepreneurship and free enterprise are flourishing. A measure of the progress: In a recent survey of leading Asian companies by the *Far Eastern Economic Review (FEER)*, India registered a higher average score than any other country in the region, including China (the survey polled over 2,500 executives and professionals in a dozen countries; respondents were asked to rate companies on a scale of one to seven for overall leadership performance). Indeed, only two Chinese firms had scores high enough to qualify for India's top 10 list. Tellingly, all of the Indian firms were wholly private initiatives, while most of the Chinese companies had significant state involvement.

Some of the leading Indian firms are true start-ups, notably Infosys, which topped *FEER*'s survey. Others are offshoots of old-line companies. Sundaram Motors, for instance, a leading manufacturer of automotive components and a principal supplier to General Motors, is part of the T.V. Sundaram group, a century-old south Indian business group.

Not only is entrepreneurship thriving in India; entrepreneurs there have become folk heroes. Nehru would surely be appalled at the adulation the Indian public now showers on captains of industry. For instance, Narayana Murthy, the 56-year-old founder of Infosys, is often compared to Microsoft's Bill Gates and has become a revered figure.

> With the help of its diaspora, China has won the race to be the world's factory. With the help of its diaspora, India could become the world's technology lab.

These success stories never would have happened if India lacked the infrastructure needed to support Murthy and other would-be moguls. But democracy, a tradition of entrepreneurship, and a decent legal system have given India the underpinnings necessary for free enterprise to flourish. Although India's courts are notoriously inefficient, they at least comprise a functioning independent judiciary. Property rights are not fully secure, but the protection of private ownership is certainly far stronger than in China. The rule of law, a legacy of British rule, generally prevails.

These traditions and institutions have proved an excellent springboard for the emergence and evolution of India's capital markets. Distortions are still commonplace, but the stock and bond markets generally allow firms with solid prospects and reputations to obtain the capital they need to grow. In a World Bank study published last year, only 52 percent of the Indian firms surveyed reported problems obtaining capital, versus 80 percent of the Chinese companies polled. As a result, the Indian firms relied much less on internally generated finances: Only 27 percent of their funding came through operating profits, versus 57 percent for the Chinese firms.

Corporate governance has improved dramatically, thanks in no small part to Murthy, who has made Infosys a paragon of honest accounting and an example for other firms. In a survey of 25 emerging market economies conducted in 2000 by Credit Lyonnais Securities Asia, India ranked sixth in corporate governance, China 19th. The advent of an investor class, coupled with the fact that capital providers, such as development banks, are themselves increasingly subject to market forces, has only bolstered the efficiency and credibility of India's markets. Apart from providing the regulatory framework, the Indian government has taken a back seat to the private sector.

In China, by contrast, bureaucrats remain the gatekeepers, tightly controlling capital allocation and severely restricting the ability of private companies to obtain stock market listings and access the money they need to grow. Indeed, Beijing has used the financial markets mainly as a way of keeping the SOEs afloat. These policies have produced enormous distortions while preventing China's markets from gaining depth and maturity. (It is widely claimed that China's stock markets have a total capitalization in excess of $400 billion, but factoring out non-tradeable shares owned by the government or by government-owned companies reduces the valuation to just around $150 billion.) Compounding the problem are poor corporate governance and the absence of an independent judiciary.

DOLLARS AND DIASPORAS

If India has so clearly surpassed China at the grassroots level, why isn't India's superiority reflected in the numbers? Why is the gap in GDP and other benchmarks still so wide? It is worth recalling that India's economic reforms only began in earnest in 1991, more than a decade after China began liberalizing. In addition to the late start, India has had to make do with a national savings rate half that of China's and 90 percent less FDI. Moreover, India is a sprawling, messy democracy riven by ethnic and religious tensions, and it has also had a longstanding, volatile dispute with Pakistan over Kashmir. China, on the other hand, has enjoyed two decades of relative tranquility; apart from Tiananmen Square, it has been able to focus almost exclusively on economic development.

That India's annual growth rate is only around 20 percent lower than China's is, then, a remarkable achievement. And, of course, whether the data for China are accurate is an open question. The speed with which India is catching up is due to its own efficient deployment of capital and China's inefficiency, symbolized by all the money that has been frittered away on SOEs. And China's misallocation of resources is likely to become a big drag on the economy in the years ahead.

In the early 1990s, when China was registering double-digit growth rates, Beijing invested massively in the state sector. Most of the investments were not commercially viable, leaving the banking sector with a huge number of nonperforming loans—possibly totaling as much as 50 percent of bank assets. At some point, the capitalization costs of these loans will have to be absorbed, either through write-downs (which means depositors bear the cost) or recapitalization of the banks by the government, which diverts money from other, more productive uses. This could well limit China's future growth trajectory.

India's banks may not be models of financial probity, but they have not made mistakes on nearly the same scale. According to a recent study by the management consulting firm Ernst & Young, about 15 percent of banking assets in India were nonperforming as of 2001. India's economy is thus anchored on more solid footing.

The real issue, of course, isn't where China and India are today but where they will be tomorrow. The answer will be determined in large measure by how well both countries utilize their resources, and on this score, India is doing a superior job. Is it pursuing a better road to development than China? We won't know the answer for many years. However, some evidence indicates that India's ground-up approach may indeed be wiser—and the evidence, ironically, comes from within China itself.

Consider the contrasting strategies of Jiangsu and Zhejiang, two coastal provinces that were at similar levels of economic development when China's reforms began. Jiangsu has relied largely on FDI to fuel its growth. Zhejiang, by contrast, has placed heavier emphasis on indigenous entrepreneurs and organic development. During the last two decades, Zhejiang's economy has grown at an annual rate of about 1 percent faster than Jiangsu's. Twenty years ago, Zhejiang was the poorer of the two provinces; now it is unquestionably more prosperous.

India may soon have the best of both worlds: It looks poised to reap significantly more FDI in the coming years than it has attracted to date. After decades of keeping the Indian diaspora at arm's length, New Delhi is now embracing it. In some circles, it used to be jokingly said that NRI, an acronym applied to members of the diaspora, stood for "not required Indians." Now, the term is back to meaning just "nonresident Indian." The change in attitude was officially signaled earlier this year when the government held a conference on the diaspora that a number of prominent NRIs attended.

China's success in attracting FDI is partly a historical accident—it has a wealthy diaspora. During the 1990s, more than half of China's FDI came from overseas Chinese sources. The money appears to have had at least one unintended consequence: The billions of dollars that came from Hong Kong, Macao, and Taiwan may have inadvertently helped Beijing postpone politically difficult internal reforms. For instance, because foreign investors were acquiring assets from loss-making SOEs, the government was able to drag its feet on privatization.

Until now, the Indian diaspora has accounted for less than 10 percent of the foreign money flowing to India. With the welcome mat now laid out, direct investment from nonresident Indians is likely to increase. And while the Indian diaspora may not be able to match the Chinese diaspora as "hard" capital goes, Indians abroad have substantially more intellectual capital to contribute, which could prove even more valuable.

The Indian diaspora has famously distinguished itself in knowledge-based industries, nowhere more so than in Silicon Valley. Now, India's brightening prospects, as well as the changing attitude vis-à-vis those who have gone abroad, are luring many nonresident Indian engineers and scientists home and are enticing many expatriate business people to open their wallets. With the help of its diaspora, China has won the race to be the world's factory. With the help of its diaspora, India could become the world's technology lab.

China and India have pursued radically different development strategies. India is not outperforming China overall, but it is doing better in certain key areas. That success may enable it to catch up with and perhaps even overtake China. Should that prove to be the case, it will not only demonstrate the importance of homegrown entrepreneurship to long-term economic development; it will also show the limits of the FDI-dependent approach China is pursuing.

Yasheng Huang is an associate professor at the Sloan School of Management at the Massachusetts Institute of Technology. Tarun Khanna is a professor at Harvard Business School.

Dangerous Neighbours

Afghanistan's power-hungry neighbours threaten to revive the ruinous civil war of the early 1990s that gave rise to the Taliban

by Ahmed Rashid/KABUL

RUSSIA IS ARMING ONE WARLORD, Iran another. Wealthy Saudis have resumed funding Islamic extremists and some Central Asian Republics are backing their ethnic allies. India and Pakistan are playing out an intense rivalry as they secretly back opposing forces. The playing field is Afghanistan, and the interference threatens to revive a multifaceted power struggle that in the early 1990s eventually gave way to a near-ruinous rule by the Taliban.

The danger is widely recognized. On December 22, under the watchful eye of Afghanistan President Hamid Karzai, the foreign ministers or ambassadors of Afghanistan's closest neighbours—China, Iran, Pakistan, Uzbekistan and Turkmenistan—signed the Kabul Declaration in which they pledged to never again interfere in the affairs of the war-ravaged country. Officials from other interested countries like Russia, India and Saudi Arabia looked on.

But the pledges of support for Karzai and the principle of non-interference, along with promises of aid for the reconstruction of Afghanistan, hid a starker reality. The very fact that such a declaration was needed, despite the all-powerful presence of U.S. forces in Afghanistan, demonstrated the apprehensions of Afghan leaders and the world.

"We are not going to be a political football for neighbours in the region as we were in the 1990s," says Karzai in an interview at the presidential palace prior to the signing of the declaration. "The soil of Afghanistan cannot be used by any country against a third country."

IN DANGER OF SPLINTERING AFGHANISTAN

Karzai's determination to keep his country free of outside disruption won't be easy to realize because many of the same neighbours who sponsored Afghan warlords in the early 1990s, prolonging the country's turmoil and eventually helping to bring the Taliban to power, are keen to resurrect their influence. According to western diplomats, many of the neighbours believe that the U.S. forces will wind down their operations in Afghanistan once a war in Iraq begins. They feel that Karzai's government is weak and Afghanistan will split along ethnic lines if the U.S. leaves.

Karzai and the U.S. strongly refute such conjecture, but many neighbours don't seem convinced. Russia backed the former Northern Alliance during the 1990s and is continuing to support the army of Gen. Mohammed Fahim, once a powerful Northern Alliance leader and now Karzai's defence minister. Last September Russian Defence Minister Sergei Ivanov declared in Kabul that Russia would provide $100 million worth of military equipment to the Afghan army.

It seems clear at this point that the aid will flow to Fahim's army. That force is separate from the one now being trained by the U.S. and France to be the official Afghan National Army—a projected body of 70,000 men of which only 3,000 have been trained so far.

Quietly U.S. officials have asked Moscow to stop the flow of arms, but to no avail. Zamir Kabulov, director for Asia at Russia's foreign ministry, says the aid to Fahim's force is permissible according to old treaty obligations between Russia and Afghanistan. Washington is unwilling to push harder for the moment because it needs Moscow's support for its potential invasion of Iraq. "We have made it clear to all of Afghanistan's neighbours that the country should be allowed to develop without interference," says Robert Finn, the U.S. ambassador to Kabul.

Western intelligence believes some Russian spare parts and even tanks are arriving in the northern city of Kunduz from Tajikistan and being transported down to the Panjshir Valley where Fahim has a large stockpile of weapons. Fahim denies the charge. "The Russians have made no promises and so far we have received no items from Russia," he says.

Afghan ministers also say Russia has run off with the country's only geological survey of oil and gas resources, which was made in the 1970s. "We have asked Russia to return these documents," says Juma Mohammed Mohammedi, the minister of mines. Russian oil companies are reportedly negotiating with

Gen. Rashid Dostum, a warlord in northern Afghanistan, to resume supplies of Afghan gas to Central Asia. The U.S. has said it will carry out a new geological survey.

Even if Karzai can straighten out these issues, his problems have only begun. Afghan officials say Iranian Revolutionary Guards are continuing to provide cash and military support to Ismail Khan, a warlord in Herat in the west. And wealthy Saudis have apparently resumed sending money to remnants of the Taliban based in Pakistan. Not to be outdone, President Islam Karimov of Uzbekistan has provided his own bodyguards to guard the Afghan Uzbek warlord Dostum, an opponent of Fahim.

Meanwhile India and Pakistan are using Afghanistan as a proxy battleground for their ongoing conflict. New Delhi has promised to deliver military vehicles and train Afghan officers. By providing civilian aircraft, buses and hospital equipment, India has quickly developed a huge presence in the country.

India is helping develop a new export route for landlocked Afghanistan through Iran—thereby avoiding Pakistan. It has opened consulates in Mazar-e-Sharif in the north, Herat in the west, and Kandahar and Jalalabad close to the Pakistan border.

"India sees Afghanistan as a means to undermine Pakistan's western border, and Pakistan is retaliating," says a European ambassador in Kabul.

"I have assurances from India that these consulates will only be for trade and consular activities," says Karzai. "We will not allow either India or Pakistan to use Afghanistan to work against each other."

Nevertheless, India's moves have irked Pakistan tremendously. An infuriated Pakistan President Pervaiz Musharraf reportedly made testy phone calls complaining to Karzai after he heard about the new Indian consulates.

Publicly, Pakistan supports Karzai and continues to hand over Al Qaeda operatives to the U.S. But according to Western diplomats and Afghan leaders, Pakistan's Inter-services Intelligence or ISI, is also giving sanctuary to top Taliban leaders, some of whom are living openly in Pakistan, and allows them to cross into Afghanistan to orchestrate rocket attacks against U.S. bases.

Moderate Pashtun leaders in Peshawar say the ISI is fuelling Pashtun radicalism, which is the ideology that led to the creation of the Taliban in the first place. Pakistan wants to retain some influence in southern Afghanistan among ethnic Pashtuns as well as to counter the influence of India in Kabul.

PAKISTAN IS SAID TO BE SUPPORTING THE U.S. ON THE SURFACE BUT ALSO ALLOWING THE TALIBAN TO OPERATE

Retired Pakistani military officers say the army is playing two sides of the Afghanistan conflict in border cities like Peshawar and Quetta. According to these sources, one senior ISI officer and his staff work with the U.S. in Pakistan to catch Al Qaeda elements, while another senior officer works separately to help protect the Taliban.

U.S. generals in Afghanistan say 90% of attacks they face are coming from groups based in Pakistan. "I think the security situation in eastern Afghanistan is going to be a problem for some time to come," said General Richard Myers, chairman of the U.S. Joint Chiefs of Staff, at Bagram in an address to troops on December 21. Myers wants Pakistan to put more troops on the border to stop the infiltrations.

A stronger Afghanistan central authority may be coming in the spring. That's when large-scale international funding is scheduled to arrive and flow to building up the Afghan army and major infrastructure construction projects such as roads and power development. Stepped up economic development would undermine the warlords and their foreign backers and strengthen Karzai.

From *Far Eastern Economic Review*, January 9 2003, pp. 18-19 by Ahmed Rashid. © 2003 by Dow Jones & Company, Inc. Reproduced with permission of Dow Jones & Company, Inc. in the format Textbook via Copyright Clearance Center.

UNIT 9
Middle East and Africa

Unit Selections
32. **"Why Do They Hate Us?"**, Peter Ford
33. **The Reluctant Nation Builders**, Alan Sorensen
34. **The Fall of the House of Saud**, Robert Baer
35. **America and Africa**, Salih Booker, William Minter, and Ann-Louise Colgan

Key Points to Consider

- Why do so many Arabs in the Middle East say that the United States is to blame for the terrorist bombings of September 11, 2001?

- Explain why you believe Bin Ladin's message is so popular with young Arabs around the world?

- Explain why you do or do not believe a U.S.-led military intervention in Iraq was justified.

- Do you agree with the prediction that the epicenter of the War on Terrorism in the future is likely to be several countries in sub-Saharan Africa?

 **Links: www.dushkin.com/online/**
These sites are annotated in the World Wide Web pages.

Africa News Online
  http://www.africanews.org
ArabNet
  http://www.arab.net
Columbia International Affairs Online
  http://www.ciaonet.org/cbr/cbr00/video/cbr_v/cbr_v.html
ei: Electronic Intifada
  http://electronicintifada.net/new.shtml
IslamiCity
  http://islamicity.com
MEMRI: The Middle East Research Institute
  http://www.memri.org/video

President Bush echoed a question on the minds of many Americans when he asked in a speech to Congress after September 11, 2001, why 19 men chose to wreck the icons of U.S. military and economic power? Peter Ford in "Why Do They Hate Us?", uses interviews with a variety of Muslims to explain why most Muslims and Arabs, including those who are sympathetic towards the United States and those who are not, understand the feelings of the hijackers. Part of Bin Laden's success in recruiting young Arabs to his cause is because he speaks in the "vivid" language of popular Islamic preachers, and builds on a deep and widespread resentment against the West and local ruling elites identified with it.

The U.S.-led military campaign in Iraq in 2003 and in Afghanistan against al Qaeda and their Taliban hosts in Afghanistan in 2002 were surprisingly brief and successful military campaigns. However, efforts to secure the peace are proving to be more difficult and costly both in terms of dollars and lost Americans lives. By the spring of 2003 more Americans had been killed since the end of the fighting in Iraq than were killed during the U.S.-led military campaign. In "The Reluctant Nation Builders," Alan Sorensen describes how the Bush administration experiment with nation building in Iraq was not based on an extensive post-war plan or lessons from other recent nation-building exercises. Sorensen concludes that America's makeshift attempt to remake Iraq could prove hard to sustain and "prove to be a costly distraction from the war on terrorism."

Continuing terrorist attacks targeting US military personnel and other Americans who worked for the American authorities in Iraq escalated after hostilities ended in 2003. Terrorists have also targeted hundreds of Iraqi intellectuals, administrators, and newly trained police and military personnel for assassination. By May of 2003 it was clear that there was a widening campaign against Iraq's professional class. The continued violence increased the difficulties involved in reconstruction efforts. Political protests by followers of the Shiite Grand Ayatollah Sistani, Kurds, and other groups in Iraq, in the face of continuing random acts of violence, pressured the Americans to modify their plan to turn over political control to a U.S.-appointed interim Governing Council on July 1, 2003.

Faced with growing political opposition, the Bush administration reversed course and turned to the United Nations to help build support for postponing the election until after a transitional government had been formed. At the beginning of March 2004 Iraqi political leaders announced that they had reached agreement on the broad terms of an interim constitution that would provide protections for individual rights, guaranteed women 25 percent of the seats in the provisional legislature, allowed Kurdish leaders to maintain their militia in northern Iraq, and recognized Islam as the official religion but not necessarily the guiding source for laws. However, the signing ceremony had to be postponed as key Shiite leaders refused to endorse the document. The delay in finalizing the interim constitution illustrates the deepening divisions between Sunni, Shiite, Kurdish, and other communities in Iraq.

According to many accounts, Bin Laden was deeply offended by the sight of thousands of American troops stationed in Saudi Arabia after the first Gulf War. In several early fatwas or announcements issued by al Qaeda, the organization called for the withdrawal of U.S. troops from Saudi Arabia and urged Saudis to overthrow the Saudi monarchy. In recent years there has been an increase in political protests in Saudi Arabia. A series of bombings during November 2003 killed 17 people and ushered in a new wave of attacks by al Qaeda sympathizers. Terrorism is just one threat facing the embattled House of Saud. The country is also struggling with economic stagnation, high rates of unemployment, and a large number of restive and increasingly cynical young Saudis who question the right of the monarchy to rule. Crown Prince Abdullah, the day-to-day ruler of the royal family, has tried to address growing popular discontent with the existing political status quo by accelerating certain political reforms and by providing televised coverage of deliberations of the consultative assembly, whose members he appoints. Robert Baer, a former CIA operative, agrees with others who feel these reforms are too little a little too late. In "The Fall of the House of Saud," Baer explains why today's Saudi Arabia can't last much longer and why the social and economic fallout of its demise could be calamitous.

For many in the Arab world the carnage of September 11th was retribution for fifty years of U.S. policies in the region. Lots of Arabs believe that America's policy towards Israel has been excessively one-sided by supporting the Sharon government financially and diplomatically as the Israeli government expanded its borders at the expense of Palestinian claims. After the violent collapse of the Israeli-Palestinian peace negotiations in 2000, the Bush administration opted not to take a lead role in efforts to promote multinational peacekeeping efforts. Instead, the U.S. remained a staunch supporter of Israel as the country struggles to cope with growing unrest in response to a comprehensive war of attrition in the Occupied Territories. Faced with overwhelming odds, many Palestinians supported the use of extreme measures such as suicide bombers to protest Israeli actions.

The security wall that Israel built in 2003 dramatically underscored the fact that Sharon's government was determined to stop suicide attacks within Israel and achieve a decisive military victory over the Palestinian Authority before entering into another round of negotiations. Some analysts warn that if Israel's current policies continue, a "two-nation" peace settlement may not be possible.

Sweeping generalizations about political and economic trends in sub-Saharan Africa are difficult to make because conditions vary widely among countries and between sectors within the same country. Most countries in sub-Saharan Africa are neither on the verge of widespread anarchy nor at the dawn of democratic and economic renewal. Economic growth in Africa is stronger today than in the disastrous 1980s but most countries were adversely affected by the downturn in many commodity prices during the late 1990s. The spread of HIV/AIDS, the return of drought in several countries, and continued rule by corrupt leaders means that talk of "a new African renaissance" at the end of the 1990s was premature. African states, however, are taking a number of steps that may eventually lead to greater economic integration and prosperity. The year 2004 promises to be a banner test year for democracy as several African countries

hold key elections. Leaders of rival factions in Sudan and Liberia reached peace agreements during 2003 that may eventually end decades of war, while conflicts in Somalia, Burundi, the Democratic Republic of the Congo, and Uganda remain unresolved. Optimists like to point to the South Africa that is in the midst of another presidential election on the 10th anniversary of the end of apartheid.

The Organization for African Union transformed itself to the African Union in 2001 with the backing of such diverse leaders as Col. Kadaffi of Libya and President Mbeki of South Africa. The continental-wide organization seeks to establish a common market and common political institutions along the lines of the European Community. Two region organizations, ECOWAS in East Africa and SADO in southern-central Africa have negotiated far-reaching agreements in recent years that are designed to create viable regional peacekeeping capabilities. Several elder African statesmen, including Nelson Mandela in Burundi, were successful in mediation efforts in inter-state conflicts. Leaders at African Union meetings in 2002 and 2003 also supported the concept of an African peacekeeping force. At the beginning of 2004 African leaders at a summit in Rwanda took the unprecedented step of adopting a unique peer review system to judge the behavior of leaders in fellow African states. While many Americans remain indifferent to the problems of African nation-states, Salih Booker, William Minter, and Ann-Louise Colgan, in "America and Africa," argue that "Africa's issues are global issues—HIV/AIDS, human development, new models for economic growth, peace, and democracy." However, Salih Booker and his colleagues also note that U.S. priorities are being set more today by the war-on-terrorism concerns.

Partly in response to terrorist concerns, the U.S. military is scaling up its presence in Africa to deal with immediate and future threats on the continent. The new focus on Africa is part of a major restructuring as U.S. forces in Europe that are being repositioned for the longer term war on terrorism and because Africa is becoming an increasingly important alternate source for oil and other energy minerals. The United States' European Command, that oversees U.S. military activities in Africa excluding the Horn, has negotiated access to several sites, including airstrips in Angola and Gabon for stopovers, refueling, or to position troops and equipment. The U.S. military is also training and equipping the military in Mali, Niger, Mauritania, and Chad to bolster border security against penetration by terrorists and other smugglers. In the recent past, joint actions by U.S. and West African militaries targeted suspected terrorist training camps in remote areas of the Sahara.

The recent focus of U.S. military actions in Africa reflects a longer term concern about the implications of the fact that Africa, rather than the Middle East, will soon become the continent with the largest numbers of Muslims. This demographic trend helps to explain why so many of Bin Laden's foot soldiers were from Africa. Many analysts now call Africa the 'soft underbelly' of International Relations. For the past several decades, the number of Africans adopting fundamentalist Islamic (and Christian) beliefs has steadily risen. Al-Qaeda and other radical Islamic movements have found many recruits among mostly unemployed youths in urban African centers. After hosting Bin Laden for several years in the 1990s, the radical Islamic government of Sudan asked him to leave in 1996. While the Sudanese cooperated with the U.S. during the military campaign in Afghanistan by sharing intelligence, many citizens in Sudan and other African countries continue to treat Bin Ladin as an icon. A battle for the hearts and minds of millions of Africans appears to be shaping up that promises to focus much more attention on the outcome of remote African conflicts.

At the same time that many African states face challenges from Islamic extremists, many governments must also cope simultaneously with problems created by fragile economies, demands for democratic reforms, ethnic tensions, and a spreading HIV/AIDS epidemic. In parts of east and southern Africa, the HIV/AIDS epidemic is reducing life expectancy, raising mortality, lowering fertility, and leaving millions of orphans in its wake. Even though the HIV/AIDS pandemic has not yet peaked in the region, the devastating AIDS pandemic is fueling poverty by reducing gross domestic product and decimating the governments of many African states because so many young adults are dying from the disease. The result is extreme social dislocation. The experiences of Uganda and Senegal with anti-AIDS programs suggest that it is never too late to act. Good science and sensible public policy can defeat this modern pandemic. What is needed now is more awareness that the pandemic is a global problem that is likely to threaten global stability.

Article 32

PROUD FATHER: In a Pakistan pharmacy, pictured above, Amirul Haq (r.), says he is 'satisfied' that his Muslim son was killed in the Kashmir. He's 'against America, because it doesn't care about those who die in Pakistan.'

'Why do they hate us?'

By Peter Ford
Staff writer

asked President Bush in his speech to Congress last Thursday night. It is a question that has ached in America's heart for the past two weeks. Why did those 19 men choose to wreck the icons of US military and economic power?

Most Arabs and Muslims knew the answer, even before they considered who was responsible. Retired Pakistani Air Commodore Sajad Haider—a friend of the US—understood why. Radical Egyptian-born cleric and US enemy Abu Hamza al-Masri understood. And Jimmy Nur Zamzamy, a devout Muslim and advertising executive in Indonesia, understood.

They all understood that this assault was more precisely targeted than an attack on "civilization." First and foremost, it was an attack on America.

In the United States, military planners are deciding how to exact retribution. To many people in the Middle East and beyond, where US policy has bred widespread anti-Americanism, the carnage of Sept. 11 *was* retribution.

And voices across the Muslim world are warning that if America doesn't wage its war on terrorism in a way that the Muslim world considers just, America risks creating even greater animosity.

Mr. Haider is a hero of Pakistan's 1965 war against India, and a sworn friend of America. But he and his neighbors in one of Islamabad's toniest districts are clear about why their warm feelings toward the US are not widely shared in Pakistan.

In his dim office in a north London mosque, Abu Hamza al-Masri sympathizes with the goals of Osama bin Laden, fingered by US officials as the prime suspect behind the Sept. 11 attacks. Abu Hamza has himself directed terrorist operations abroad, according to the British police, although for lack of evidence, they have never brought him to trial.

Mr. Zamzamy, a 30-something advertising executive in Jakarta, knew what was behind the attack, too. Trying to give his ads some zip and still stay within the bounds of his Muslim faith, he is keenly aware of the tensions between Islam and American-style global capitalism.

The 19 men—who US officials say hijacked four American passenger jets and flew them on suicide missions that left more than 7,000 people dead or missing—were all from the Middle East. Most of the hijackers have been identified as Muslims.

The vast majority of Muslims in the Middle East were as shocked and horrified as any American by what they saw happening on their TV screens. And they are frightened of being lumped together in the popular American imagination with the perpetrators of the attack.

But from Jakarta to Cairo, Muslims and Arabs say that on reflection, they are not surprised by it. And they do not share Mr. Bush's view that the perpetrators did what they did because "they hate our freedoms."

Rather, they say, a mood of resentment toward America and its behavior around the world has become so commonplace in their countries that it was bound to breed hostility, and even hatred.

And the buttons that Mr. bin Laden pushes in his statements and interviews—the injustice done to the Palestinians, the cruelty of continued sanctions against Iraq, the presence of US troops in Saudi Arabia, the repressive and corrupt nature of US-backed Gulf governments—win a good deal of popular sympathy.

The resentment of the US has spread through societies demoralized by their recent history. In few of the world's 50 or so Muslim countries have governments offered their citizens either prosperity or democracy. Arab nations have lost three wars against their arch-foe—and America's closest ally—Israel. A sense of failure and injustice is rising in the throats of millions.

Three weeks ago, a leading Arabic newspaper, Al-Hayat, published a poem on its front page. A long lament about the plight of the Arabs, addressed to a dead Syrian poet, it ended:

"Children are dying, but no one makes a move.
Houses are demolished, but no one makes a move.
Holy places are desecrated, but no one makes a move....
I am fed up with life in the world of mortals.
Find me a hole near you. For a life of dignity is in those holes."

It sounds as if it could have been written by a desperate and hopeless man, driven by frustration to seek death, perhaps martyrdom. A young Palestinian refugee planning a suicide bomb attack, maybe. In fact, it was written by the Saudi Arabian ambassador to London, a member of one of the wealthiest and most influential families in the kingdom that is Washington's closest Arab ally.

Against the background of that humiliated mood, America's unchallenged military, economic, and cultural might be seen as an affront even if its policies in the Middle East were neutral. And nobody voices that view.

From one end of the region to the other, the perception is that Israel can get away with murder—literally—and that Washington will turn a blind eye. Clearly, the US and Israel have compelling reasons for their actions. But little that US diplomats have done in recent years to broker a peace deal between Israel and the Palestinians has persuaded Arabs that the US is a fair-minded and equitable judge of Middle Eastern affairs.

Over the past year, Arab TV stations have broadcast countless pictures of Israeli soldiers shooting at Palestinian youths, Israeli tanks plowing into Palestinian homes, Israeli helicopters rocketing Palestinian streets. And they know that the US sends more than $3 billion a year in military and economic aid to Israel.

"You see this every day, and what do you feel?" asks Rafiq Hariri, the portly prime minister of Lebanon, who is not an excitable man. "It hurts me a lot. But for hundreds of thousands of Arabs and Muslims, it drives them crazy. They feel humiliated."

RESENTMENT RISES, AND A RADICAL IS BORN

Ask Sheikh Abdul Majeed Atta why Palestinians may not like the United States, and he does not immediately answer. Instead, he pads barefoot across the red swirls of his living room carpet and reaches for three framed photographs on the floor beside a couch.

The black-and-white prints show dusty, rock-strewn hills dotted with tiny tents and cinderblock houses: the early days of Duheisheh refugee camp, south of Bethlehem in the West Bank. It was where Mr. Atta was born, and where his family has lived for more than half a century. Atta's family village was destroyed in the struggle between Palestinian Arabs and Jews after Britain divided Palestine between them in 1948. For 10 years his family of

13 lived in a tent. The year Atta was born, the United Nations gave them a one-room house.

It doesn't matter to Atta that the United States was not directly involved in "the catastrophe," as Palestinians refer to the events of 1948. Washington averted its eyes when it could have helped, he says, and since then has been firmly on Israel's side.

Heavyset, solid, with a neatly trimmed full beard, Atta is the preacher at a nearby mosque. He looks the part of the community leader, always meticulously turned out in crisp shirts and pressed trousers, gold-rimmed reading glasses tucked into a pocket.

In the past year of the Palestinian-Israeli conflict, Atta has joined Hamas, the radical group responsible for recently sending most of the suicide bombers into Israeli towns. Frustration at watching the rising Palestinian death toll at the hands of the Israeli army played a large part in his decision, he says.

His resentment at Israel, though, dates back to his infancy, and the stories he heard of his village, Ras Abu Amar, which he never knew. That village is still alive for him, just as millions of Palestinians in the West Bank and Gaza Strip, and throughout the Middle East cherish photos, house keys, and deeds to homes that no longer exist or which have housed Israelis for generations.

Today he lives in his own house in Duheisheh, a sprawling tangle of densely packed concrete buildings that crowd snaking, narrow alleys. But he still dreams of the home he never knew, and recalls who took it from him, and remembers who they rely on for their strength.

What happened on Sept. 11 "was an awful thing, a tragedy, and since we live a continuous tragedy, we felt like this touched us," he adds. "But when we see something like this in Israel or the US, we feel a contradiction. We see it's a tragedy, but we remember that these are the people behind our tragedy."

"Even small children know that Israel is nothing without America," says Atta. "And here America means F-16, M-16, Apache helicopters, the tools Israelis use to kill us and destroy our homes."

SUPERPOWER SWAGGER

Such weapons are very much the visible face of American policy in the Middle East, where military might has held the balance of power for 50 years. Thousands of US soldiers stationed in the Gulf, and billions of US dollars each year in military aid to Israel, Egypt, and other allies, have shored up Washington's interests in the strategically crucial, oil-rich region.

That military presence and power looks like swagger to some in the Muslim world, even far from the flashpoints. "Now America is ready with its airplanes to bomb this poor nation [Afghanistan], and most people in Indonesia don't like arrogance," says Imam Budi Prasodjo, an Indonesian sociologist and talk-show host.

"You are a superpower, you are a military superpower, and you can do whatever you want. People don't like that, and this is dangerous," he adds.

"America should spread its culture, rather than weapons or tanks," adds Mohammed el-Sayed Said, deputy director of Cairo's influential Al Ahram think tank. "They need to act like any respectable commander or leader of an army. They can't just project an image of contempt for those they wish to lead."

Ten years ago, at the head of a broad coalition of Western and Arab countries, the United States used its superpower status to kick the Iraqi army out of Kuwait. Since then, however, Washington has found itself alone—save for loyal ally Britain—in its determination to keep bombing Iraq, and to keep imposing strict economic sanctions that the United Nations says are partly responsible for the deaths of half a million Iraqi children.

'We wish the American people could see what their governments are doing in the rest of the world.'

—Saniya Ghussein, whose daughter, Raafat, was killed in the 1986 US bombing raid against Libya

---

Those deaths, and those bombs (which US and British planes drop regularly, but without fanfare), are felt keenly among fellow Arabs. And Saniya Ghussein knows all about bombs.

A DAUGHTER DIES, AND PARENTS WAIT FOR US APOLOGY

In the middle of the night of April 16, 1986, the deafening sound of anti-aircraft guns woke Saniya Ghussein with a sudden start. "My God," she thought, "there's a war being fought above my house."

She slipped out of bed and ran into the bedroom where her husband Bassem and their 7-year-old daughter Kinda had fallen asleep earlier in the evening. "Bassem, the Americans are here," she said urgently. "It looks like they're going to hit us."

She checked on her other daughter, Raafat. She had been suffering from her annual bout of hay fever, and the 18-year-old art student was in the television room next to the humidifier so she could breathe easier.

Raafat was still sleeping, completely oblivious of all the commotion going on around her, due to the medication she had taken earlier. There was little Saniya felt she could do. She climbed back into bed and pulled the sheets tight around her.

Bassem lay awake on the bed, listening to the appalling noise in the night sky above.

A Palestinian-born Lebanese national, Bassem had worked in Libya as an engineer for Occidental, the American oil giant, for 20 years, helping exploit the country's

'When Bush talked of a crusade...it was not a slip of the tongue.'
—Sajad Haider, retired Pakistani air force officer

massive oil reserves. He and his family lived in the upmarket Ben Ashour neighborhood of Tripoli, the Libyan capital, on the ground floor of a two-story apartment block.

Bassem never heard the explosion. Instead, he watched in astonishment as the window frame suddenly flew into the room, and the roof collapsed on top of him and his daughter.

Kinda was screaming in the darkness near him. Bassem tried to move, but was pinned by the rubble. He groped in the blackness for Kinda. "Don't worry," he said, squeezing his daughter's hand. "Daddy's here, don't cry, it will be okay."

The blast had knocked Saniya unconscious. She woke to hear Bassem calling from the next room and Kinda screaming. She stumbled in the darkness, barefoot across the rubble and glass shards, choking on the fumes from the missile blast, as she called her daughter's name "Raafat! Raafat!" for several minutes. But there was no response, and Saniya knew with a terrible certainty that her daughter was dead.

"Bassem," she cried. "Raafat has gone."

Pinned beneath the rubble, Bassem heard his wife's words, and he felt a deep sense of anger and resentment well up inside him. His life and that of his family had been shattered, and nothing would ever be the same again.

It took them eight hours to dig Raafat out from under the ruins of the house. "Our pain and agony, which I cannot describe, started at that moment," Saniya says.

Raafat was one of an estimated 55 victims of an air raid mounted by US warplanes against a series of targets in Tripoli and another Libyan city, Benghazi.

The attacks were in retaliation for the bombing of a disco in Berlin, Germany, 10 days earlier in which 200 people were injured, 63 of them US soldiers; one soldier and one civilian were killed. The Reagan administration blamed Libyan leader Muammar Qaddafi.

Bassem and Saniya Ghussein are not natural anti-Americans. Bassem studied in the US before going to work for Esso and then Occidental. He sent Raafat to an American Catholic school, and on family trips to the US, Saniya would take Raafat to Disney World in Florida. "We did all the typical American things," she says.

But since that terrible night 16 years ago, neither Bassem nor Saniya have stepped foot in America. They returned to Beirut in 1994 when Bassem retired.

In 1989, the Libyan government enlisted the help of Ramsey Clark, an attorney general during the Carter administration, to file a lawsuit against President Ronald Reagan and British Prime Minister Margaret Thatcher for the civilian deaths during the air raids. "When Clark came to collect our documents and evidence, I asked him if he thought we had a case," Bassem recalls. "He said 'Oh, definitely. This was murder.' "

But US district court judge Thomas Penfield Jackson disagreed. He dismissed the suit, and fined Clark for presenting a "frivolous" case that "offered no hope whatsoever of success."

Twelve years later, the court's decision still rankles with Bassem. "I will only return to America when I know someone will listen to me and say: 'yes, it was our fault your daughter died, and I am sorry.' So long as they think my daughter's death is 'frivolous,' I won't go back," Bassem says.

The Ghusseins have no sympathy for religious extremism and thoroughly condemn the Sept. 11 suicide bombings in New York and Washington. Yet they both maintain that the devastating attack was a result of America's "arrogant" policies in the Middle East and elsewhere. "We wish the American people could see what their governments are doing in the rest of the world," Saniya says.

A FEELING OF BETRAYAL AMONG FRIENDS

On the other side of Asia, in Pakistan, Air Commodore Haider would sympathize with the Ghusseins' wish. He has always been a friend of the United States, and not just because he enjoyed the 10 years he spent in Washington as his country's military attaché. Like most other members of the ruling elite in Pakistan, in the armed forces, in business, and in the political parties, he sees America as a natural ally.

But not a reliable one. The prevailing mood in Pakistan of anger and suspicion toward the United States springs from a deeply rooted perception that the US has been a fickle friend, Haider says, and not just to Pakistan, but to other nations in the Muslim world.

If there was a moment of betrayal for Haider, it was the 1965 war between India and Pakistan, largely over the future of Kashmir. As Indian tanks advanced on the Pakistani metropolis of Lahore, Haider was head of a squadron of F-86 Sabre jets sent to destroy them. India's Soviet allies helped with money, arms, and diplomatic support. But at a crucial moment, Pakistan's ally, the US, refused to send more weapons. As it turned out, Pakistan was able to defeat the Indian attack on Lahore and elsewhere without US help. Haider's squadron decimated the column of Indian tanks that had reached to within six miles of Lahore. But the lesson lingered: America cannot be trusted.

"There is a feeling of being betrayed, it's a feeling of being let down, and you can only be let down by somebody you care for," says Haider, out for an evening stroll in a tony Islamabad neighborhood.

"They said you will be the bulwark of America and of the free world against Communism. But then they dropped a friend for no good reason."

Today, Haider sees a "convergence of interests" between the United States and Pakistan in the fight against terrorism. But he says that President Bush will need to watch his language when he talks about the Muslim world. "When Bush talked of a Crusade… it was not a slip of the tongue. It was a mindset. When they talk of terrorism, the only thing they have in mind is Islam."

Ultimately, Haider does see a way for America and Muslim nations to become lasting friends, but only if the US begins to give as much weight to the interests of Muslim nations as it does to Israel.

"When you deny justice to people, which you have been doing for several decades in Palestine, and they are intelligent, sensitive people, they are going to find something to do," warns Haider. "They might take shelter in Islam, in fatalism, and some will come to despise you."

AN EGYPTIAN 'INSPIRED' TO JOIN AFGHAN FIGHTERS

Sheikh Abu Hamza al-Masri, the radical Muslim cleric who runs a mosque in a shabby district of north London, has certainly come to despise America.

Abu Hamza says he used to admire the West when he was a young man—so much so that he dropped out of university in his native Alexandria, Egypt, to study in Britain. And he clearly had nothing against the British government when he took a job as a civil engineer at Sandhurst, the British equivalent of West Point, after he graduated.

But as he immersed himself more and more in religious studies, and came into contact with more and more Arab mujahideen, who had travelled from the mountains of Afghanistan to England for medical treatment, he began to change his outlook.

"When you see how happy they are, how anxious to just have a new limb so they can run again and fight again, not thinking of retiring, their main ambition is to get killed in the cause of God… you see another dimension in the verses of the Koran," says Abu Hamza.

How the world views a US military response

In your opinion, once the identity of the terrorists is known, should the American government launch a military attack on the country or countries where the terrorists are based, or should the American government seek to extradite the terrorists to stand trial?

	Launch attack	Try the terrorists	Don't know
Israel	77%	19%	4%
India	72	28	0
United States	54	30	16
Korea	38	54	9
France	29	67	4
Czech Republic	22	64	14
Italy	21	71	8
South Africa	18	75	7
United Kingdom (excluding N. Ireland)	18	75	7
Germany	17	77	6
Bosnia	14	80	6
Colombia	11	85	4
Pakistan	9	69	22
Greece	6	88	6
Mexico	2	94	3

Source: Gallup International surveys Sept. 14 to 17.

Inspired by their example, he took his family to Afghanistan in 1990, to work there as a civil engineer, building roads, tunnels, and "anything I could do." And he also fought with the mujahideen against Afghan President Mohammad Najibullah (seen as a Russian stand-in supported by the Soviets), until he blew both his hands off and lost the sight in his left eye, in a mine explosion.

What transformed him and his comrades-in-arms from anti-Soviet to anti-American militants, he says, was the way Washington abandoned them at the end of the war in Afghanistan, and sought to disarm and disperse them.

"It was when the Americans took the knife out of the Russians and stabbed it in our back, it's as simple as that," says Abu Hamza. "It was a natural turn, not a theoretical one.

"In the meantime, they were bombarding Iraq and occupying the [Arabian] peninsula," he says, referring to the US troops stationed in Saudi Arabia after the Gulf War, "and then with the witch-hunt against the mujahideen, all of it came together, that was a full-scale war, it was very clear."

Abu Hamza would rather see Islamic militants fight corrupt or secular Arab governments before they take on America (indeed, the Yemeni government has sought his extradition from Britain for plotting to overthrow the government in Sana). But he is in no doubt that the American government brought the events of Sept. 11 on its own head.

"The Americans wanted to fight the Russians with Muslim blood, and they could only justify that by triggering the word 'jihad,' " he argues. "Unfortunately for everybody except the Muslims, when that button is pushed, it does not come back that easy. It only keeps going on and on until the Muslim empire swallows every empire existing."

Can he understand the motivation behind the assault on New York and Washington? "The motivation is everywhere," he says, with the current US administration. "When a president stands up before the planet and says an American comes first, he is only preaching hatred. When a president stands up and says we don't honor our missile treaty with the Russians, he is only preaching arrogance. When he refuses to condemn what's happening in Palestine, he is only preaching tyranny.

"American foreign policy has invited everybody, actually, to try to humiliate America, and to give it a bloody nose," he adds.

IN JAKARTA, COUNTERING AMERICAN CULTURE WITHOUT VIOLENCE

You wouldn't catch Rizky "Jimmy" Nur Zamzamy justifying violence that way, though he professes just as deep an attachment to Islam as Abu Hamza.

Mr. Zamzamy, a rangy young Indonesian advertising executive in a pink shirt, is sitting in a Western-style cafe in Jakarta, his cellphone at the ready, and his fried chicken growing cold as he explains how he tries to be a good Muslim by right action, not fighting.

That, he feels, is the best way of countering what he sees as the corrupting influence of American culture and morals on traditional Indonesian ways of life in the largest Muslim country in the world.

Until a few years ago, Zamzamy led a regular secular life, hanging out in bars and dating women. Then he met a Muslim teacher who became his spiritual guide. Now he follows Islamic teachings and donates most of his $1,300 monthly salary to his "guru" to be spent on building mosques and helping the poor.

He says he has made sure that none of the money goes to extremist groups that use violence in the name of Islam, such as the Laskar Jihad group, locked in bloody battle with Christians in the Maluku region of Indonesia.

Two years ago, in line with his growing religious beliefs, he quit the advertising agency he had worked for and set up his own company along Islamic lines: He won't take banks or alcoholic-beverage producers as clients, for example, and he does no business on Friday, the Muslim holy day.

But he is relaxed about those who don't share his beliefs: He does not insist that his wife wear a headscarf, for example, and he is not uncomfortable sitting alongside the rich young Jakartans in the cafe who are flirting and drinking. They must make their own choices, he says.

And though he does not like the sexual overtones of American pop culture, he knows that "you can't hide from American culture." By living his life according to Islamic precepts, he says, "I am fighting America in my own way. But I don't agree with violence."

AMBIVALENCE ABOUT AMERICA

All over the Muslim world, young people like Zamzamy are juggling their sense of Islamic identity with the trappings of a globalized, secular society.

In a classroom of Al Khair University, set in a concrete office park in Islamabad, Nabil Ahmed, a business student, and his classmates are fuming over their president's betrayal of the Pakistani people by pledging to support what they fear will turn into a crusade against Muslims.

Ahmed and his friends are well-dressed, middle-class boys, and represent neither the old-money security of Pakistan's elite nor dirt-poor peasants who make up the bulk of Pakistan's angry conservative masses. They are the silent majority of Pakistan, with their feet firmly planted in both the East and the West. On weekdays, they listen to Whitney Houston and Michael Bolton, wear Dockers and Van Heusen shirts. On weekends, many switch to traditional salwar kameez outfits and go with their fathers to the mosque to pray.

'It is [the] double standard that creates hatred.'

—Nabil Ahmed, a business student at Al Khair University in Islamabad

---

They have much to gain from a Western style of life, and most have plans to move to the United States for a few years to make some money before returning home to Pakistan. Yet despite their attraction to the West, they are wary of it too.

"Most of us here like it both ways, we like American fashion, American music, American movies, but in the end, we are Muslims," says Ahmed. "The Holy Prophet said that all Muslims are like one body, and if one part of the body gets injured, then all parts feel that pain. If one Muslim is injured by non-Muslims in Afghanistan, it is the duty of all Muslims of the world to help him."

Like his friends, Ahmed feels that America has double standards toward its friends and enemies. America attacks Iraq if it invades Kuwait, but allows Israel to bulldoze Palestinian homes in the West Bank and Gaza Strip.

ROBERT HARBISON—STAFF

AFGHAN REFUGEES: These boys are among some 60,000 displaced Afghans at Jalozai refugee camp near Peshawar, Pakistan, along its border with Afghanistan. The camp is crowded, and Pakistan has recently forbidden the UNHCR to register any more refugees.

It ostracizes a Muslim nation like Sudan for oppressing its Christian minority, but allows Russia to bomb its Muslim minority into submission in Chechnya.

And while the US supported many "freedom fighter" movements in the past few decades, including the contra movement in Nicaragua, America labels Pakistan and Afghanistan as terrorist states because they support militant Muslim groups fighting in the Indian state of Kashmir and elsewhere.

'The Americans wanted to fight the Russians with Muslim blood.'

—Sheikh Abu Hamza al-Masri, a radical Muslim cleric who runs a mosque in London

---

"There is only one way for America to be a friend of Islam," says Ahmed. "And that is if they consider our lives to be as precious as their own. "If Americans are concerned about the 6,500 deaths in the World Trade Center, let them talk also about the deaths in Kashmir, in Palestine, in Chechnya, in Bosnia. It is this double standard that creates hatred."

Ahmed's ambivalence about America—his desire to live and work there, his admiration for its values, but his anger at its behavior around the world—is broadly shared across the Muslim world and Arab world.

"I think they hate us because of what we do, and it seems to contradict who we say we are," says Bruce Lawrence, a professor of religion at Duke University, referring to people in the Middle East. "The major issue is that our policy seems to contradict our own basic values."

That seems clear enough to Muslims who sympathize with the Palestinians, and who say that Washington should force Israel to abide by United Nations resolutions to withdraw from the occupied territories. "The Americans say September 11th was an attack on civilization," says Mr Hariri, the Lebanese prime minister. "But what does civilized society mean if not a society that lives according to the law?"

It also seems clear to citizens of monarchical states in the Gulf, where elections are unknown and women's rights severely restricted. "Since the Cold War ended, America has talked about promoting democracy," says John Esposito, head of the Center for Muslim-Christian Understanding at Georgetown University in Washington. "But we don't do anything about it in repressive regimes in the Middle East, so you can understand widespread anti-Americanism there."

At the same time, the state-run media—which is all the media there is across much of the Middle East—often fan the flames of anti-American and anti-Israel sentiment because that helps focus citizens' minds on something other than their own government's shortcomings.

In Sana, the Yemeni capital, where queues of visa-seekers line up daily outside the US embassy, the ambivalence about America is clear. "When you go there, you really

50 YEARS OF US POLICY IN THE MIDDLE EAST

1947–48
UN votes to partition Palestine into two states—one for Jews, one for Palestinian Arabs. Arab states invade; 300,000 Palestinians flee Jewish-controlled areas. Jewish forces prevail, declaring Israeli independence. US recognizes Israel.

1953
CIA helps Iran's military stage a coup, deposing elected PM Mohammad Mossadeq, whom US sees as communist threat. US oversees installation of Shah Mohammad Reza Pavlavi as ruler of Iran.

1956
Israel attacks Egypt for control of Suez Canal. Britain and France veto US-sponsored UN resolution calling for halt to military action. British forces attack Egypt.

1960
Iran, Iraq, Kuwait, Saudi Arabia, and Venezuela form Organization of Petroleum Exporting Nations (OPEC).

1966
US sells its firs jet bombers to Israel, breaking with a 1956 decision not to sell arms to the Jewish state.

1967
Six-Day War. Israel launches preemptive strike against Arab neighbors, capturing Jerusalem, the Sinai Peninsula, the Gaza Strip, and the Golan Heights. Kuwait and Iraq cut oil supplies to US, UN adopts Resolution 242, calling on Israel to withdraw from captured territory. Israel refuses.

1968
First major hijacking by Arab militants occurs on El Al flight from Rome to Tel Aviv, marking decades of hostage-takings, hijackings, and assassinations as a strategy by Arab militant groups.

1969
Mummar Qaddafi comes to power in Libyan coup and orders US Air Force to evacuate Tripoli.

1972
Eight Arab commandos of Palestinian group Black September kill 11 Israeli athletes at the Munich Olympic Games.

1973
Egypt and Syria attack Israel over its occupation of the Golan Heights and the Sinai Peninsula. US gives $2.2 billion in emergency aid to Israel, turning tide of battle to Israel's favor. Arab states cut US oil shipments.

1974
UN General Assembly recognizes right of Palestinians to independence.

1976
The UN votes on a resolution accusing Israel of war crimes in occupied Arab territories. US casts lone "no" vote. US Ambassador to Lebanon Francis Meloy and an adviser are shot to death in Beirut. US closes Embassy there.

1978
Egypt and Israel sign US-brokered Camp David peace treaty. Eighteen Arab countries impose an economic boycott on Egypt. Egyptian president Anwar Sadat and Israeli Prime Minister Menachem Begin receive Nobel Peace Prize.

1979
Ayatollah Ruhollah Khomeini leads grass-roots Islamic revolution in Iran, deriding the US as "the great Satan." Iranian students storm US Embassy in Tehran, taking 66 Americans hostage for next 15 months. US imposes sanctions. Protesters attack US Embassies in Libya and Pakistan.

1981
Israel bombs Iraqi nuclear reactor. Muslim militants opposed to Egypt's peace treaty with Israel assassinate Egyptian President Sadat.

1982
Israel invades Lebanon to expel the Palestine Liberation Organization, facilitate election of friendly government, and form 25-mile security zone along Israel's border. Defense Minister Ariel Sharon permits Lebanese Christian militiamen to enter the Sabra and Shatila refugee camps outside Beirut. The ensuing three-day massacre kills 600 or more civilian refugees. US and other nations deploy peacekeeping troops in Lebanon.

1983
A truck bomb explodes in US Marines' barracks in Beirut, Lebanon, killing 241 soldiers. US forces withdraw.

1986
Us bombs Libya in retaliation for the bombing of a Berlin nightclub frequented by US servicemen. The airstrike kills 15 people, including the infant daughter of leader Muammar Qaddafi. All Arab nations condemn the attack.

1987
Start of the Palestinian intifada, or uprising, in the West Bank and Gaza Strip.

1990
Iraq invades Kuwait. Saddam Hussein links pullout to Israel's withdrawal from occupied territories. UN imposes sanctions that continue to hobble Iraq's economy in effort to force Iraqi compliance with weapons resolutions.

(continued)

50 YEARS (continued)

1991

US and coalition launch attacks against Iraq from Saudi Arabia. Gulf War ends after some three months, but US deployment continues even now, with 17,000 to 24,000 US troops in region at any time.

1993

World Trade Center in New York is bombed, killing six. US Special Forces, deployed as peacekeepers in Somalia, attempt to capture warlord Mohamed Farah Aidid. Eighteen US servicemen are killed. Israeli PM Yitzhak Rabin and Palestinian leader Yasser Arafat sign historic peace declaration in White House ceremony with President Clinton.

1994

Jordan and Israel sign peace treaty. Yasser Arafat, Yitzhak Rabin, and Foreign Minister Shimon Peres receive Nobel Peace Prize for 1993 agreement.

1995

US announces trade ban against Iran, reinforcing sanctions in effect since 1979. Rabin is assassinated, two years after peace deal with Palestinians. In Riyadh, Saudi Arabia, a car bomb explodes outside an office housing US military personnel. Seven are killed, including five Americans. Three Islamist groups claim responsibility.

1996

A truck bomb explodes outside a US military barracks in Khobar, Saudi Arabia, killing 19 US airmen. UN reports that sanctions cause 4,500 Iraqi children under 5 to die each month.

1997

Egyptian Islamic Group massacres 62 people, mostly foreign tourists, in Luxor, Egypt. The group claims it is retaliation for US imprisonment of Sheikh Omar Abdel al-Rahman, who is later convicted in 1993 World Trade Center bombing.

1998

Bombs explode outside US Embassies in Kenya and Tanzania, killing 224 people. US launches cruise-missile attacks on sites in Sudan and Afghanistan allegedly linked to Osama bin Laden. US indicts bin Laden for committing acts of terrorism against Americans abroad.

1999

Islamic militants, traced to bin Laden, are arrested for plot to bomb tourist sites during millennium celebrations.

2000

Camp David negotiations fail. Sharon visits Temple Mount in Jerusalem, sparking current Palestinian uprising. USS Cole bombing in Yemen's Aden harbor kills 17 American sailors. Bin Laden denies responsibility, but applauds the act.

2001

Hijackers crash two planes into World Trade Center in New York, one into Pentagon, and one in Pennsylvania. More than 7,000 people are dead or missing.

Compiled by Julie Finnin Day

SOURCES: "THE MIDDLE EAST" (CONGRESSIONAL QUARTERLY), NEWS REPORTS.

---

love the United States," says Murad al-Murayri, a US-trained physicist. "You are treated like a human being, much better than in your own country. But when you go back home, you find the US applies justice and fairness to its own people, but not abroad. In this era of globalization, that cannot stand."

Nor has the mood that has gripped Washington over the past two weeks done much to reassure skeptics, says François Burgat, a French social scientist in Yemen.

"When Bush says 'crusade', or that he wants bin Laden 'dead or alive', that is a *fatwa* (religious edict) without any judicial review," he cautions. "It denies all the principles that America is supposed to be."

A *fatwa* is something Amirul Haq, a Pakistani shopkeeper whose son died two years ago in a jihad in Kashmir, understands better than judicial review. "When I heard that my son died, I was satisfied," he says.

It's a sentiment shared by Azad Khan, too. On a hot Sunday afternoon in Mardan, Pakistan, Mr. Khan and his family have laid out a feast in a small guesthouse next to the local mosque. They are celebrating because they have just heard that Mr. Khan's 20-year-old son, Saeed, has been killed in a gun battle with Indian troops in the part of Jammu and Kashmir state that is under Indian control. With his death, Saeed has become another *shahid*, a martyr and heroic defender of the Muslims against the enemies of Islam. According to the Koran, *shahideen* are not actually dead; they are still alive, they just can't be seen. And through acts of bravery, a *shahid* guarantees that his whole family will go to heaven.

"It is not a thing to be mourned. We are happy," says Khan, sitting down to a meal of chicken and mutton, rice and bread, along with leaders of the group with which Saeed had fought. "I told him to take part in jihad [holy war] because he is the son of a Muslim," Khan says. "And just as we fight in Kashmir, if we need to fight against the United States in Afghanistan we are ready, because we are Muslims. It is our duty to fight against any infidels who are threatening our Muslim brothers."

It's not likely that many Pakistanis, or other Muslims, will actually go to Afghanistan to fight the Americans—assuming American soldiers land there. Khan's militant views are not shared by most of his countrymen.

But in a broader sense, and in the longer term, many people in the Middle East fear that the coming war against terrorism—unless it is waged with the utmost caution—could unleash new waves of anti-American sentiment.

Jamal al-Adimi, a US-educated Yemeni lawyer, speaks for many when he warns that "if violence escalates, you bring seeds and water for terrorism. You kill someone's brother or mother, and you will just get more crazy people."

Trying to root out terrorism without re-plowing the soil in which it grows—which means rethinking the policies that breed anti-American sentiment—is unlikely to succeed, say ordinary Middle Easterners and some of their leaders.

On the practical level, Hariri points out, "launching a war is in the hands of the Americans, but winning it needs everybody. And that means everybody should see that he has an interest in joining the coalition" that Washington is building.

On a higher level, argues Bassam Tibi, a professor of international relations at Gottingen University in Germany, and an expert on political Islam, "we need value consensus between the West and Islam on democracy and human rights to combat Islamic fundamentalism. We can't do it with bombs and shooting—that will only exacerbate the problem."

Reported by staff writers Scott Baldauf in Islamabad, Pakistan; Cameron W. Barr in Amman, Jordan; Peter Ford in London; Nicole Gaouette in Jerusalem; Robert Marquand in Beijing; Scott Peterson in Sana, Yemen; Ilene R. Prusher in Tokyo; as well as contributors Nicholas Blanford in Beirut, Lebanon; Sarah Gauch in Cairo; and Simon Montlake in Jakarta, Indonesia.

This article first appeared in *The Christian Science Monitor,* September 27, 2001, and is reproduced with permission. © 2001 by The Christian Science Monitor (www.csmonitor.com). All rights reserved.

The Reluctant Nation Builders

ALAN SORENSEN

George W. Bush now calls Iraq "the central front" in America's war on terror. Absent the discovery there of either weapons of mass destruction or ties between Saddam Hussein and Al Qaeda, what remains of this front—besides daily assaults on US troops and Iraqis who help them—is the project of remaking the country into a democratically governed, peaceable model for the region.

The mantle of nation builder sits uncomfortably on this president's shoulders. It is one of those compelling ironies of history: that the war on terror should spawn among foreign-policy hardliners a newfound interest in state building and alleviating the "root causes" of conflict. Entering office full of disdain for such activities, Bush found himself after 9-11 overseeing a national security strategy that worries about failed states as potential terrorist nests. This fall found him asking Congress for $87 billion in emergency spending, essentially for nation building in Iraq and Afghanistan.

And yet, as a portent for renewed forays into nation building, after several such efforts failed or were discredited during the 1990s, America's latest conversion to the cause remains problematic for two reasons.

First, reluctant preparation has reduced the odds of success. The Bush administration planned with great care and enthusiasm for the removal of regimes in Afghanistan and Iraq, less so for their replacement. And no wonder: the latter amounts to "a long, hard slog," as Defense Secretary Donald Rumsfeld put it recently. Postwar Germany and Japan notwithstanding, the history of nation building warns against confidence. The endeavor involves enormous complexity, expense, and danger. The aim of establishing not a friendly autocrat but true democracy compounds the difficulty, as does the presence of rivalrous factions, guerrilla resisters, and suicidal bombers. Tortuously defined as part of the war on terror, America's makeshift attempt to remake Iraq could prove hard to sustain, much less replicate, especially if it continues to provoke more terrorism than it puts down.

Second, the Bush administration's leadership in the promotion of nation building, however energetic and unanticipated, has been misdirected. The largely unilateral, dubiously rationalized, and defiantly prosecuted occupation of Iraq has distracted from the need to develop international consensus and capacity for nation building and other benevolent interventions. What the world needs now, besides a better set of rules guiding intrusion into countries, is possibly a new multilateral organization or two specialized in peacekeeping and state-building operations. The sole superpower would benefit from these developments as much as anyone, but they will not happen without the US leadership's rededication to global governance.

America's latest experiment in nation building may yet succeed, despite the bloodshed dominating news from Iraq. Security may improve. A majority of Iraqis share with their occupiers an interest in economic reconstruction and the orderly transfer of sovereignty to a government that respects individual rights and the rule of law. Democracy may take time as a salve against terrorism, but it seems likely to prove the ultimate destination for all societies, even those accustomed to tyranny and torn by tribalism.

The stakes in Iraq, moreover, have perhaps grown high enough to render failure unacceptable. The outcome will affect prospects not just for Iraqis but for their region and America's influence in the world. The importance of the task, however, only underscores the need to engage the best efforts of policy makers and the international community. They should heed the lessons of past state-building efforts. They should appreciate the hazards that face future efforts lacking globally institutionalized legitimacy and resources.

THE LONG, HARD SLOG

Nation building is nothing new. As Marina Ottaway of the Carnegie Endowment for International Peace has noted, colonial nation-building bequeathed to the world most of today's failed and quasi-states, such as Somalia and Afghanistan. Countries formed by the internal mobilization of people and resources have tended to fare better than those created by outsiders' decisions. Yet, in the past century, the US occupation of postwar Germany and Japan set a standard for successful reconstruction that remains unmatched. The results prove that democracy can be transferred to defeated nations.

During the cold war, the United States made few attempts to replicate these successes. It generally applied military power for containment and deterrence: to preserve rather than alter the global status quo, to manage crises rather than address their causes. It did initiate regime changes, but not as part of nation-building exercises. The environment changed, however, with the Soviet Union's breakup. As a recent RAND report, *America's Role in Nation-Building: From Germany to Iraq*, points out, the end of the cold war occasioned the collapse of numerous states. It created opportunities to intervene without fear of superpower conflict. And it reduced constraints on United Nations action. Since 1945, the UN has conducted 55 peace operations. Fortyone of them started after 1989.

America in combination with others has undertaken six major nation-building projects since the end of the cold war—in Somalia, Haiti, Bosnia, Kosovo, Afghanistan, and now Iraq—but it has done so only reluctantly, and with mixed results. The United States quit Somalia at the first sign of trouble in 1993. It restored a deposed president in Haiti in 1995, but left without installing self-sustaining democratic or economic reforms. It resisted for four years European entreaties to help end the ethnic repression and bloodshed in the Balkans, before finally aiding efforts to force a political settlement and initiate nation building in Bosnia and Kosovo. After routing the Taliban and Al Qaeda in Afghanistan, it installed a new government in Kabul, but abandoned the rest of the country to warlords and drug traffickers.

In preparing for Iraq, the United States might have consulted studies analyzing past UN and US nation-building efforts—all of which point to a lengthy, difficult task. As the RAND report found: "The most important determinant [of success] seems to be the level of effort—measured in time, manpower, and money." And: "To date, no effort at enforced democratization has taken hold in less than five years."

Experience yields lessons relevant to the early stages of nation building: Start planning long before reconstruction begins. Commit and retain forces sufficient to win the peace as well as the war. Amass military police ready to deploy immediately after conflict ceases so as to avoid chaos. Quickly install a highly visible civilian occupation leader with undisputed clout. Take care not to announce an exit strategy. Avoid prematurely declaring the end of combat.

Every one of these lessons US officials ignored. Unlike postwar Germany or Japan, Iraq does not constitute an ethnically homogeneous state with strong national identity, parliamentary traditions, or experience with the rule of law. So American planners had to know it would prove no "cakewalk." They might have been expected to take great pains to prepare carefully for the challenge. But this is what happened in Iraq:

The US military failed to deploy enough force to establish security, permitting looting and lawlessness to continue unchecked. It initially appointed (then dismissed) a low-key, low-profile coordinator to oversee reconstruction. It grossly underestimated the costs of restoring services and rebuilding infrastructure. It attempted to promote an émigré political figure with little experience in his native country. It failed to secure critical facilities, including arms caches, many of them still unguarded. It diverted significant resources and manpower to a failed attempt to find weapons of mass destruction. It consigned the Iraqi army to resentful unemployment. It emptied the government of knowledgeable technocrats. It invited Iraq's former imperial masters from Turkey to join the occupation. It favored select American businesses in the distribution of no-bid contracts. It failed miserably to engage in effective public diplomacy. It ignored a pre-invasion State Department report that had laid out with startling precision many of the challenges now bedeviling authorities.

> *Tortuously defined as part of the war on terror, America's makeshift attempt to remake Iraq could prove hard to sustain, much less replicate.*

None of these failures precludes ultimate success. Yet the question remains: Why did they happen? The likeliest answer seems a combination of wishful thinking, willful ignorance, and general ambivalence about the enterprise. It was a telling note that the Army announced it would close its tiny peacekeeping institute, the only school of its kind in the US military, before changing its mind later this year. In many ways, the military's reluctance to embrace nation building is understandable. As National Security Adviser Condoleezza Rice once remarked, the 82nd Airborne has more important work than "escorting kids to kindergarten." Soldiers riding in armored vehicles, wary of the next roadside bomb, make less than ideal nation builders.

The answer, however, is not to ignore postconflict challenges—shrinking from all intervention, ousting regimes without consideration for their replacement, or performing only halfhearted reconstruction planning. Part of the answer is to invest more resources in the needed capabilities and place them under civilian management. The other part is to begin developing more robust multilateral means for sharing the burdens and shoring up the effectiveness of nation building wherever it occurs.

THE MULTILATERAL IMPERATIVE

President Bush's 2002 *National Security Strategy* describes the dangers posed by the intersection of failed states, terrorism, weapons proliferation, and political turmoil. Conflict today more likely will occur within countries than between them. Sudden threats can emerge out of nowhere, as state weakness rather than strength incubates and spreads all manner of afflictions. Such an environment may have outgrown cold-war institutions and policies designed to deter different kinds of dangers.

From this analysis, the administration has inferred the need to emphasize preemptive and unilateral responses. Yet the same analysis could argue for collective efforts to build new international consensus and institutions around strategies to address emerging threats, as happened after World War II. In this light, the nation-building project in Iraq might be weighed not only on its own terms, but also as an outcome of the preemptive doctrine and, as such, a lost opportunity to strengthen and update international law.

The need for state-building efforts around the globe seems unlikely to diminish. Not only in Africa but in places like Georgia, Colombia, Burma, the Balkans, and across much of southwestern Asia, turbulence continues to spread. Greedy dictators divide their spoils with criminal networks. War lords and drug traffickers compete over lawless territory. As the contagion of instability creeps across borders, weapons proliferate and terrorists organize.

Many states could use help bolstering their institutions—legal codes and court systems, police forces and administrative bureaucracies, central banks and transparent financial procedures—while decentralizing political power and fostering responsive, accountable government. These are the bulwarks against state collapse. Yet the international community lacks the capacity to move quickly and effectively to help states develop or restore them. As Chester Crocker, a former American diplomat, puts it, "The concept of military readiness is well understood, but readiness for what happens after the fighting stops is just as important."

Who are you going to call? Not the United States. The prospect of America, embroiled in Iraq and stretched thin everywhere else, taking on more nation building anytime soon seems decidedly slim—unless, perhaps, a state can claim a credible terrorist threat. Meanwhile, in the absence of settled legal and moral criteria or authority for humanitarian interventions, including nation building, the task is left to ad hoc coalitions of interested and capable parties. Issues of legitimacy and accountability arise. The United Nations struggles to cope. And the possibility looms that more countries will become, as Afghanistan did, hothouses for radicalism and terror.

A superpower as intent on leadership as independence would strive to build the political will, technical expertise, and military and economic capacity of the United Nations, regional security organizations, and other groups to assist states in developing their governments, economies, and civil society. It would help organize interventions and trusteeships where necessary, and support courts that prosecute humanrights abusers. It would recognize the congruence between the war on terror and transnational lawenforcement strategies that recover looted state assets, punish corrupt officials, and combat the predators who cause state failure and thrive off it.

If the United Nations proves too unwieldy an instrument for consistently applied nation building—its record so far is mixed but better than most believe—America could support efforts to found a new multilateral institution, perhaps along the lines of a World Bank or International Monetary Fund. Credibly international but aligned with US values and leadership, able to avoid the frustrations that attend national patronage in UN agencies and diplomatic quirks in the Security Council's structure, such an organization could encompass both a reconstruction fund financed by rich nations and a standing all-volunteer military force ready and experienced for deployment. The alternative, as Michael Ignatieff has suggested, is empire: "a muddled, lurching America policing an ever more resistant world alone, with former allies sabotaging it at every turn."

DEFINING SUCCESS

The Iraq experiment's impact on terrorism will take time to sort out. In the short term, occupation of an Arab country may produce more fury than friends. Terrorists retain a global capacity to conduct horrifying attacks. In Iraq, postcombat deaths suffered by American forces already exceed the combined postconflict toll incurred during US occupations of Germany, Japan, Haiti, Bosnia, Kosovo, and Afghanistan. If Iraq descends into chaos or comes to be regarded as a costly distraction from the war on terror instead of its central front, American resolve could falter.

An Iraq success story in the long term almost certainly would promote positive change in the region. The Bush administration in any event deserves credit for questioning traditional approaches, according to which Arabs might assume that the United States stands for democracy everywhere except the

Middle East. People of good will the world over will hope the nation building succeeds in Iraq. But America's policy makers should know that the example will not be often or easily replicated, and that even success will not vindicate the lost chance to promote an international framework. They should understand that a foreign policy inspiring confidence and support would not mistake regime change for democratization or unilateralism for leadership.

ALAN SORENSEN is the associate editor of Current History.

Article 34

The Fall of the House of Saud

Americans have long considered Saudi Arabia the one constant in the Arab Middle East—a source of cheap oil, political stability, and lucrative business relationships. But the country is run by an increasingly dysfunctional royal family that has been funding militant Islamic movements abroad in an attempt to protect itself from them at home. A former CIA operative argues, in an article drawn from his new book, Sleeping With the Devil, *that today's Saudi Arabia can't last much longer—and the social and economic fallout of its demise could be calamitous.*

Robert Baer

In the decades after World War II the United States and the rest of the industrialized world developed a deep and irrevocable dependence on oil from Saudi Arabia, the world's largest and most important producer. But by the mid-1980s—with the Iran-Iraq war raging, and the OPEC oil embargo a recent and traumatic memory—the supply, which had until that embargo been taken for granted, suddenly seemed at risk. Disaster planners in and out of government began to ask uncomfortable questions. What points of the Saudi oil infrastructure were most vulnerable to terrorist attack? And by what means? What sorts of disruption to the flow of oil, short-term and long-term, could be expected? These were critical concerns. Underlying them all was the fear that a major attack on the Saudi system could cause the global economy to collapse.

The Saudi system seemed—and still seems—frighteningly vulnerable to attack. Although Saudi Arabia has more than eighty active oil and natural-gas fields, and more than a thousand working wells, half its proven oil reserves are contained in only eight fields—including Ghawar, the world's largest onshore oil field, and Safaniya, the world's largest offshore oil field. Various confidential scenarios have suggested that if terrorists were simultaneously to hit only a few sensitive points "downstream" in the oil system from these eight fields—points that control more than 10,000 miles of pipe, both onshore and offshore, in which oil moves from wells to refineries and from refineries to ports, within the kingdom and without—they could effectively put the Saudis out of the oil business for about two years. And it just would not be that hard to do.

The most vulnerable point and the most spectacular target in the Saudi oil system is the Abqaiq complex—the world's largest oil-processing facility, which sits about twenty-four miles inland from the northern end of the Gulf of Bahrain. All petroleum originating in the south is pumped to Abqaiq for processing. For the first two months after a moderate to severe attack on Abqaiq, production there would slow from an average of 6.8 million barrels a day to one million barrels, a loss equivalent to one third of America's daily consumption of crude oil. For seven months following the attack, daily production would remain as much as four million barrels below normal—a reduction roughly equal to what *all* of the OPEC partners were able to effect during their 1973 embargo.

Oil is pumped from Abqaiq to loading terminals at Ras Tanura and Ju'aymah, both on Saudi Arabia's east coast. Ras Tanura moves only slightly more oil than Ju'aymah does (4.5 million barrels per day as opposed to 4.3 million barrels), but it offers a greater variety of targets and more avenues of attack. Nearly all of Ras Tanura's export oil is handled by an offshore facility known as The Sea Island, and the facility's Platform No. 4 handles half of that. A commando attack on Platform 4 by surface boat or even by a Kilo-class submarine—available in the global arms bazaar—would be devastating. Such an attack would also be easy, as was made abundantly clear in 2000 by the attack on the USS *Cole*, carded out with lethal effectiveness by suicide bombers piloting nothing more than a Zodiac loaded with plastic explosives.

Another point of vulnerability is Pump Station No. 1, the station closest to Abqaiq, which sends oil uphill, into the Aramah Mountains, so that it can begin its long journey across the peninsula to the Red Sea port of Yanbu. If Pump No. 1 were taken out, the 900,000 barrels of Arabian light and superlight crude that are pumped daily to Yanbu would suddenly stop arriving, and Yanbu would be out of business.

Even the short pipe run from Abqaiq to the Gulf terminals at Ju'aymah and Ras Tanura is not without opportunity. If heavy damage were inflicted on the Qatif Junction manifold complex, which directs the flow of oil for all of eastern Saudi Arabia, the flow would be stopped for months. The pipes that connect the terminals and processing facilities can be replaced off the shelf, but those at Qatif require custom fabrication.

Promoters of Alaskan, Mexican Gulf, Caspian, and Siberian oil all like to point out that the United States has been weaning itself from Saudi Arabian oil, for protection against the effects of just such an attack on the Saudi oil system. Saudi Arabia may sit on 25 percent of the world's known oil reserves, they argue, but it provides somewhere around 18 percent of the crude oil consumed by the United States—and that is down from 28 percent in only a decade. What these people fail to mention is that Saudi Arabia has the world's only important surplus production capacity—two million barrels a day. This keeps the world market liquid. Not only that, but because the Saudis more or less determine the price of oil globally by deciding how much oil to produce, even countries that don't buy Saudi oil would be vulnerable if the flow of that oil were disrupted.

The Saudis have repeatedly used their surplus production capacity to stabilize the international oil market. They used it to break the OPEC embargo (but not before they had enriched themselves by tens of billions of dollars), in 1974. They used it again during the protracted Iran-Iraq war, to keep oil flowing to the industrialized West. They used it during the Gulf War, in 1990-1991; with help from a couple of other Gulf states, they produced an extra five million barrels a day, making up for the loss of Iraqi and Kuwait oil.

And they used it again on September 12, 2001. Less than twenty-four hours after the attacks on the World Trade Center and the Pentagon, the Saudis decided to send nine million barrels of oil to the United States over the next two weeks. The result was that the United States experienced only a slight inflation spike in the wake of the most devastating terrorist attack in history. Had that same surplus capacity been taken out of play with twenty pounds of Semtex, all bets would have been off. The U.S. Strategic Petroleum Reserve can support the domestic market for only about seventy days. And if Saudi Arabia's contribution to the world's oil supply were cut off, crude petroleum could quite realistically rise from around $40 a barrel today to as much as $150 a barrel. It wouldn't take long for other economic and social calamities to follow.

Americans have long considered Saudi Arabia the one constant in the Arab Middle East. The Saudis banked our oil under their sand, and losing Saudi Arabia would be like losing the Federal Reserve. Even if the Saudi rulers one day did turn anti-American, the argument went, they would never stop pumping oil, because that would mean cutting their own throats. This, at any rate, is the way we looked at the matter before fifteen Saudis and four other terrorists launched their suicide attacks on September 11; before Osama bin Laden suddenly became for the Arab world the most popular Saudi in history; before *USA Today* reported last summer that nearly four out of five hits on a clandestine al Qaeda Web site came from inside Saudi Arabia; and before a recent report commissioned by the UN Security Council indicated that Saudi Arabia has transferred $500 million to al Qaeda over the past decade.

Five extended families in the Middle East own about 60 percent of the world's oil. The Saud family, which rules Saudi Arabia, controls more than a third of that amount. This is the fulcrum on which the global economy teeters, and the House of Saud knows what the West is only beginning to learn: that it presides over a kingdom dangerously at war with itself. In the air in Riyadh and Jidda is the conviction that oil money has corrupted the ruling family beyond redemption, even as the general population has grown and gotten poorer; that the country's leaders have failed to protect fellow Muslims in Palestine and elsewhere; and that the House of Saud has let Islam be humiliated—that, in short, the country needs a radical "purification."

We can try to wish this away all we want. But the reality is getting harder and harder to ignore. Per capita income in Saudi Arabia fell from $28,600 in 1981 to $6,800 in 2001. The country's birth rate has soared, becoming one of the highest in the world. Its police force is corrupt, and the rule of law is a sham. Saudi Arabia almost certainly leads the world in public beheadings, the venue for which is often a Riyadh plaza popularly known as Chop-Chop Square. Illegal arms routinely flow into and out of the country. Taking into account its murky "off-budget" defense spending, Saudi Arabia may spend more per capita on defense than any other country in the world (some estimates put the figure at 50 percent of its total revenues), and the House of Saud believes this is necessary for its personal protection. The regime is threatened by increasingly hostile neighbors—and by determined enemies within the country's borders. Popular preachers all over Saudi Arabia call openly for a *jihad* against the West—a designation that clearly includes the royal family itself—in terms as vitriolic as anything heard in Iran at the height of the Islamic revolution there. The kingdom's mosque schools have become a breeding ground for militant Islam. Recent attacks in Bali, Bosnia, Chechnya, Kenya, and the United States, not to mention those against U.S. military personnel within Saudi Arabia, all point back to these schools—and to the House of Saud itself, which, terrified at the prospect of a militant uprising against it, shovels protection money at the fundamentalists and tries to divert their attention abroad.

Recent examples of Saudi support for the fundamentalists abound. In 1997 a high-ranking member of the royal family coordinated a $100 million aid package for the Taliban. In Los Angeles two of the 9/11 hijackers met with a Saudi working for a company contracted to the Ministry of Defense. A raid on the Hamburg apartment of a suspected accomplice of the hijackers turned up the business card of a Saudi diplomat attached to the religious-affairs section of the embassy in Berlin. Most of the more than 650 al Qaeda prisoners being held at the Guantanamo Bay Naval Base in Cuba—"the worst of the worst," according to Secretary of Defense Donald Rumsfeld—are rumored to be Saudis.

I served for twenty-one years with the CIA's Directorate of Operations in the Middle East, and during all my years there I accepted on faith my government's easy assumption that the money the House of Saud was dumping into weaponry and national security meant that the family's armed forces and bodyguards could keep its members—and their oil—safe. "The royal family is like the fingers of a hand," my colleagues at the State Department liked to say. "Threaten it, and they become a fist." I no longer believe this. Saudi Arabia is more and more a breathtakingly irrational state. For a surprising number of Saudis, including some members of the royal family, taking the kingdom's oil off the world market—even for years, and at the risk of destroying their own economy—is an acceptable alternative to the status quo.

Saudi Arabia has existed as a formal nation only since 1932, when the tribal leader Abdul Aziz ibn Saud gained control of much of the Arabian Peninsula, named the territory after his clan, and proclaimed himself king. But the House of Saud had been powerful in the region ever since the eighteenth century, when the radical cleric Muhammad ibn Abdul Wahhab, the founder of the puritanical Wahhabi movement, wandered into Dar'iya, near present-day Riyadh, and made a bargain with its ruler, Muhammad ibn Saud. The Saud family would provide the generals, and the Wahhabis would provide the foot soldiers. Until recently it was a marriage made in heaven.

An attack on Saudi Arabia's largest oil-processing facility could create a reduction in the flow of oil roughly equal to what *all* of OPEC effected in its 1973 embargo.

If I had to pick a single moment when the House of Saud truly began to fall apart, it would be when Abdul Aziz ibn Saud's son Fahd, who has been king since 1982, suffered a near fatal stroke, in 1995. As soon as the royal family heard about Fahd's stroke, it went on high alert. From all over Riyadh came the *thump-thump* of helicopters and the sirens of convoys converging on the hospital where Fahd had been taken.

Among the first to arrive were Jawhara al-Ibrahim, Fahd's fourth and favorite wife, and their spoiled, megalomaniac twenty-nine-year-old son Abdul Aziz—or "Azouzi" ("Dearie"), as Fahd called him. Anyone who knew how Fahd's court ran knew the extent to which Fahd had come to depend on Jawhara, who helped him with everything from remembering his medicine to handling intricate problems of foreign policy. If a prince wanted a matter immediately brought to Fahd's attention, he called Jawhara. As for Abdul Aziz, he was the youngest of Fahd's children and the apple of his father's eye. Fahd indulged him in everything. Stories circulated widely about Abdul Aziz's riding a Harley-Davidson inside his father's palace, chasing servants and smashing furniture. Most of the royal family found the king's indulgence strange. Abdul Aziz was pimply, craven, a bit slow. Apparently, though, he was regarded as the king's good-luck charm. Fahd's favorite soothsayer had once told him that as long as Abdul Aziz was by his side, the king would have a long, fulfilling life. So Fahd did not complain when Abdul Aziz spent $4.6 billion on a sprawling palace and theme park outside Riyadh, because Abdul Aziz was "interested" in history. The property includes a scale model of old Mecca, with actors attending mosque and chanting prayers twenty-four hours a day, and also replicas of the Alhambra, Medina, and half a dozen other Islamic landmarks.

Next to arrive at the hospital, in a great show of solidarity, were Fahd's full brothers—Sultan, the Defense Minister; Nayef, the Interior Minister; and Salman, the governor of Riyadh province. To outsiders, they were a tight bunch. Their mother, from the Sudayri clan, had taught them from an early age to stick together or risk being elbowed out by the forty-odd other children of their father.

Other princes—the children and grandchildren of Ibn Saud's children—hurried to the hospital too, from all over the kingdom and the rest of the world. Private executive jets were lined up wing to wing at Riyadh's airport. These princes couldn't get anywhere near Fahd, but by being close at hand they could pick up more-reliable news and, just as important, demonstrate their fealty. Most of them lived off his largesse—royal stipends, which ran from $800 to $270,000 a month. The princes knew they were breaking the treasury—all told, their brethren numbered 10,000 to 12,000. Would Crown Prince Abdullah—Fahd's half brother, a seventy-one-year-old reformer who was next in line for the throne—cut back on their stipends, or even eliminate them if Fahd died? They had to stick around to find out.

A recent report commissioned by the UN Security Council indicated that Saudi Arabia has transferred $500 million to al Qaeda over the past decade.

At this point Fahd's brothers were calling doctors in the United States and Europe. They wanted to know not whether Fahd would ever recover his mental capacities, or what kind of life he would be able to live, but what it would take to keep his heart beating and his body warm. Money, of course, wasn't a problem. They told the doctors they were prepared to lease as many Boeing 747 cargo jets as needed to bring in mobile hospitals and medical teams. The doctors couldn't understand the reasoning behind the questions—but only because they didn't understand the politics of the kingdom. What the family knew and the doctors didn't was that Crown Prince Abdullah had long been eager to take power. The only way to keep him at bay was to keep Fahd alive—God willing, until Abdullah died.

Abdullah had always been the odd prince out. To begin with, his mother was from the Rashid tribe, traditional enemies of the Saud. Ibn Saud had married her to cement a truce with the Rashid, and although the Rashid were now loyal subjects, Abdullah was still mistrusted by Fahd's full brothers. Almost alone among the top members of the royal family, Abdullah had chosen the way of the desert, turning his back on the luxuries of Riyadh, Jidda, and Ta'if. He never vacationed lavishly in Europe, unlike King Fahd and his entourage, who typically spent $5 million a day during visits to the palace at Marbella, on the Spanish Riviera. Abdullah preferred to spend his time in a tent, drinking camel's milk and eating dates. He interspersed his conversation with Bedouin aphorisms and turns of phrase. All his children were raised according to the customs of the desert. It is Abdullah who has recently called publicly for democratic reforms, the reining in of the conservative clergy, and military disengagement from the United States.

The royal family hated being reminded that they had abandoned their Bedouin roots, but they hated still more that Abdullah was trying to cut back royal corruption and entitlements. Aping the senior members of the family, the lesser princes had fantastic financial expectations, and their stipends didn't suffice. The third-generation princes were getting only about $19,000 a month—a fraction of what they needed for the lifestyles they sought. To keep even a modest yacht on the French Riviera requires a million dollars a year. What were they supposed to do? In order to make ends meet they had been getting into nastier and nastier business, taking bribes from construction firms (mostly the bin Laden family's) seeking government contracts, getting involved in arms deals, expropriating property from commoners, and selling Saudi visas to guest workers. Another trick they'd discovered was borrowing money from private banks and simply refusing to pay it back. It wasn't as if the larger family could somehow discipline or shame them. There were so many princes that they didn't even all know one another.

Abdullah had made no secret of his intention to put an end to the thievery when he became king—and for a while it looked as if he might get his way even before becoming king. In the mid-1990s, as Saudi Arabia was facing increasingly dire financial difficulties, he persuaded King Fahd to appoint a handful of reformist ministers. Abdullah first had them zero in on expropriations. The practice had become so widespread among the lesser princes that it was completely alienating Saudi Arabia's traditional merchant class and fledgling middle class. A prince might walk into a restaurant, see that it was doing well, and write out a check to buy the place, usually well below market price. There was nothing the owner could do. He knew that if he resisted, he'd end up in jail on trumped-up charges.

The senior princes used their government positions to do the same thing, but on a much grander scale. One of them would pick out a valuable piece of property—maybe a particularly good location for a shopping mall or a new road—and then order a court to condemn it in the name of the state, which would clear the way for the king to award it to him. The money to be earned was staggering, and senior princes had started to rely on the practice to maintain their ever more bloated personal budgets. In Abdullah's view, however, crooked property deals and the like were only a small part of the problem. The off-budget deals were a much bigger part. In off-budget spending, revenue from oil sales goes directly to special accounts, bypassing the Saudi treasury altogether. The money is then used to pay for pet projects, from defense procurement to construction, with no government audits or accountability of any sort. Commissions and bribes are enormous.

As a reformer, Abdullah was kept out of the tight circle that gathered around Fahd after his stroke. Bitterness against Abdullah within the family was so deep that he was in fact blamed for the stroke. One version had it that Fahd and Abdullah had been on the telephone, arguing about who would attend a meeting of the Gulf Cooperation Council in Oman. It was a fundamentally unimportant decision, but relations between the two men had become so toxic, it was said, that Fahd's anger brought on the event. Another rumor in circulation held that Fahd and Abdullah had been arguing about what they always argued about—looming financial collapse. There were even whispers that Abdullah had intentionally provoked Fahd, knowing his health wouldn't withstand a shouting match.

It eventually became clear that Fahd would live, but the extent of his impairment also became clear—embarrassingly so when, during a therapy session not long after the stroke, Fahd defecated in his pool in front of his family. His mind was affected too. Those close to him knew that he would never truly rule again, though he is still led out for ceremonial appearances.

A year and a half after Fahd's stroke Sultan had come to so despise Abdullah that he stopped attending cabinet meetings chaired by him. For Abdullah, the feeling was mutual. In July of 1997 he simply bypassed the Council of Ministers, which was heavily stacked in favor of the Sudayri, and tried to get Fahd to sign off on decrees and laws he thought needed passing. Jawhara and Abdul Aziz teamed up to thwart him.

Mind you, it is not as if the rest of the Fahd clan is united. Sultan, Salman, and Nayef may have arrived at the hospital together in a show of solidarity, but they got a rude shock once they pushed through the front doors. Jawhara and Abdul Aziz blocked them from seeing their brother. The two had set up camp outside Fahd's hospital room and were deciding who and what would or would not get in. That included ministers, senior princes, and doctors, along with petitions, decrees, and everything else.

Saudi succession doesn't operate according to primogeniture. By tradition, senior princes come to a consensus on succession, usually choosing one from their ranks who is thought to have the necessary experience and wisdom. So far the system had served the royal family well, even though Abdullah had become a gadfly, but now Fahd's brothers were afraid that Abdul Aziz was trying to circumvent custom and place himself higher in the line of succession. For one thing, he had started getting more and more involved in national security, from foreign affairs to intelligence. Even the Americans noticed it. When the commander of U.S. forces in the Middle East, General J. H. Binford Peay, came to Riyadh to meet with Fahd, in July of 1997, he was surprised to find Abdul Aziz at Fahd's side, whispering in his father's ear. Where was Abdullah? What had become of Sultan? Peay had to meet with Abdullah separately, and even then Abdullah didn't talk about the issues at hand.

What really worried some members of his family was that Abdul Aziz was funding radical Wahhabi causes and was gaining strength and popularity as a result. They had little doubt that money was going to clerics and causes that were associated with Osama bin Laden. Abdul Aziz hadn't rediscovered his faith, of course: he was courting favor with the Wahhabis because he knew he would need their support to become king. In September of 1997 he helped to coordinate that $100 million aid package for the Taliban, even though the Taliban were protecting bin Laden—a man who not only had vowed to overthrow the House of Saud but also seemed increasingly capable of doing so. Abdul Aziz was buying support wherever he could find it. In December of 1993 Abdul Aziz authorized $100,000 for a Kansas City mosque. On September 15, 1995, he opened the King Fahd Academy, in Bonn, and two days later he dedicated a new mosque there. Nine days after that he invited the head of the Islamic Society of Spain, Mansur Abdul Salam, to Riyadh. In May of 1996 he and Jawhara arranged for King Fahd to release Muhammad al Fasi from prison. Al Fasi had been imprisoned for opposing the Gulf War and the presence of U.S. troops in Saudi Arabia; in other words, he shared some of bin Laden's chief grievances. In December of 1999 the press finally caught wind of Abdul Aziz's penchant for backing radical Islamic causes. One regional account made available by U.S. translation services noted that he was believed to have been funding an associate of bin Laden's, Sa'd al Burayk, who in turn was giving the money to Islamic groups dedicated to killing Russian soldiers and civilians in Chechnya. Nayef promised to put a stop to Abdul Aziz and bring his charity back under control—but he appears to have done nothing.

All the while, throughout the 1990s, the royal family kept growing and growing. A prince might sire forty to seventy children during a lifetime of healthy copulation; however, the resources to support the growing population of the entitled were shrinking, not just in relative terms but in absolute ones. Young royals were pushing up from below, chafing at leaders who were slipping into their late seventies and eighties. The incapacitated King Fahd will turn eighty this year; Crown Prince Abdullah will turn seventy-nine. Many of the most active court intriguers are also in their seventies.

The Saudi royals number 30,000 and may grow to 60,000 in a generation. What would oil have to cost to support even the most basic privileges they now enjoy?

The House of Saud currently has some 30,000 members. The number will be 60,000 in a generation, maybe much higher. According to reliable sources, anecdotal evidence, and the Saudi gossip machine, the royal family is obsessed with gambling, alcohol, prostitution, and parties. And the commissions and other outlays to fund their vices are constant. What would the price of oil have to be in 2025 to support even the most basic privileges—for example, free air travel anywhere in the world on Saudia, the Saudi national airline—that the Saudi royals have come to enjoy? Once the family numbers 60,000, or 100,000, will there even be a spare seat for a mere commoner who wants to fly out of Riyadh or Jidda? Reformers among the royal family talk about cutting back the perks, but that's a hard package to sell.

Saudi Arabia operates the world's most advanced welfare state, a kind of anti-Marxian non-workers paradise. Saudis get free health care and interest-free home and business loans. College education is free within the kingdom, and heavily subsidized for those who study abroad. In one of the world's driest spots water is almost free. Electricity, domestic air travel, gasoline, and telephone service are available at far below cost. Many of the kingdom's best and brightest—the most well-educated and in theory, the best prepared for the work world—have little motivation to do any work at all.

About a quarter of Saudi Arabia's population, and more than a third of all residents aged fifteen to sixty-four, are foreign nationals, allowed into the kingdom to do the dirty work in the oil fields and to provide domestic help, but also to program the computers and manage the refineries. Seventy percent of all jobs in Saudi Arabia—and close to 90 percent of all private-sector jobs—are filled by foreigners.

Among men, at least, the Saudis have an admirably high literacy rate, especially for a place that only three generations back was inhabited mostly by nomadic tribesmen. About 85 percent of Saudi men aged fifteen and older can read and write, as opposed to less than 70 percent of Saudi women of the same age. But because in recent years the Saudi education system has been largely entrusted to Wahhabi fundamentalists, as a form of appeasement that many in the royal family hope will direct the fundamentalists' animus at foreign targets, its products are generally ill prepared to compete in a technological age or a global economy. Today two out of every three Ph.D.s earned in Saudi Arabia are in Islamic studies. Doctorates are only very rarely granted in computer sciences, engineering, and other worldly vocations. Younger Saudis are being educated to take part in a world that will exist only if the Wahhabi *jihad*ists succeed in turning back the clock not just a few decades but a few centuries.

Then there's the demographic problem. Saudi Arabia has one of the highest birth rates in the world outside Africa—37.25 births for every 1,000 citizens last year, compared with 14.5 per 1,000 in the United States. Ninety-seven percent of all Saudis are sixty-four or younger, and half the population is under eighteen. The simple presence of so many people of working age, and especially so many just now ready to enter the work force, places enormous pressure on an economy—particularly one designed less to accommodate those who want to work than to provide sustenance for those who would rather contemplate original intent in the Koran. A middle class stabilizes society. Saudi Arabia's middle class is imploding.

The functioning of the world's most advanced welfare state is influenced overwhelmingly by fluctuations in the price of oil. In 1981, when the entire kingdom was in effect being put on the dole, oil was selling at nearly $40 a barrel, and the annual per capita income was $28,600. A decade later, just before Iraq invaded Kuwait, refiners were able to buy oil for about $15 a barrel. The Gulf War sent prices back up to about $36 a barrel before they quickly fell. Today a barrel of oil once again fetches around $40, but twenty years' worth of inflation, combined with a population explosion, has brought per capita income down to below $7,000. Because roughly 85 percent of Saudi Arabia's total revenues are oil-based, every dollar increase in the price of a barrel of oil means a gain of about $3 billion to the Saudi treasury. In the early 1980s the kingdom boasted cash reserves on the order of $120 billion; today the figure is estimated to be $21 billion.

Given all these threatening forces, one might think that every map in official Washington would have a red flag sticking out of Riyadh, as a reminder that Saudi Arabia is on life support. The truth is quite the opposite. Before 9/11 the United States never issued an advisory indicating the obvious security problems for Americans traveling to Saudi Arabia. Dependents of U.S. citizens residing there were never advised to leave. According to official Washington, even today the country is stable: its government is in undisputed control of its borders; its police force and army are efficient and loyal; its people are well clothed, well fed, and well educated.

Consider the way the State Department has handled visas for Saudi nationals. Until 9/11, Saudis were not even required to appear at the U.S. embassy in Riyadh or the consulate in Jidda for a visa interview. Under a system called Visa Express a Saudi had only to send his passport, an application, and the application fee to a travel agent. The Saudi travel agent, in other words, stood in for the U.S. government. Just about any Saudi who had the money could book a flight to New York after a mere twenty-hour wait. Until recently Saudis were exempt from the new anti-terrorism entry regulations that apply to citizens of other Middle Eastern countries, despite the fact that most of the 9/11 terrorists were Saudis.

"The Saudi Arabian Government, at all levels, continued to reaffirm its commitment to combating terrorism," the State Department's 1999 report *Patterns of Global Terrorism* soberly asserted. The report went on to state, "The Government of Saudi Arabia continued to investigate the bombing in June 1996 of the Khobar Towers." This was false; Prince Nayef, Saudi Arabia's grim Interior Minister, had been stalling the investigation for years. Nayef told the kingdom's other senior princes that he was reluctant to help the United States with the Khobar investigation. In one heated meeting Nayef ignored Defense Minister Sultan when Sultan warned that stonewalling the FBI would end up causing a rift with the United States. To make his point Nayef went out of his way to avoid meeting the FBI's director, Louis Freeh, when Freeh showed up in Saudi Arabia to see what he could do to get the Khobar investigation going. Nayef put himself out of reach—on his yacht, anchored off the coast near Jidda, in the Red Sea—and turned the chore over to two low-ranking officials in the internal-security service, neither of whom knew anything about the Khobar investigation.

Even after the 1998 attacks on the U.S. embassies in Kenya and Tanzania, which were organized by Osama bin Laden from his bases in Afghanistan, the Saudi royals continued to aid the Taliban and its main supporter in the region, Pakistan. This was hardly a secret: in July of 2000 *Petroleum Intelligence Weekly*, which calls itself the "bible" of the international petroleum industry, reported that Saudi Arabia was sending as many as 150,000 barrels of oil a day to Afghanistan and Pakistan in off-budget foreign aid that had a value of something like $2 million a day. Furthermore, the United States had known since 1994 that the Saudis were supporting Pakistan's nuclear development program, ultimately contributing upwards of a billion dollars. More recently, because Saudi law does not allow foreign agencies to directly question Saudi citizens, the FBI has not been allowed to interview Saudi suspects, including the families of the fifteen Saudi hijackers, about the 9/11 attacks. For more than a year after September 11 Saudi Arabia refused to provide advance manifests for flights coming into the United States—which could have led to a basic and potentially fatal breach of security. Although there are plenty of possible al Qaeda members awaiting trial, as of this writing there hasn't been a single Saudi arrest related to 9/11—not even of a material witness.

As for the CIA, the Agency let the State Department take the lead and decided simply to ignore Saudi Arabia. The CIA recruited no Saudi diplomats to tell us, for instance, what the religious-affairs sections of Saudi embassies were up to. The CIA's Directorate of Intelligence avoided writing national intelligence estimates—appraisals, drawn from various U.S. intelligence services, about areas of potential crisis—on Saudi Arabia, knowing that such estimates, especially when negative, have a tendency to find their way onto the front pages of U.S. newspapers, where they might have an undesired effect on public opinion. The CIA's line became the same as State's: There's no need to worry about Saudi Arabia and its oil reserves.

No need to worry, of course, means business as usual—and for decades now that's meant that almost every Washington figure worth mentioning has been involved with companies doing major deals with Saudi Arabia. Spending a lot of money was a tacit part of the U.S.-Saudi relationship practically from the very beginning: the Americans would buy Saudi Arabia's oil and would provide the Saudis with protection and security; the Saudis would buy American weapons, construction services, communications systems, and drilling rigs. In the global-economics game this is known as "recycling," and in this case it worked well: two-way trade between Saudi Arabia and the United States grew from $56.2 million in 1950 to $19.3 billion in 2000.

Consider the case of the Carlyle Group—a private investment company, founded in 1987, that almost since its inception has been conducting immensely profitable business with Saudi Arabia. From 1993 to 2002 the chairman of Carlyle was Frank Carlucci, who served first as Ronald Reagan's National Security Adviser and then as his Secretary of Defense. Carlyle's senior counselor is James Baker, who served as Secretary of State under George H.W. Bush—who in his post-presidency also happens to be a Carlyle adviser. Others who hang their hats at Carlyle include Arthur Levitt, the head of the Securities and Exchange Commission under Bill Clinton, and now Carlyle's senior adviser; John Major, a former Prime Minister of Great Britain and the current chairman of Carlyle Europe; William Kennard, who chaired the Federal Communications Commission during the second Clinton Administration; Afsaneh Mashayekhi Beschloss, a former treasurer and chief investment officer of the World Bank; and Richard Darman, who ran the Office of Management and Budget under the first President Bush and also served as deputy secretary of the treasury under Reagan.

Carlyle isn't the only company in this business. Halliburton, run by Dick Cheney between his stints as Secretary of Defense under the first George Bush and Vice President under the second, has been a frequent beneficiary of Saudi money. In late 2001 Halliburton landed a $140 million contract to develop a new Saudi oil field. For many years Condoleezza Rice, now President Bush's National Security Adviser, served on the board of Chevron, which merged in 2001 with Texaco. The new corporation, ChevronTexaco, is a partner with Saudi Aramco in

Article 34. The Fall of the House of Saud

several ventures and has recently joined forces with Nimir Petroleum to develop oil fields in Kazakhstan. Currently on the board of ChevronTexaco are Carla Hills, who served as the Secretary of Housing and Urban Development under Gerald Ford and as a U.S. trade representative under George H.W. Bush; the former Louisiana senator J. Bennett Johnston, who made a specialty of energy issues while in Congress; and the former Georgia senator Sam Nunn, who served most notably as head of the Senate Armed Services Committee.

Elsewhere, Nicholas Brady, the Secretary of the Treasury under the first President Bush, and Edith Holiday, a former assistant to the first President Bush, serve on the board of Amerada Hess, which has teamed with some of Saudi Arabia's most powerful royal-family members to exploit the rich oil resources of Azerbaijan. In 1998 Amerada Hess formed a joint venture, Delta Hess, with the Saudi-owned Delta Oil to explore and exploit petroleum resources in Azerbaijan. The Houston-based Frontera Resources Corporation joined the Azerbaijan hunt the same year, teaming with the newly created Delta Hess. Among the members of Frontera's board of advisers: the former Texas senator former Secretary of the Treasury, and 1988 Democratic vice-presidential candidate Lloyd Bentsen; and John Deutch, a former CIA director.

Just to make sure that no one upsets the workings of this system, perhaps by meddling in internal Saudi affairs, Saudi Arabia now keeps possibly as much as a trillion dollars on deposit in U.S. banks—an agreement worked out in the early eighties by the Reagan Administration, in an effort to get the Saudis to offset U.S. government budget deficits. The Saudis hold another trillion dollars or so in the U.S. stock market. This gives them a remarkable degree of leverage in Washington. If they were suddenly to withdraw all their holdings in this country, the effect, though perhaps not as catastrophic as having a major source of oil shut down, would still be devastating.

The U.S.-Saudi relationship would not be as cozy as it is without there being someone well connected on both sides who can move comfortably between them. That someone is the fifty-four-year-old Prince Bandar. Although he ranks low on the royal bloodline (his father is King Fahd's brother Sultan, the Saudi Defense Minister, but his mother was a house servant), Prince Bandar has been the Saudi ambassador to the United States since 1983. He is the only foreign ambassador to have a security detail assigned to him by the State Department. A daredevil fighter pilot in his younger years, a Muslim with a taste for single-malt Scotch, and an envoy with a perpetually open wallet, Bandar has proved adept at working both the public and the private sides of diplomacy. As the Saudi military attache to the United States, he scored a stunning coup in 1981 by persuading Congress to approve the sale of AWACS air-defense technology to his country, over the objections of AIPAC, the pro-Israeli Washington lobby. Later, as ambassador, Bandar conveyed the kingdom's thanks by secretly placing $10 million in a Vatican City bank, as reported last year in *The Washington Post*; the money, deposited at the request of William Casey, then the director of the CIA, was to be used by Italy's Christian Democratic Party in a campaign against Italian Communists. Later still, in June of 1984, Bandar started paying out $30 million from the royal family so that Lieutenant Colonel Oliver North could buy arms for the Nicaraguan contras.

Leading U.S. corporations hire and rehire known Saudi crooks and known financiers of terrorism, so that they can land deals back in Saudi Arabia.

It is on the personal front, however, where Bandar shines. A visit in the early nineties to the summer home of George H.W. Bush, in Kennebunkport, Maine, earned the prince the affectionate family sobriquet "Bandar Bush." Bandar reciprocated by inviting Bush to hunt pheasant on his estate in England. For good measure he also contributed a million dollars to the construction of the Bush Presidential Library, in College Station, Texas. King Fahd sent another million to Barbara Bush's campaign against illiteracy. (He had donated a million dollars to Nancy Reagan's "Just Say No" campaign against drugs four years earlier.) Bandar was once Colin Powell's racquetball partner.

Press accounts portrayed Bandar as largely on the outside during the Clinton years, passing melancholy weeks at his mountain compound in Aspen, Colorado (more than 50,000 square feet, thirty-two rooms, sixteen bathrooms). If Bandar was less physically present, however, he was his usual useful self. In 1992 he persuaded King Fahd to donate $20 million to the University of Arkansas's new Center for Middle Eastern Studies, a gesture of respect for the Arkansas governor who had just been elected President. He is said to have played a role in persuading the Libyans, in 1999, to turn over two intelligence operatives suspected in the 1988 bombing of Pan Am Flight 103, over Lockerbie, Scotland. As he reportedly does at the end of every administration, whether he is perceived as friend or foe, Bandar also invited each of the Clinton Cabinet members out to dinner, at a restaurant of their choice, in a private room or a public one, depending on their willingness to be seen with him.

Prince Bandar once told associates that he is very careful to look after U.S. government officials when they return to private life. "If the reputation then builds that the Saudis take care of friends when they leave office," Bandar has observed, according to a source cited in *The Washington Post*, "you'd be surprised how much better friends you have who are just coming into office." Practically every deal with the Saudis eventually becomes hard to trace, lost in some desert sandstorm back near the wellheads where the money sprang from in the first place. Many of Washington's lobbyists, PR firms, and lawyers live off Saudi money. Just about every Washington think tank has taken it. So have the John F. Kennedy Center for the Performing Arts, the Children's National Medical Center, and every presidential library built in the past thirty years.

Bandar hurried back to prominence after the election of George W. Bush, occupying a spot somewhere between ambassador and permanently enthroned visiting head of state. But after 9/11 he began to experience some difficulty in maintaining a positive Saudi image. In March of last year agents of the Treasury Department raided the northern-Virginia offices of four Saudi-based charities: the SAAR Foundation, the Safa Trust, the International Institute for Islamic Thought (IIIT), and the International Islamic Relief Organization (IIRO). Also raided was the local headquarters for the Muslim World League, an umbrella group funded by the Saudi government. All five organizations are located only a few miles from Bandar's mansion overlooking the Potomac River. The organizations can point to a long list of genuinely humanitarian causes they have aided and supported; but they also have a long list of alarming associations. Testifying before Congress in August of 2002, Matthew Levitt, a senior fellow with the Washington Institute for Near East Policy, noted that Tarik Hamdi, an IIIT employee, had personally provided Osama bin Laden with batteries for his satellite phone—a critical link in the stateless world that bin Laden inhabits. IIIT and the SAAR Foundation are suspected of helping to finance Hamas and the Palestinian Islamic Jihad, the sponsors of some of the most lethal suicide bombers in the Middle East. From 1986 to 1994 Muhammad Jamal Khalifa, a brother-in-law to Osama bin Laden, ran the IIRO's Philippine office, from which he channeled funds to al Qaeda. Only excellent work by the Indian police prevented another IIRO employee, Sayed Abu Nasir, from bombing the U.S. consulates in Calcutta and Madras.

In mid-2002 word leaked to the press that the semi-official Defense Policy Board, chaired by the notorious cold warrior Richard Perle, had sponsored a report declaring Saudi Arabia to be part of the problem of international terrorism rather than part of the solution. Saudi Arabia, the report stated, was "central to the self-destruction of the Arab world and the chief vector of the Arab crisis and its outwardly-directed aggression." It went on to say, "The Saudis are active at every level of the terror chain, from planners to financiers, from cadre to foot-soldier, from ideologist to cheerleader." Within hours Colin Powell was on the phone to the Saudi Foreign Minister, assuring him—and through him, the royal family—that such apostasy was not and never would be the official stance of the Bush Administration. To reinforce the message, President Bush invited Bandar down to the family ranch at Crawford, Texas.

And yet the image problems have continued. In October of 2001, NATO forces raided the offices of the Saudi High Commission for Aid to Bosnia, founded by Prince Salman, and discovered, among other items, photos of the U.S. embassies in Kenya and Tanzania, before and after they were bombed; photos of the World Trade Center and the USS *Cole*; information on the use of crop-duster planes; and materials for forging U.S. State Department badges. His job wasn't made any easier when, in the fall of last year, Bandar found himself having to explain away the fact that about $130,000 in charitable contributions from his wife, Princess Haifa, might have ended up with two of the 9/11 hijackers.

In the wake of these revelations a U.S. delegation headed by Alan Larson, President Bush's undersecretary of state for economic affairs, traveled to Riyadh last November, ostensibly to prod the Saudis toward increasing the surveillance of their charities and financial networks. But U.S. and Saudi sources say that one of the main reasons for Larson's trip was to ensure that if the United States invaded Iraq, the Saudis would guarantee the flow of extra oil into the World market. The U.S. embrace of the House of Saud was as tight as ever.

Washington's answer for Saudi Arabia—apart from repeating that nothing is wrong—is to suggest that a little democracy will cure everything. Talk the royal family into ceding at least part of its authority; support the reform-minded princes; set up a model parliament; co-opt the firebrands with a cabinet position or two, a minor political party, and some outright bribery; send Jimmy Carter in to monitor the first election; and in a few generations Riyadh will be Ankara, maybe even London. The governmental mechanism may be faulty, the Washington view maintains, but the people who administer the government are for the most part committed to rooting out corruption, rounding up terrorists, and recognizing the right of the people to self-government.

It's utter nonsense, of course. If an election were held in Saudi Arabia today, if anyone who wanted to could run for the office of president, and if people could vote their hearts without fear of having their heads cut off afterward in Chop-Chop Square, Osama bin Laden would be elected in a landslide—not because the Saudi people want to wash their hands in the blood of the dead of September 11, but simply because bin Laden has dared to do what even the mighty United States of America won't do: stand up to the thieves who rule the country.

Saudi Arabia today is a mess, and it is our mess. We made it the private storage tank for our oil reserves. We reaped the benefits of a steady petroleum supply at a discounted price, and we grabbed at every available Saudi petrodollar. We taught the Saudis exactly what was expected of them. We cannot walk away morally from the consequences of this behavior—and we *really* can't walk away economically. So we crow about democracy and talk about someday weaning ourselves from our dependence on foreign oil, despite the fact that as long as America has been dependent on foreign oil there has never been an honest, sustained effort at the senior governmental level to reduce long-term U.S. petroleum consumption.

Not all the wishing in the world will change the basic reality of the situation.

- Saudi Arabia controls the largest share of the world's oil and serves as the market regulator for the global petroleum industry.
- No country consumes more oil, and is more dependent on Saudi oil, than the United States.
- The United States and the rest of the industrialized world are therefore absolutely dependent on Saudi Arabia's oil reserves, and will be for decades to come.
- If the Saudi oil spigot is shut off, by terrorism or by political revolution, the effect on the global economy, and particularly on the economy of the United States, will be devastating.
- Saudi oil is controlled by an increasingly bankrupt, criminal, dysfunctional, and out-of-touch royal family that is hated by the people it rules and by the nations that surround its kingdom.

Signs of impending disaster are everywhere, but the House of Saud has chosen to pray that the moment of reckoning will not come soon—and the United States has chosen to look away. So nothing changes: the royal family continues to exhaust the Saudi treasury, buying more and more arms and funneling more and more "charity" money to the *jihad*ists, all in a desperate and self-destructive effort to protect itself.

The fact is that the West, especially the United States, has left the Saudis little choice. Leading U.S. corporations hire and re-hire known Saudi crooks and known financiers of terrorism to represent their interests, so that they can land the deals that will pay the commissions back in Saudi Arabia—commissions that will further erode the budget and thus further divide the ruling class from everyone else. Former CIA directors serve on boards whose members have to hold their noses to cut deals with Saudi companies—because that's business, that's the price of entry, that's the way it's done. Ex-Presidents, former prime ministers, onetime senators and congressmen, and Cabinet members walk around with their hands out, acting as if they're doing something else but rarely slowing down, because most of them know it's an endgame too. But sometime soon, one way or another, the House of Saud is coming down.

Robert Baer served for twenty-one years with the CIA, primarily as a field officer in the Middle East. He resigned from the agency in 1997 and was awarded its Career Intelligence Medal in 1998. This article is adapted from the forthcoming book Sleeping With the Devil *(Crown Publishers), to be published in June.*

From *The Atlantic Monthly*, May 2003, pp. 53-62. © 2003 by Robert Baer. Reprinted by permission of the author.

"Africa's issues are global issues—HIV/AIDS, human development, new models for economic growth, peace, and democracy. Worldwide consciousness of the HIV/AIDS pandemic has even forced its way into the pages of a United States president's State of the Union address. In practice, however, priorities are being set by another agenda, a war agenda."

America and Africa

SALIH BOOKER, WILLIAM MINTER, AND ANN-LOUISE COLGAN

In a dangerous replay of the cold war, the United States is likely to ignore Africa's priorities, placing military basing rights above human rights. The war against AIDS, by far the most important global effort and an especially urgent priority for Africa, will continue to suffer from a lack of resources. The American war on Iraq will also have a major negative impact on the global economy, with dire consequences for African development. In addition, this year will likely see United States unilateralism directly at odds with African interests in building multilateral approaches to the continent's greatest challenges, which range from HIV/AIDS to international trade rules and peacekeeping.

Last year saw African efforts toward building greater political and economic unity often offset by a failure to provide collective leadership for the continent's most pressing problems. The African Union replaced the 39-year-old Organization of African Unity as a framework for stepped-up cooperation across the continent. The new union, expected to evolve out of a process of accelerated integration, is seen as more ambitious than the European Union. (The African Union has plans for continent-wide institutions, including a pan-African parliament, plus phased economic integration of member countries at both subcontinental and continental levels.)

Africans welcomed the prospect of new commitments to unity. But many have raised the fear that African leaders are still unwilling to act more decisively as a bloc within international affairs or to promote regional economic and political integration at the expense of nationalist interests.

Under South African leadership, the African Union embraced the New Partnership for African Development (NEPAD), designed to increase cooperation among African states, donor countries, and multilateral organizations. But critics charged that NEPAD adopted the failed economic policies and programs of the World Bank and rich-country governments while neglecting to lay the basis for the democratic participation of African people. Moreover, the framework initially failed in its major objective of winning substantially increased resources from the Group of Eight industrialized donor countries in terms of new economic aid, debt relief, or increased investment. NEPAD avoids any mention of Western obligations to support development in Africa and thus does not make reference to reparations.

The most dramatic failure for African governments and world leaders last year was in combating HIV/AIDS. Despite the ever-louder chorus of warnings and promises, the rich countries and most African governments moved at a snail's pace in responding to the emergency. The Global Fund to Fight AIDS, Tuberculosis, and Malaria received only a fraction of the resources needed.[1] The South African government stalled on providing antiretroviral drugs to people living with HIV/AIDS (one in nine South Africans is living with HIV/AIDS). Grassroots and government health programs throughout Africa continued to be crippled by a lack of resources.

As the new year began, it was not AIDS but Iraq that cast a looming shadow over every other issue. In January former South African President Nelson Mandela called on the world to "condemn both Blair and Bush and let them know in no uncertain terms that what they are doing [by pushing for war] is wrong." At the meeting of the African Union in early February, President Thabo Mbeki of South Africa warned that war in the Persian Gulf could trigger an economic meltdown in Africa and set development back more than three decades. But Washington's lack of regard for African opinion was illustrated earlier by the perfunctory cancellation of President George W. Bush's projected January visit to five African countries.

At the end of January, President Bush surprised many by accepting, for the first time, the need to supply antiretroviral drugs and by promising additional resources for Africa to fight AIDS. But if the United States fails to at least triple its spending on AIDS this year, the gesture will be seen in retrospect as simply a public relations adjunct to the push for war on Iraq. Early signs have not been encouraging.

WHAT POLICY DIRECTION?

African countries must cope with the HIV/AIDS pandemic and its root causes while dealing with a myriad of other problems greatly exacerbated by this health crisis. It is a staggering prospect. The chances of success will be fundamentally affected by how much attention the world pays to Africa, and by whether rich countries contribute their share to addressing these global issues centered there. The level of world attention, as well as Africa's internal capacity to act, will in turn be affected by the length of the United States war on Iraq.

Africa's priorities have been fairly consistent in recent years, defined in part by the huge and unavoidable challenges of HIV/AIDS, poverty, and conflicts. Broad consensus exists among African and international nongovernmental organizations, most international agencies, and many African governments on what needs to be done.[2]

The Africa policy agenda for the United States government encompasses the following:

- provide adequate resources for a serious war against AIDS, including affordable life-saving drugs;
- cancel Africa's unsustainable and largely illegitimate debts;
- back attempts by African diplomats and civil society to resolve conflicts, manage peace negotiations, and achieve lasting peace;
- support efforts to move beyond formal elections toward increased participation and accountability for national and multinational institutions;
- invest development resources not only in health, education, and other sectors that build African human resources, but also in communications and transportation infrastructure necessary to make human resources economically productive.

Although the details may be debated, African and international civil society groups are virtually unanimous in favoring such commonsense proposals. Similar views have significant support among opinion makers in international agencies and African governments. Breakthroughs on some issues have even occurred, such as President Bush's belated acknowledgment that affordable antiretroviral treatment is imperative.

Yet the resources to implement this agenda are not available, and rich countries—particularly the United States—do not accept that providing these resources is both an obligation and a necessity for building common security in today's world. For their part, African governments still give more weight to pleasing donors and preserving their power than to meeting popular demands to address critical needs.

In Bush's Washington, when priorities are measured by resource allocation, war on Iraq comes first. This is followed by the United States push for free trade, designed to promote the interests of American corporations. Africa's urgent needs to fight AIDS and promote human development are far down the list. Other rich-country governments and African governments tried to slow Washington's rush to war, and they criticized the Bush administration's refusal to engage with multilateral responses to global problems. But with varying degrees of nuance, they have also bought into the imposed "Washington consensus" that places faith in free-market fundamentalism as the key to development (the Washington consensus tests the discredited belief that opening borders for the free flow of trade and investment will eventually provide countries with the resources to meet their people's needs).

Popular pressure has forced African leaders and even President Bush to make new promises. The pressure to deliver on these will continue, regardless of events in Iraq. It is impossible, however, to dispute the stubborn reality: as long as distraction, denial, and dogmatism prevail, constructive efforts to address Africa's needs will be diminished or crippled. On issue after issue, the real test is not rhetoric but action.

BEYOND DENIAL AND DISREGARD

In January 2002 Africa Action director Salih Booker noted that Presidents Thabo Mbeki and George Bush epitomized the "two greatest impediments to the fight against AIDS: denial and disregard."[3] The South African government continues to vacillate on its commitment to provide treatment for HIV/AIDS through the public sector. And, despite President Bush's pledge this January of $10 billion in new "emergency" funds to fight AIDS in Africa, in February the administration tried to block a congressional decision to provide an additional $150 million from this year's budget to the Global Fund to Fight AIDS, Tuberculosis, and Malaria.

A closer examination of the president's plan reveals even more gaps. As Africa Action and other activist groups quickly pointed out, the president's proposal, despite its "emergency" label, provided no new money for HIV/AIDS this year. The $10 billion in new money would start small, with much less than $1 billion disbursed in 2004 and with no guarantee that it would not be edged out of future budgets by rising costs of war in Iraq or other priorities pushed by powerful lobbyists. Also *The New York Times* noted that the increase in AIDS funds comes partly by cutting nearly $500 million from international child health programs. "The White House should not be forcing the babies of Africa to pay for their parents' AIDS drugs," the February 17 *Times* editorial concluded. And although the president's promise was for additional funds for "Africa," his budget proposals count all money spent worldwide for HIV/AIDS toward his pledge.

An equally important question is how the money will be spent. While the Global Fund is facing bankruptcy, the president's proposal calls for only $200 million for the fund annually, essentially freezing United States contributions at the level of previous years. The bulk of the new funds apparently will be channeled either through the notoriously cumbersome bureaucracy of the United States Agency for International Development or through some new bilateral mechanism yet to be established—a further reflection of this administration's preference for "going it alone."

To the extent that American unilateralism prevails this year, the global response to AIDS will be further weakened and delayed. Some activists also fear that distribution of the funds may be used to advance right-wing religious agendas (for example, the international "gag rule" preventing support for many reproductive health programs) or as leverage for other United States diplomatic aims.

Although President Bush acknowledged that affordable antiretroviral drugs are necessary, as of early March 2003 it remained highly doubtful that United States policy would in fact help countries import these drugs or build manufacturing capability. A practical approach would be for rich countries and the Global Fund to work closely together to help African countries import generic versions of these drugs from Brazil, India, and Thailand and begin treating people immediately. Instead, United States trade representatives were still blocking even an agreement in principle on implementation of the 2001 Doha agreement of the World Trade Organization that called for loosened patent rules to facilitate such exports.

Negotiators freely admitted to journalists that the United States resistance to generics was being driven by the American pharmaceutical company lobby. The response to the AIDS pandemic will be the most telling indicator of United States and global response to African priorities. If past patterns prevail, congressional debate on the United States budget for fiscal year 2004 may well continue early into that year. Yet the Global Fund does not have sufficient assets for a third round of proposals in October 2003. The next summit of the G-8 will be held in June in France. The question is whether AIDS will even be on the agenda, and, if so, whether it will bring only new promises or real resources.

The level of response to AIDS also reveals the priority given by the international community to health and human development more generally. Countless international conferences have affirmed the need for additional resources, for partnership and participation, and for independent evaluation of results not dominated by bilateral political agendas. Yet all signs indicate that United States policy is moving in the opposite direction.

THE DEVELOPMENT DEFICIT

The United Nations estimated Africa's economic growth rate in 2002 at 2.9 percent, higher than the world average of 1.7 percent. But with more than 38 million people facing famine at year's end, and the continued escalation of the AIDS pandemic undermining the capacity to respond at all levels, these growth figures were deceptive. Families struggling to survive and governments hard-pressed to meet the minimum demands of their societies could find little comfort in such a report.

In recent years, United States economic policy toward Africa has revolved around the African Growth and Opportunity Act (AGOA), a program that may now be supplemented with plans for a new worldwide aid program called the Millennium Challenge Account (MCA). While AGOA centers on expanding trade opportunities, and the MCA focuses on development assistance, both initiatives feature implementation through bilateral agreements between the United States and selected countries, and unilateral determination by Washington of procedures, criteria, and evaluation of results. Although they may deliver some benefits to a small subset of recipients, neither initiative responds to civil society demands for reform of the international aid system.

President Bush has asked for $1.3 billion in 2004 for the MCA—$300 million less than the amount originally promised when he announced the program in March 2002. The 10 to 15 countries expected to qualify include only 3 from Africa: Uganda, Senegal, and Ghana. Eligibility will be based on sixteen indicators: six for "governing justly," four for "investing in people," and six for "promoting economic freedom." Inclusion of indicators for "investing in people" is a concession to the need for human development, but all indicators are based on data provided by a narrow range of institutions: ten indicators draw on data from the World Bank, two from the International Monetary Fund, two from Freedom House, one from *Institutional Investor* magazine, and one from the Heritage Foundation. No African institutions are even considered relevant for assessing African conditions.

Washington will no doubt continue to pursue this unilateral stance, minimizing its participation in multilateral efforts to deal with African development issues. Despite its concessions to donor perspectives, the NEPAD approach to rich countries pursued by African leaders—with the support of Canada, Britain, and some other European countries—finds little backing from the White House or other United States agencies.

The United States approach forces each African country to compete with its neighbors in negotiating its relationship with Washington and its access to resources or trade concessions under different United States government programs. President Bush may or may not fulfill his pledge to reschedule a visit to the continent for later this year. Regardless, high-level attention in Washington to African priorities such as further debt cancellation, adequate funding for multilateral institutions at the global and African levels, and reduction in rich-country agricultural trade subsidies is likely to be minimal.

WAR, PEACE, AND HUMAN RIGHTS

Despite inadequate international support, African countries made several significant advances toward peace in 2002, reducing the overall level of conflict from the previous year. Angola and Sierra Leone moved beyond war to reconstruction. African leaders and the United Nations kept fragile peace processes alive in Sudan, Burundi, and Congo. The cease-fire held on the Ethiopia–Eritrea border. In September, however, Ivory Coast erupted into a conflict that has continued into the new year.

On the democracy front, Kenyans celebrated as the 24-year reign of President Daniel arap Moi ended with December elections that were largely peaceful. In Zimbabwe, however, President Robert Mugabe stayed in power through elections that were widely criticized as not free and fair. Despite the suspension of Zimbabwe from the Commonwealth, political and economic crisis in that country continued to escalate during the year. Seventeen other African countries held presidential or parliamentary elections in 2002. In almost all countries, however, civil society and opposition groups pointed to huge gaps between the promise and the practice of democracy.

The principal factors in resolving or aggravating unrest will be the actions of African parties to conflict, neighboring states, and the pressures of African civil society and public opinion demanding peace. Similar factors will determine the extent to which elections are free and fair, and whether human rights are defended against abusive rulers or other violent forces.

Particularly influential—both for their own sake and for their effect on neighboring countries—will be the outcome of elections in Nigeria; the fate of peace processes in Sudan, Congo, Burundi, and Ivory Coast; and the extent to which the new Kenyan government can begin to meet voters' high expectations. Angola too faces enormous challenges in delivering on expected peace dividends. And Zimbabwe's stability is threatened not only by AIDS and famine, but by the escalation of internal repression and the failure of outside parties to force the government to moderate its stance on issues such as the government's controversial land redistribution program, which has increased tensions and exacerbated the political and economic crisis.

Yet United States engagement with security and democracy issues in African countries is driven more by geopolitical considerations, in a dangerous replay of cold war disregard for African concerns. Increased United States interest in projecting military force into the Persian Gulf has led to a massive increase in the American military presence in the Horn of Africa and efforts to form alliances with African governments according to their perceived value in the framework of the war on terrorism. In West Africa and Central Africa, United States policymakers are focusing on the strategic value of oil. This raises the same question of whether issues of human rights and resolving internal conflict will be neglected in key oil-producing countries such as Nigeria and Angola.

Resolution of conflicts in Sudan and Congo and support for democracy in Nigeria, Zimbabwe, and Kenya should be among the highest priorities for United States diplomatic efforts in Africa. Rhetorical support for these goals already forms part of American policy. It is unknown, however, whether this will be accompanied by additional diplomatic attention and resources for multilateral and civil society actors engaged with these issues.

As in the economic sphere, there may be positive United States contributions in these areas, at the initiative of some American officials specializing in African affairs. But the chances that these contributions will grow and bear fruit while war rages in Iraq will be slim indeed.

AFRICAN ISSUES, GLOBAL ISSUES

Africa's issues are global issues—HIV/AIDS, human development, new models for economic growth, peace, and democracy. Worldwide consciousness of the HIV/AIDS pandemic has even forced its way into the pages of a United States president's State of the Union address. In practice, however, priorities are being set by another agenda, a war agenda. This year will prove particularly decisive in determining whether Africa and the world can build momentum for a change of course.

NOTES

1. The fund is an independent body directed by representatives of government and civil society from both rich and poor countries, and from the private sector, the nonprofit sector, and organizations representing people living with HIV/AIDS.
2. In January Africa Action released "Africa Policy for a New Era: Ending Segregation in U.S. Foreign Relations." This report (available at <http://www.africaaction.org/featdocs/afr2003.htm>) reflects and summarizes this emerging consensus and provides a formulation of positive directions for policy.
3. "AIDS: Another World War," *The Nation*, January 7, 2002.

SALIH BOOKER *is executive director,* WILLIAM MINTER *a senior research fellow, and* ANN-LOUISE COLGAN *a research associate at Africa Action. This essay is adapted from "Policy Report: Africa Policy Outlook 2003," published in March 2003 by Foreign Policy in Focus* <www.fpif.org>.

UNIT 10
International Organizations and Global Issues

Unit Selections
36. **United Nations**, Madeleine K. Albright
37. **The Five Wars of Globalization**, Moises Naim

Key Points to Consider

- What specific actions would you recommend the UN take to diffuse the current conflict in Iraq and ensure that the future Iraqi government has the widespread support of all Iraqis-Sunnis, Shiites, Kurds, and other nationalities?

- Can the UN do anything to help ensure that the future government of Iraq does not attempt to resume covert weapons of mass destruction research?

- Why is the United Nations such a useful organization for dealing with global problems such as disease, poverty, global crime, and war?

- Which one of the networks described by Naim related to terrorism, drugs, arms, intellectual property, people, and money is the most destabilizing for world stability?

- What is your reaction to the following statement? "Infectious diseases in the 21st century, much like the black plague in the Middle Ages, will fundamentally alter political systems and change the way international relations are conducted during my life time." Be specific.

 **Links: www.dushkin.com/online/**
These sites are annotated in the World Wide Web pages.

United Nations
http://untreaty.un.org

HIV/AIDS
http://www.unaids.org

Commission on Global Governance
http://www.sovereignty.net/p/gov/gganalysis.htm

Global Trends 2005 Project
http://www.csis.org/gt2005/sumreport.html

InterAction
http://www.interaction.org

IRIN
http://www.irinnews.org

The North-South Institute
http://www.nsi-ins.ca/ensi/index.html

Uniited Nations Home Page
http://www.un.org

The most visible international institution since World War II has been the United Nations. Membership grew from the original 50 in 1945 to 185 in 1995. The UN, across a variety of fronts, achieved noteworthy results—eradication of disease, immunization, provision of food and shelter to refugees and victims of natural disasters, and help to dozens of countries that have moved from colonial status to self-rule.

After the first Gulf War in the early 1990s, the UN guided enforcement of economic sanctions against Iraq, sent peacekeeping forces to former Yugoslavia and to Somalia, monitored an unprecedented number of elections and cease-fire agreements, and played an active peacekeeping role in almost every region of the world. However, the withdrawal of the UN mission in Somalia, the near-collapse of the UN peacekeeping mission in Bosnia prior to the intervention of NATO-sponsored troops, and the delayed response of the UN in sending troops to monitor cease-fire agreements in East Timor and Sierra Leone in 1999 raised doubts about the ability of the organization to continue to be involved in peacekeeping worldwide. Some observers now call for the United Nations to scale back its current level of peacekeeping in order to focus more effectively on global problems that nobody else can or will tackle.

Pressures to reduce UN peacekeeping efforts are also fueled by the realities of scarce resources created largely by the refusal of many member states, including the United States, to pay back dues. The United States withheld payments of back dues for several years as part of a campaign led by conservative members of the U.S. Congress. The congressmen demanded that the UN undertake extensive internal reforms and reduce the amount of the United States' contribution before the United States pay its back dues. The United States narrowly averted losing the right to vote in the General Assembly by agreeing at the end of 1999 to pay a portion of its $1.02 billion debt over a 3-year period. At the end of 2000, UN and U.S. negotiators reached agreement on a deal that includes a large reduction in the dues that the United States pays. How much the terrorist attacks in September 2001 changed the political landscape in the United States was evidence at the end of 2001. As the U.S. made the second of three installments to pay off the U.S. debt to the UN there was no criticism to the U.S. $582 million dollar outlay. Congressional opposition to paying for the UN melted away in the face of the U.S. need for international support in its war on terrorism.

The first four years of George W. Bush's administration in the United States proved to be a difficult period for the United Nations. The Bush administration's emphasis on unilateral action and its doctrine of preventive war have posed a profound challenge to the U.N.'s founding principle of collective security and threatens the organization's continued relevance. The war in Iraq brought these conflicts to a new height. Washington's rush to invade Iraq split the Security Council in ways that have still not healed. Yet the months since the Iraq invasion showed how

much the United States still needs the U.N.'s unparalleled ability to confer international legitimacy and its growing experience in nation building. Even after the UN presence left Iraq, the U.S. had to turn to the organization at the end of 2003 in an effort to build legitimacy and support for delaying elections in Iraq. The policy reversal on the part of the Bush administration underscored the fact that the UN is a useful body even for a unilateralist-orientated regime such as the Bush administration. In "United Nations," the former Secretary of State under Clinton, Madeleine K. Albright, explains why the UN, despite all of its problems, remains the world's best hope against disease, poverty, global crime, and war—and all at a reasonable price.

Since the Bush administration came to power it has been terminating U.S. agreements designed to support the multinational diplomatic system and develop new international norms. During the Bush administration, the U.S. announced that it would not support negotiations on how to implement the Kyoto Protocol on global warming and would not ratify a Protocol Agreement designed to give the 1972 Conventional on the Prohibition of Biological Warfare inspection powers.

Tensions among member states continue over dues allocation, the role of the UN in the Iraqi conflict, and the proper role of United Nation's peacekeeping forces in conflicts worldwide. For many, the UN has not lived up to expectations in securing a disarmed and peaceful world. Especially since the genocide in Rwanda in 1994, there have been several efforts to obtain justice at the international level for the worst war crimes.

The lack of a global consensus on what a broader definition of security would entail and who should be responsible for guaranteeing universal human rights remains a core dispute in international relations and law. Disagreements also extend to other issue areas where analysts, policy makers of nation-states, and international and non-governmental organizations disagree about the most appropriate factor to tackle a variety of broader transnational problems often called "human security" issues. In "Five Wars of Globalization," Moises Naim describes how, in addition to terrorism, governments today are fighting several other networks involved in drugs, arms, intellectual property, people, and money. Naim argues that governments will lose these wars until they adopt new strategies to deal with a larger, unprecedented struggle that now shapes the world as much as confrontations between nation-states once did.

Southern Africa reflects the fact that the HIV/AIDs pandemic is the worst infectious disease pandemic since the black plague swept through feudal Europe. Both pandemics created a series of interrelated and unfamiliar problems. Today many experts warn that the global HIV/AIDS pandemic is shifting from Africa to Eurasia. The death toll in three pivotal countries in the region—Russia, India, and China—could be staggering. The spreading disease will cause a humanitarian tragedy, change the economic potential of the region's major states, and may alter the global balance of power. However, most governments throughout Eurasia much like their counterparts in Africa are ignoring the implications of the shifting global HIV/AIDS crisis. This lack of attention may lead to several governments being swept away much like the old political order in Europe was in the aftermath of the disruptions caused by the black plague. As HIV/AIDS spreads worldwide, it is becoming increasingly evident that transnational trends and problems will increasingly do more to reshape our collective future than more familiar security problems and threats.

Article 36

United Nations

By Madeleine K. Albright

Bureaucratic. Ineffective. Undemocratic. Anti-United States. And after the bitter debate over the use of force in Iraq, critics might add "useless" to the list of adjectives describing the United Nations. So why was the United Nations the first place the Bush administration went for approval after winning the war? Because for $1.25 billion a year—roughly what the Pentagon spends every 32 hours—the United Nations is still the best investment that the world can make in stopping AIDS and SARS, feeding the poor, helping refugees, and fighting global crime and the spread of nuclear weapons.

"The United Nations Has Become Irrelevant"

NO. The second Gulf War battered the U.N. Security Council's already shaky prestige. Hawks condemned the council for failing to bless the war; opponents for failing to block it. Nevertheless, when major combat stopped, the United States and Great Britain rushed to seek council authorization for their joint occupation of Iraq, the lifting of sanctions, and the right to market Iraqi oil.

What lessons will emerge from the wrangle over Iraq? Will France, Russia, and China grudgingly concede U.S. dominance and cooperate sufficiently to keep the United States from routinely bypassing the Security Council? Or might they form an opposition bloc that paralyzes the body? Will the United States and United Kingdom proceed triumphantly? Or will they suffer so many headaches in Iraq that they conclude, in hindsight, that initiating the war without council support was a mistake?

Both sides have reason to move toward cooperation. The French, Russians, and Chinese all derive outsized influence from their status as permanent Security Council members; they see the panel as a means to mitigate U.S. hegemony and do not want the White House to pronounce it dead. And despite their unilateralist tendencies, Bush administration officials will welcome council support when battling terrorists and rogue states in the future. Although the council is not and never has been the preeminent arbiter of war and peace that its supporters wish it were, it remains the most widely accepted source of international legitimacy—and legitimacy still has meaning, even for empires. That is why U.S. President George W. Bush and U.S. Secretary of State Colin Powell both made their major prewar, pro-war presentations before a U.N. audience.

Beyond the council itself, the United Nations' ongoing relevance is evident in the work of the more than two dozen organizations comprising the U.N. system. In 2003 alone, the International Atomic Energy Agency reported that Iran had processed nuclear materials in violation of its Nuclear Nonproliferation Treaty obligations; the International Criminal Tribunal for the Former Yugoslavia tried deposed Yugoslav leader Slobodan Milosevic for genocide; and the World Health Organization successfully coordinated the global response to severe acute respiratory syndrome (SARS). Meanwhile, the World Food Programme has fed more than 70 million people annually for the last five years; the U.N. High Commissioner for Refugees maintains a lifeline to the international homeless; the U.N. Children's Fund has launched a campaign to end forced childhood marriage; the Joint U.N. Programme on HIV/AIDS remains a focal point for global efforts to defeat HIV/AIDS; and the U.N. Population Fund helps families plan, mothers survive, and children grow up healthy in the most impoverished places on earth. The United Nations may seem useless to the self-satisfied, narrow-minded, and micro-hearted minority, but to most of the world's population, it remains highly relevant indeed.

"Relations Between the United States and the United Nations are at an All-Time Low"

NOT EVEN CLOSE. One day before the U.N. General Assembly convened in 1952, Republican Sen. Joseph McCarthy of Wisconsin began hearings in New York on the loyalty of U.S. citizens employed by the United Nations. A federal grand jury then opened a competing inquiry in the same city on the same subject. (Some U.N. employees called to testify even invoked their constitutional right against self-incrimination.) The furor generated massive indignation and mutual U.S.-U.N. distrust. J.B. Mathews, chief investigator for the House Un-American Activities Committee, declared that the United Nations "could not be less of a cruel hoax if it had been organized in Hell for the sole purpose of aiding and abetting the destruction of the United States."

East-West and North-South tensions transformed the General Assembly into hostile territory through much of the 1970s and 1980s. U.S. ambassadors such as Daniel P. Moynihan and Jeane Kirkpatrick earned combat pay rebutting the verbal pyrotechnics of delegates in the throes of anti-Semitic passions and Marxist moonbeams. The low point was the passage in 1975 of a resolution equating Zionism with racism.

In the 1990s, supporters of the Contract With America, led by Republican Rep. Newt Gingrich of Georgia, lambasted U.N. peacekeeping, blocked payment of U.N. dues, and ridiculed U.N. programs. Similarly, Republican Sen. Jesse Helms of North Carolina spoke for many of the far-right-minded but wrong-headed when he termed the United Nations "the nemesis of millions of Americans."

Today according to the Pew Global Attitudes Project, U.S. citizens consider U.N. Secretary-General Kofi Annan the fourth most respected world leader (trailing, in order, British Prime Minister Tony Blair, U.S. President George W. Bush, and Israeli Prime Minister Ariel Sharon). The United States has paid back most of its acknowledged U.N. arrears. The United Nations' agenda and core U.S. security interests have gradually converged. For example, the U.N. Charter says nothing about the importance of elected government, yet U.N. missions routinely sponsor democratic transitions, monitor elections, and promote free institutions. The charter explicitly prohibits U.N. intervention in the internal affairs of any government (save for enforcement actions), yet the U.N. High Commissioner for Human Rights, created in 1993 at the United States' urging, exists solely to nudge governments to do the right thing by their own people. The United Nations' founders never mentioned terrorism, yet today the United Nations encourages governments to ratify anti-terrorist conventions, freeze terrorist assets, and tighten security on land, in air, and at sea. Polls continue to show that a significant majority of U.S. citizens believe the United States should seek U.N. authorization before using force and should cooperate with other nations through the world body.

"The Bush Administration's Doctrine of Preemption Is Not Authorized by the U.N. Charter"

SO? The charter calls upon states to attempt to settle disputes peacefully and, failing that, to refer matters to the Security Council for appropriate action. Article 51 provides that nothing in the charter "shall impair the inherent right of individual or collective self-defense if an armed attack occurs against a Member of the United Nations, until the Security Council has taken measures necessary to maintain international peace and security."

Compare that to this passage from President Bush's 2002 National Security Strategy: "Given the goals of rogue states and terrorists, the United States can no longer solely rely on a reactive posture as we have in the past. The inability to deter a potential attacker, the immediacy of today's threats, and the magnitude of potential harm that could be caused by our adversaries' choice of weapons, do not permit that option. We cannot let our enemies strike first."

The mystery here is not what the administration said, but rather why it chose to arouse global controversy by elevating what has always been a residual option into a highly publicized doctrine. In reality, no U.S. president would allow an international treaty to prevent actions genuinely necessary to deter or preempt imminent attack upon the United States. The notion that the United States has relied solely "on a reactive posture" in the past is not true. In the name of self-defense, U.S. administrations of both parties initiated actions during the Cold War that violated the sovereignty of other nations. In 1994, the Clinton administration considered military strikes against nuclear facilities in North Korea. In 1998, U.S. President Bill Clinton launched cruise missiles into Afghanistan and Sudan in retaliation for the terrorist bombing of two U.S. embassies in Africa and in an effort to prevent al Qaeda leader Osama bin Laden from striking again.

Whether tracking the language of Article 51 or not, the Bush administration's preemption doctrine will prove a departure from past practice only if it is implemented in a manner that is aggressive, indifferent to precedent, and careless of the information used to justify military action. Calibrated and effective actions taken against real enemies posing an imminent danger should not overturn the international legal apple cart. Measures wide of that standard would indeed raise troubling questions about whether the United States is setting itself above the law or tacitly acknowledging the right of every nation to act militarily against threats that are merely potential and suspected. The administration approached that line by invading Iraq, but the issue was blurred by the multiple rationales given for the conflict—enforcement of Security Council resolutions (relatively strong legal grounds), self-defense (in this case, shaky), and liberation (shakier still). The issue now is whether the administration intends to strike first against nuclear aspirants North Korea and Iran (and, if so, on what evidence) and whether it will exhaust other options first. Thus far, the administration is traveling the diplomatic route.

"Political Correctness Often Trumps Substance at the United Nations"

CORRECT. The Cold War and the rapid growth in U.N. membership following decolonization shaped the United Nations' civil service, requiring the distribution of jobs on the basis of geography rather than qualifications. The U.S. Congress did not help over the years by buying in to the notion that the United States was entitled to many jobs and then filling them with defeated politicians.

While at the United Nations, I used to joke that managing the global institution was like trying to run a business with 184 chief executive officers—each with a different language, a distinct set of priorities, and an unemployed brother-in-law seeking a paycheck. Secretary-General Kofi Annan has done about everything possible within the system to reward high achievers and improve recruitment, but the pressure to satisfy members from Afghanistan to Zimbabwe remains a management nightmare.

Another long-standing problem is that decisions on U.N. committee chairs and memberships are most often made on a regional and rotating basis, with equal weight given to, for example, South Africa and Swaziland. By tradition, these decisions are sacrosanct, leading to the recent spectacle of Libya chairing the U.N. Commission on Human Rights. To prevent such an outcome, one must be willing to break some diplomatic china. Former President Clinton did so in blocking the reelection of Secretary-General Boutros Boutros-Ghali in 1996 and defeating Sudan's regionally endorsed nomination for Security Council membership in 2000. Both initiatives prompted resentment toward the United States, but both enhanced the standing and credibility of the United Nations.

"U.N. Peacekeeping Has Failed"

UNTRUE. U.N. peacekeeping has maintained order in such diverse places as Namibia, El Salvador, Cambodia, eastern Slavonia, Mozambique, and Cyprus. The traditional U.N. mission is a confidence-building exercise, conducted in strict neutrality between parties that seek international help in preserving or implementing peace. U.N. peacemaking, however, is quite another matter. During my years as the U.S. permanent representative to the United Nations, the tragic experiences in Bosnia and Herzegovina, Somalia, and Rwanda showed that traditional U.N. peacekeepers lack the mandate, command structure, unity of purpose, and military might to succeed in the more urgent and nasty cases—where the fighting is hot, the innocent are dying, and the combatants oppose an international presence. Such weaknesses, sadly, are inherent in the voluntary and collective nature of the United Nations. When the going gets tough, the tough tend to go wherever they want, notwithstanding the wishes of U.N. commanders.

One possible solution: peace-enforcement missions authorized by the United Nations, in which the Security Council deputies an appropriate major power to organize a coalition and enforce the world's collective will. The council sets the overall mandate, but the lead nation calls the shots—literally and figuratively—necessary to achieve the mission. The U.S.-led intervention in Haiti (1994), the Australian-led rescue of East Timor (1999), and the British action in Sierra Leone (2000) were largely successful and provide a model for the future.

Peacemaking is a hard, dangerous, and often thankless task. To deter people with guns, other people with more and bigger guns are necessary, and finding such people is not easy. It is one thing to expect a soldier to risk life and limb defending his or her homeland. It is another to expect that same soldier to travel halfway around the world and perhaps to die while trying to quell a struggle over diamonds, oil, or ethnic dominance on someone else's home turf. Most people are simply not that altruistic, especially when they see many intervention forces blamed for what such forces fail to accomplish rather than credited for the burdens they assume. As a result, the world is left with an international system of crisis response that is pragmatic, episodic, and incremental rather than principled, reliable, and decisive.

Without any expectation of perfection or even consistency, the international community can nevertheless make the best of things by doing more to equip and train selected military units willing to volunteer in advance for peace enforcement; by recruiting personnel to fill the gap between lightly armed police and heavily armed conventional military; by prosecuting war criminals; and by attacking the roots of conflicts such as arms peddling and economic desperation.

"The U.N. Security Council Should Be Enlarged"

INDUBITABLY, but don't hold your breath. Probably no U.N. issue has been studied more with less to show for the effort than Security Council enlargement.

To ensure the council's strength as a guardian of international security and peace, the United Nations' founders assigned permanent membership and veto authority to the five leading nations on the winning side of World War II: the United States, Great Britain, France, the Soviet Union, and China. (Other countries compete for election to fill the 10 remaining council seats, with the winners serving a two-year term.) Obviously, the world has changed a bit since 1945: U.N. membership has more than tripled, and three of the eight most populous nations in the world can now be found in South Asia. Despite an apparent consensus to enlarge the council, its members have been tied up in knots trying to decide how. Major debates include fair regional representation (if India deserves a permanent seat, what about Pakistan?) and reluctance to extend veto power to additional countries.

During my years at the United Nations in the mid-1990s, the United States supported expanding the council to no more than 21 members and granting permanent seats to Japan and Germany. This position outraged Italian Ambassador F Paolo Fulci, whose country opposed the addition of more permanent members. By that logic, he argued, if Japan and Germany joined the Security Council, Italy should be included as a permanent member, too. "After all," he argued, "Italians also lost World War II."

"The United Nations Is a Threat to the Sovereignty of the United States"

BALDERDASH. The United Nations' authority flows from its members; it is servant, not master. The United Nations has no armed forces of its own, no power of arrest, no authority to tax, no right to confiscate, no ability to regulate, no capacity to override treaties, and—despite the paranoia of some—no black helicopters poised to swoop down upon innocent homes in the middle of the night and steal lawn furniture. The U.N. General Assembly has little power, except to approve the U.N. budget, which it does by consensus. Meanwhile, the Security Council, which does have power, cannot act without the acquiescence of the United States and the other four permanent members. That means that no secretary-general can be elected, no U.N. peacekeeping operation initiated, and no U.N. tribunal established without the approval of the United States. Questions about the efficiency of the United Nations and many of its specific actions are legitimate, but worries about U.S. sovereignty are misplaced and appear to come primarily from people aggrieved to find the United Nations so full of foreigners. That, I am constrained to say, simply cannot be helped.

"The United Nations Is a Huge Bureaucracy"

NOPE. A bureaucracy certainly, but not huge. The annual budget for core U.N. functions-based in New York City, Geneva, Nairobi, Vienna, and five regional commissions—is about $1.25 billion, or roughly what the Pentagon spends every 32 hours. The U.N. Secretariat has reduced its staff by just under 25 percent over the last 20 years and has had a zero-growth budget since 1996. The entire U.N. system, composed of the secretariat and 29 other organizations, employs a little more than 50,000 people, or just 2,000 more than work for the city of Stockholm. Total annual expenditures by all U.N. funds, programs, and specialized agencies equal about one fourth the municipal budget of New York City.

Madeleine K. Albright served as U.S. secretary of state from 1997 to 2001 and as U.S. permanent representative to the United Nations from 1993 to 1996. She is the author of Madam Secretary: A Memoir *(New York: Miramax Books, 2003).*

THE Five Wars OF GLOBALIZATION

The illegal trade in drugs, arms, intellectual property, people, and money is booming. Like the war on terrorism, the fight to control these illicit markets pits governments against agile, stateless, and resourceful networks empowered by globalization. Governments will continue to lose these wars until they adopt new strategies to deal with a larger, unprecedented struggle that now shapes the world as much as confrontations between nation–states once did.

By Moisés Naím

The persistence of al Qaeda underscores how hard it is for governments to stamp out stateless, decentralized networks that move freely, quickly, and stealthily across national borders to engage in terror. The intense media coverage devoted to the war on terrorism, however, obscures five other similar global wars that pit governments against agile, well-financed networks of highly dedicated individuals. These are the fights against the illegal international trade in drugs, arms, intellectual property, people, and money. Religious zeal or political goals drive terrorists, but the promise of enormous financial gain motivates those who battle governments in these five wars. Tragically, profit is no less a motivator for murder, mayhem, and global insecurity than religious fanaticism.

In one form or another, governments have been fighting these five wars for centuries. And losing them. Indeed, thanks to the changes spurred by globalization over the last decade, their losing streak has become even more pronounced. To be sure, nation-states have benefited from the information revolution, stronger political and economic linkages, and the shrinking importance of geographic distance. Unfortunately, criminal networks have benefited even more. Never fettered by the niceties of sovereignty, they are now increasingly free of geographic constraints. Moreover, globalization has not only expanded illegal markets and boosted the size and the resources of criminal networks, it has also imposed more burdens on governments: Tighter public budgets, decentralization, privatization, deregulation, and a more open environment for international trade and investment all make the task of fighting global criminals more difficult. Governments are made up of cumbersome bureaucracies that generally cooperate with difficulty, but drug traffickers, arms dealers, alien smugglers, counterfeiters, and money launderers have refined networking to a high science, entering into complex and improbable strategic alliances that span cultures and continents.

Defeating these foes may prove impossible. But the first steps to reversing their recent dramatic gains must be to recognize the fundamental similarities among the five wars and to treat these conflicts not as law enforcement problems but as a new global trend that shapes the world as much as confrontations between nation-states did in the past. Customs officials, police officers, lawyers, and judges alone will never win these wars. Governments must recruit and deploy more spies, soldiers, diplomats, and economists who understand how to use incentives and regulations to steer markets away from bad social outcomes. But changing the skill set of government combatants alone will not end these wars. Their doctrines and institutions also need a major overhaul.

THE FIVE WARS

Pick up any newspaper anywhere in the world, any day, and you will find news about illegal migrants, drug busts, smuggled weapons, laundered money, or counterfeit goods. The global nature of these five wars was unimaginable just a decade ago. The resources—financial, human, institutional, technological—deployed by the combatants have reached unfathomable orders of magnitude. So have the numbers of victims. The tactics and tricks of both sides boggle the mind. Yet if you cut through the fog of daily headlines and orchestrated photo ops, one inescapable truth emerges: The world's governments are fighting a qualitatively new phenomenon with obsolete tools, inadequate laws, inefficient bureaucratic arrangements, and ineffective strategies. Not surprisingly, the evidence shows that governments are losing.

Drugs

The best known of the five wars is, of course, the war on drugs. In 1999, the United Nations' "Human Development Report" calculated the annual trade in illicit drugs at $400 billion, roughly the size of the Spanish economy and about 8 percent of world trade. Many countries are reporting an increase in drug use. Feeding this habit is a global supply chain that uses everything from passenger jets that can carry shipments of cocaine worth $500 million in a single trip to custom-built submarines that ply the waters between Colombia and Puerto Rico. To foil eavesdroppers, drug smugglers use "cloned" cell phones and broadband radio receivers while also relying on complex financial structures that blend legitimate and illegitimate enterprises with elaborate fronts and structures of cross-ownership.

The United States spends between $35 billion and $40 billion each year on the war on drugs; most of this money is spent on interdiction and intelligence. But the creativity and boldness of drug cartels has routinely outstripped steady increases in government resources. Responding to tighter security at the U.S.-Mexican border, drug smugglers built a tunnel to move tons of drugs and billions of dollars in cash until authorities discovered it in March 2002. Over the last decade, the success of the Bolivian and Peruvian governments in eradicating coca plantations has shifted production to Colombia. Now, the U.S.-supported Plan Colombia is displacing coca production and processing labs back to other Andean countries. Despite the heroic efforts of these Andean countries and the massive financial and technical support of the United States, the total acreage of coca plantations in Peru, Colombia, and Bolivia has increased in the last decade from 206,200 hectares in 1991 to 210,939 in 2001. Between 1990 and 2000, according to economist Jeff DeSimone, the median price of a gram of cocaine in the United States fell from $152 to $112.

Even when top leaders of drug cartels are captured or killed, former rivals take their place. Authorities have acknowledged, for example, that the recent arrest of Benjamin Arellano Felix, accused of running Mexico's most ruthless drug cartel, has done little to stop the flow of drugs to the United States. As Arellano said in a recent interview from jail, "They talk about a war against the Arellano brothers. They haven't won. I'm here, and nothing has changed."

Arms Trafficking

Drugs and arms often go together. In 1999, the Peruvian military parachuted 10,000 AK-47s to the Revolutionary Armed Forces of Colombia, a guerrilla group closely allied to drug growers and traffickers. The group purchased the weapons in Jordan. Most of the roughly 80 million AK-47s in circulation today are in the wrong hands. According to the United Nations, only 18 million (or about 3 percent) of the 550 million small arms and light weapons in circulation today are used by government, military, or police forces. Illicit trade accounts for almost 20 percent of the total small arms trade and generates more than $1 billion a year. Small arms helped fuel 46 of the 49 largest conflicts of the last decade and in 2001 were estimated to be responsible for 1,000 deaths a day; more than 80 percent of those victims were women and children.

Small arms are just a small part of the problem. The illegal market for munitions encompasses top-of-the-line tanks, radar systems that detect Stealth aircraft, and the makings of the deadliest weapons of mass destruction. The International Atomic Energy Agency has confirmed more than a dozen cases of smuggled nuclear-weapons-usable material, and hundreds more cases have been reported and investigated over the last decade. The actual supply of stolen nuclear-, biological-, or chemical-weapons materials and technology may still be small. But the potential demand is strong and growing from both would-be nuclear powers and terrorists. Constrained supply and increasing demand cause prices to rise and create enormous incentives for illegal activities. More than one fifth of the 120,000 workers in Russia's former "nuclear cities"—where more than half of all employees earn less than $50 a month—say they would be willing to work in the military complex of another country.

Governments have been largely ineffective in curbing either supply or demand. In recent years, two countries, Pakistan and India, joined the declared nuclear power club. A U.N. arms embargo failed to prevent the reported sale to Iraq of jet fighter engine parts from Yugoslavia and the Kolchuga anti-Stealth radar system from Ukraine. Multilateral efforts to curb the manufacture and distribution of weapons are faltering, not least because some powers are unwilling to accept curbs on their own activities. In 2001, for example, the United States blocked a legally binding global treaty to control small arms in part because it worried about restrictions on its own citizens' rights to own guns. In the absence of effective international legislation and enforcement, the laws of economics dictate the sale of more weapons at cheaper prices: In 1986, an AK-47 in Kolowa, Kenya, cost 15 cows. Today, it costs just four.

Other Fronts

Drugs, arms, intellectual property, people, and money are not the only commodities traded illegally for huge profits by international networks. They also trade in human organs, endangered species, stolen art, and toxic waste. The illegal global trades in all these goods share several fundamental characteristics: Technological innovations and political changes open new markets, globalization is increasing both the geographical reach and the profit opportunities for criminal networks, and governments are on the losing end of the fight to stop them. Some examples:

Human organs: Corneas, kidneys, and livers are the most commonly traded human parts in a market that has boomed thanks to technology, which has improved preservation techniques and made transplants less risky. In the United States, 70,000 patients are on the waiting list for major organ transplants while only 20,000 of them succeed in getting the organ they need. Unscrupulous "organ brokers" partly meet this demand by providing, for a fee, organs and transplant services. Some of the donors, especially of kidneys, are desperately poor. In India, an estimated 2,000 people a year sell their organs. Many organs, however, come from nonconsenting donors forced to undergo operations or from cadavers in police morgues. For example, medical centers in Germany and Austria were recently found to have used human heart valves taken without consent from the cadavers of poor South Africans.

Endangered species: From sturgeon for caviar in gourmet delicatessens to tigers or elephants for private zoos, the trade in endangered animals and plants is worth billions of dollars and includes hundreds of millions of plant and animal types. This trade ranges from live animals and plants to all kinds of wildlife products derived from them, including food products, exotic leather goods, wooden musical instruments, timber, tourist curiosities, and medicines.

Stolen art: Paintings and sculptures taken from museums, galleries, and private homes, from Holocaust victims, or from "cultural artifacts" poached from archeological digs and other ancient ruins are also illegally traded internationally in a market worth an estimated $2 billion to $6 billion each year. The growing use of art-based transactions in money laundering has spurred demand over the last decade. The supply has boomed because the Soviet Union's collapse flooded the world's market with art that had been under state control. The Czech Republic, Poland, and Russia are three of the five countries most affected by art crime worldwide.

Toxic waste: Innovations in maritime transport, tighter environmental regulations in industrialized countries coupled with increased integration of poor countries to the global economy and better telecommunications have created a market where waste is traded internationally. Greenpeace estimates that during the 20 years prior to 1989, just 3.6 million tons of hazardous waste were exported; in the five years after 1989, the trade soared to about 6.7 billion tons. The environmental organization also reckons that 86 to 90 percent of all hazardous waste shipments destined for developing countries—purportedly for recycling, reuse, recovery, or humanitarian uses—are toxic waste.

—M.N.

Intellectual Property

In 2001, two days after recording the voice track of a movie in Hollywood, actor Dennis Hopper was in Shanghai where a street vendor sold him an excellent pirated copy of the movie with his voice already on it. "I don't know how they got my voice into the country before I got here," he wondered. Hopper's experience is one tiny slice of an illicit trade that cost the United States an estimated $9.4 billion in 2001. The piracy rate of business software in Japan and France is 40 percent, in Greece and South Korea it is about 60 percent, and in Germany and Britain it hovers around 30 percent. Forty percent of Procter & Gamble shampoos and 60 percent of Honda motorbikes sold in China in 2001 were pirated. Up to 50 percent of medical drugs in Nigeria and Thailand are bootleg copies. This problem is not limited to consumer products: Italian makers of industrial valves worry that their $2 billion a year export market is eroded by counterfeit Chinese valves sold in world markets at prices that are 40 percent cheaper.

The drivers of this bootlegging boom are complex. Technology is obviously boosting both the demand and the supply of illegally copied products. Users of Napster, the now defunct Internet company that allowed anyone, anywhere to download and reproduce copyrighted music for free, grew from zero to 20 million in just one year. Some 500,000 film files are traded daily through file-sharing services such as Kazaa and Morpheus; and in late 2002, some 900 million music files could be downloaded for free on the Internet—that is, almost two and a half times more files than those available when Napster reached its peak in February 2001.

Global marketing and branding are also playing a part, as more people are attracted to products bearing a well-known brand like Prada or Cartier. And thanks to the rapid growth and integration into the global economy of countries, such as China, with weak central governments and ineffective laws, producing and exporting near perfect knockoffs are both less expensive and less risky. In the words of the CEO of one of the best known Swiss watchmakers: "We now compete with a product manufactured by Chinese prisoners. The business is run by the Chinese military, their families and friends, using roughly the same machines we have, which they purchased at the same industrial fairs we go to. The way we have rationalized this problem is by assuming that their customers and ours are different. The person that buys a pirated copy of one of our $5,000 watches for less than $100 is not a client we are losing. Perhaps it is a future client that some day will want to own the real thing instead of a fake. We may be wrong and we do spend money to fight the piracy of our products. But given that our efforts do not seem to protect us much, we close our eyes and hope for the better." This posture stands in contrast to that of companies that sell cheaper products such as garments, music, or videos, whose revenues are directly affected by piracy.

Governments have attempted to protect intellectual property rights through various means, most notably the World Trade Organization's Agreement on Trade-Related Aspects of Intellectual Property Rights (TRIPS). Several other organizations such as the World Intellectual Property Organization, the World Customs Union, and Interpol are also involved. Yet the large and growing volume of this trade, or a simple stroll in the streets of Manhattan or Madrid, show that governments are far from winning this fight.

Alien Smuggling

The man or woman who sells a bogus Hermes scarf or a Rolex watch in the streets of Milan is likely to be an illegal alien. Just as likely, he or she was transported across several continents by a trafficking network allied with another network that specializes in the illegal copying, manufacturing, and distributing of high-end, brand-name products.

Alien smuggling is a $7 billion a year enterprise and according to the United Nations is the fastest growing business of organized crime. Roughly 500,000 people enter the United States illegally each year—about the same number as illegally enter the European Union, and part of the approximately 150 million who live outside their countries of origin. Many of these backdoor travelers are voluntary migrants who pay smugglers up to $35,000, the top-dollar fee for passage from China to New York. Others, instead, are trafficked—that is, bought and sold internationally—as commodities. The U.S. Congressional Research Service reckons that each year between 1 million and 2 million people are trafficked across borders, the majority of whom are women and children. A woman can be "bought" in Timisoara, Romania, for between $50 and $200 and "resold" in Western Europe for 10 times that price. The United Nations Children's Fund estimates that cross-border smugglers in Central and Western Africa enslave 200,000 children a year. Traffickers initially tempt victims with job offers or, in the case of children, with offers of adoption in wealthier countries, and then keep the victims in subservience through physical violence, debt bondage, passport confiscation, and threats of arrest, deportation, or violence against their families back home.

Governments everywhere are enacting tougher immigration laws and devoting more time, money, and technology to fight the flow of illegal aliens. But the plight of the United Kingdom's government illustrates how tough that fight is. The British government throws money at the problem, plans to use the Royal Navy and Royal Air Force to intercept illegal immigrants, and imposes large fines on truck drivers who (generally unwittingly) transport stowaways. Still, 42,000 of the 50,000 refugees who have passed through the Sangatte camp (a main entry point for illegal immigration to the United Kingdom) over the last three years have made it to Britain. At current rates, it will take 43 years for Britain to clear its asylum backlog. And that country is an island. Continental nations such as Spain, Italy, or the United States face an even greater challenge as immigration pressures overwhelm their ability to control the inflow of illegal aliens.

Money Laundering

The Cayman Islands has a population of 36,000. It also has more than 2,200 mutual funds, 500 insurance companies, 60,000 businesses, and 600 banks and trust companies with almost $800 billion in assets. Not surprisingly, it figures prominently in any discussion of money laundering. So does the United States, several of whose major banks have been caught up in investigations of money laundering, tax evasion, and fraud. Few, if any, countries can claim to be free of the practice of helping individuals and companies hide funds from governments, creditors, business partners, or even family members, including the proceeds of tax evasion, gambling, and other crimes. Estimates of the volume of global money laundering range between 2 and 5 percent of the world's annual gross national product, or between $800 billion and $2 trillion.

Smuggling money, gold coins, and other valuables is an ancient trade. Yet in the last two decades, new political and economic trends coincided with technological changes to make this ancient trade easier, cheaper, and less risky. Political changes led to the deregulation of financial markets that now facilitate cross-border money transfers, and technological changes made distance less of a factor and money less "physical." Suitcases full of banknotes are still a key tool for money launderers, but computers, the Internet, and complex financial schemes that combine legal and illegal practices and institutions are more common. The sophistication of technology, the complex web of financial institutions that crisscross the globe, and the ease with which "dirty" funds can be electronically morphed into legitimate assets make the regulation of international flows of money a daunting task. In Russia, for example, it is estimated that by the mid-1990s organized crime groups had set up 700 legal and financial institutions to launder their money.

Faced with this growing tide, governments have stepped up their efforts to clamp down on rogue international banking, tax havens, and money laundering. The imminent, large-scale introduction of e-money—cards with microchips that can store large amounts of money and thus can be easily transported outside regular channels or simply exchanged among individuals—will only magnify this challenge.

WHY GOVERNMENTS CAN'T WIN

The fundamental changes that have given the five wars new intensity over the last decade are likely to persist. Technology will continue to spread widely; criminal networks will be able to exploit these technologies more quickly than governments that must cope with tight budgets, bureaucracies, media scrutiny, and electorates. International trade will continue to grow, providing more cover for the expansion of illicit trade. International migration will likewise grow, with much the same effect, offering ethnically based gangs an ever growing supply of recruits and victims. The spread of democracy may also help criminal cartels, which can manipulate weak public institutions by corrupting police officers or tempting politicians with offers of cash for their increasingly expensive election campaigns. And ironically, even the spread of international law—with its growing web of embargoes, sanctions, and conventions—will offer criminals new opportunities for providing forbidden goods to those on the wrong side of the international community.

> **Even the spread of international law will offer criminals new opportunities for providing forbidden goods to those on the wrong side of the international community.**

These changes may affect each of the five wars in different ways, but these conflicts will continue to share four common characteristics:

They are not bound by geography.

Some forms of crime have always had an international component: The Mafia was born in Sicily and exported to the United States, and smuggling has always been by definition international. But the five wars are truly global. Where is the theater or front line of the war on drugs? Is it Colombia or Miami? Myanmar (Burma) or Milan? Where are the battles against money launderers being fought? In Nauru or in London? Is China the main theater in the war against the infringement of intellectual property, or are the trenches of that war on the Internet?

They defy traditional notions of sovereignty.

Al Qaeda's members have passports and nationalities—and often more than one—but they are truly stateless. Their allegiance is to their cause, not to any nation. The same is also true of the criminal networks engaged in the five wars. The same, however, is patently *not* true of government employees—police officers, customs agents, and judges—who fight them. This asymmetry is a crippling disadvantage for governments waging these wars. Highly paid, hypermotivated, and resource-rich combatants on one side of the wars (the criminal gangs) can seek refuge in and take advantage of national borders, but combatants of the other side (the governments) have fewer resources and are hampered by traditional notions of sovereignty. A former senior CIA official reported that international criminal gangs are able to move people, money, and weapons globally faster than he can move resources inside his own agency, let alone worldwide. Coordination and information sharing among government agencies in different countries has certainly improved, especially after September 11. Yet these tactics fall short of what is needed to combat agile organizations that can exploit every nook and cranny of an evolving but imperfect body of international law and multilateral treaties.

They pit governments against market forces.

In each of the five wars, one or more government bureaucracies fight to contain the disparate, uncoordinated actions of thousands of independent, stateless organizations. These groups are motivated by large profits obtained by exploiting international price differentials, an unsatisfied demand, or the cost advantages produced by theft. Hourly wages for a Chinese cook are far higher in Manhattan than in Fujian. A gram of cocaine in Kansas City is 17,000 percent more expensive than in Bogotá. Fake Italian valves are 40 percent cheaper because counterfeiters don't have to cover the costs of developing the product. A well-funded guerrilla group will pay anything to get the weapons it needs. In each of these five wars, the incentives to successfully overcome government-imposed limits to trade are simply enormous.

They pit bureaucracies against networks.

The same network that smuggles East European women to Berlin may be involved in distributing opium there. The proceeds of the latter fund the purchase of counterfeit Bulgari watches made in China and often sold on the streets of Manhattan by illegal African immigrants. Colombian drug cartels make deals with Ukrainian arms traffickers, while Wall Street brokers controlled by the U.S.-based Mafia have been known to front for Russian money launderers. These highly decentralized groups and individuals are bound by strong ties of loyalty and common purpose and organized around semiautonomous clusters or "nodes" capable of operating swiftly and flexibly. John Arquilla and David Ronfeldt, two of the best known experts on these types of organizations, observe that networks often lack central leadership, command, or headquarters, thus "no precise heart or head that can be targeted. The network as a whole (but not necessarily each node) has little to no hierarchy; there may be multiple leaders.... Thus the [organization's] design may sometimes appear acephalous (headless), and at other times polycephalous (Hydra-headed)." Typically, governments respond to these challenges by forming interagency task forces or creating new bureaucracies. Consider the creation of the new Department of Homeland Security in the United States, which encompasses 22 former federal agencies and their 170,000 employees and is responsible for, among other things, fighting the war on drugs.

RETHINKING THE PROBLEM

Governments may never be able to completely eradicate the kind of international trade involved in the five wars. But they can and should do better. There are at least four areas where efforts can yield better ideas on how to tackle the problems posed by these wars:

Develop more flexible notions of sovereignty.

Governments need to recognize that restricting the scope of multilateral action for the sake of protecting their sovereignty is often a moot point. Their sovereignty is compromised daily, not by nation-states but by stateless networks that break laws and cross borders in pursuit of trade. In May 1999, for example, the Venezuelan government denied U.S. planes authorization to fly over Venezuelan territory to monitor air routes commonly used by narcotraffickers. Venezuelan authorities placed more importance on the symbolic value of asserting sovereignty over air space than on the fact that drug traffickers' planes regularly violate Venezuelan territory. Without new forms of codifying and "managing" sovereignty, governments will continue to face a large disadvantage while fighting the five wars.

Strengthen existing multilateral institutions.

The global nature of these wars means no government, regardless of its economic, political, or military power, will make much progress acting alone. If this seems obvious, then why does Interpol, the multilateral agency in charge of fighting international crime, have a staff of 384, only 112 of whom are police officers, and an annual budget of $28 million, less than the price of some boats or planes used by drug traffickers? Similarly, Europol, Europe's Interpol equivalent, has a staff of 240 and a budget of $51 million.

One reason Interpol is poorly funded and staffed is because its 181 member governments don't trust each other. Many assume, and perhaps rightly so, that the criminal networks they are fighting have penetrated the police departments of other countries and that sharing information with such compromised officials would not be prudent. Others fear today's allies will become tomorrow's enemies. Still others face legal impediments to sharing intelligence with fellow nation-states or have intelligence services and law enforcement agencies with organizational cultures that make effective collaboration almost impossible. Progress will only be made if the world's governments unite behind stronger, more effective multilateral organizations.

Devise new mechanisms and institutions.

These five wars stretch and even render obsolete many of the existing institutions, legal frameworks, military doctrines, weapons systems, and law enforcement techniques on which governments have relied for years. Analysts need to rethink the concept of war "fronts" defined by geography and the definition of "combatants" according to the Geneva Convention. The functions of intelligence agents, soldiers, police officers, customs agents, or immigration officers need rethinking and adaptation to the new realities. Policymakers also need to reconsider the notion that ownership is essentially a physical reality and not a "virtual" one or that only sovereign nations can issue money when thinking about ways to fight the five wars.

Move from repression to regulation.

Beating market forces is next to impossible. In some cases, this reality may force governments to move from repressing the market to regulating it. In others, creating market incentives may be better than using bureaucracies to curb the excesses of these markets. Technology can often accomplish more than government policies can. For example, powerful encryption techniques can better protect software or CDs from being copied in Ukraine than would making the country enforce patents and copyrights and trademarks.

In all of the five wars, government agencies fight against networks motivated by the enormous profit opportunities created by other government agencies. In all cases, these profits can be traced to some form of government intervention that creates a major imbalance between demand and supply and makes prices and profit margins skyrocket. In some cases, these government interventions are often justified and it would be imprudent to eliminate them—governments can't simply walk away from the fight against trafficking in heroin, human beings, or weapons of mass destruction. But society can better deal with other segments of these kinds of illegal trade through regulation, not prohibition. Policymakers must focus on opportunities where market regulation can ameliorate problems that have defied approaches based on prohibition and armed interdiction of international trade.

Ultimately, governments, politicians, and voters need to realize that the way in which the world is conducting these five wars is doomed to fail—not for lack of effort, resources, or political will but because the collective thinking that guides government strategies in the five wars is rooted in wrong ideas, false assumptions, and obsolete institutions. Recognizing that governments have no chance of winning unless they change the ways they wage these wars is an indispensable first step in the search for solutions.

Moisés Naím is editor of FOREIGN POLICY *magazine.*

From *Foreign Policy*, January/February 2003, pp. 28-37 © 2003 by the Carnegie Endowment for International Peace (www.foreignpolicy.com). Permission conveyed through Copyright Clearance Center, Inc.

Index

A

Abdul Aziz, prince of Saudi Arabia, 205, 206–207
Abdullah, prince of Saudi Arabia, 205–206, 207
Acheson, Dean, 133
Afghanistan, 128–129, 153, 184–185
Africa, U.S. and, 212–215
African Growth and Opportunity Act (AGOA), 214
African Union, 212
Agreed Framework, 48, 49, 50, 57–58, 175, 176, 177
agriculture: biobased economy and, 33–41; NAFTA and, 114–115; North Korea and, 54–55; Polish entry in EU and, 146
aid organizations, 20–21
AIDS, Africa and, 212, 213, 214
airpower, 10–11
al Qaeda, 12–16, 68, 152, 153, 210
Albania, 19
Albright, David, 71–72, 76
alien smuggling, illegal trafficking in, 226
Allende, Salvador, 96
American Civil Liberties Union (ACLU), 101, 102
Angola, 29, 30
Animal and Plant Health Inspection Service (APHIS), bioterrorism and, 37
Annan, Kofi, 21
anti-Americanism, Arabs and, 189–198
Anti-Ballistic Missile (ABM) Treaty, 152
anti-Semitism, terrorist training and, 161–163
anti-war demonstrators, 99; surveillance and, 102
Arabs, anti-Americanism and, 189–198
Argentina, 18, 116–117, 120; credit cards in, 26
armed forces: of Europe, 142–144; U.S., and biobased economy, 39–40; stealth operations of, 91–100
arms trafficking, 224
art, illegal trafficking in, 226
Aryan Nations, 14
Asia, use of credit cards in, 24, 25, 26, 27
assassinations, political, 97
assertive nationalists, doctrine of preemption and, 85
Aziz, Tariq, 45

B

Baker, James, 208
Bandar, prince of Saudi Arabia, 209–210
banks. See financial industry
Beam, Louis, 14
Belarus, 73

Belgium, armed forces of, 142, 143–144
Bell for Adano, A (Hershey), 93
bin Laden, Osama, 12, 13, 190
biobased economy, 33–41
biodiversity, biobased economy and, 34–35
Biodiversity Treaty, 35–36
biological weapons: Internet civil defense against terrorism and, 77–79; Iraq and, 45, 46, 47
biorefineries, 34
Biosafety Protocol to the Biodiversity Treaty, 36
bioterrorism, 37–38
Bolivia, 119, 225
Boot, Max, 98
Border Patrol, U.S., 106, 107
Botswana, 30
Brazil, 116, 117, 121
Britain, immigrants in, 139–141
bureaucracy, UN and, 222–223
Bush, George W., 69, 75, 129, 135; China and, 169–174; doctrine of preemption and, 83–90, 220–221; as revolutionary, 83–90; Russia and, 152, 153, 154

C

Callard, James, 99
Cambodia, 29
Canada: NAFTA and, 113–115; security issues of terrorism and, 104–108, 109
Cardoso, Fernando Henrique, 116
Carlucci, Frank, 208
Carlyle Group, 208
Cartegena Protocol on Biosafety, 36
cartography, terrorist training and, 158
Cayman Islands, 227
Central Intelligence Agency (CIA), 97, 103, 208, 209; weapons of mass destruction program in Iraq and, 45–47
Chávez, Hugo, 117–118, 121
Chechnya, 152, 153
chemical weapons, Iraq and, 45–46, 47
Cheney, Dick, 85, 209
ChevronTexaco, 208–209
children, monitoring of, and bioterrorism, 78
Chile, 96, 116, 121
China, 18, 136; credit cards in, 25, 26; economic growth in, versus India, 180–183; North Korea and, 48, 55–56, 172, 178, 179; U.S. and, 169–174
Christian missionaries, terrorist training and, 163–164
citizenship: globalization and, 5; immigrants in Europe and, 141
civil society, globalization and, 5

civil war, 28–32; ethnic conflict and, 28–29; natural resources and, 29, 30, 31
Clinton, Bill, 132
Cochran, Thomas, 68, 69, 72
Colombia, 32, 91, 95–96, 99, 225
colonialism, 94
combination warfare, 99
Cooperative Threat Reduction (CTR) program, 154
credit cards, increasing worldwide use of, 24–27
Croatia, 17, 19, 97
cultural artifacts, illegal trafficking in, 226

D

Darman, Richard, 208
de Gualle, Charles, 134–135
democracy: civil war and, 29; promotion of, by military, 95–96; war and, 9
demolitions, terrorist training and, 159
demonstrators, 99; surveillance and, 102
Denmark, immigrants in, 140
desalination, biobased economy and, 38
diamonds, civil war and, 29, 30, 31
diasporas. See emigrants
direct-current stabilizers, 63
dirty bombs, 68, 73, 74, 75
drug trafficking, 225
Duhalde, Eduardo, 120
Dulles, John Foster, 133
DuPont, 34

E

Ecuador, 118
Egyptian Islamic Jihad (EIJ), 12
El Salvador, 95, 96
electronic matrix, 78–79
emigrants: civil wars and, 31; economic growth in China versus India and, 180, 181, 182–183; in Europe, 139–141; impact of, on home countries, 17–19; NAFTA and, 114–115
employment, NAFTA and, 113–114
"End of History" thesis, of Francis Fukuyama, 3
endangered species, illegal trafficking in, 226
entrepreneurs, economic growth in India and, 180–183
EP-3 incident, 170, 174
Eritrea, 17, 19
Estonia, 19, 32
ethnic conflict, civil war and, 28–29
Europe: armed forces of, 142–144; U.S. and, 125–130, 131–138

European Coal and Steel Community (ECSC), 134
European Commission, 145
European Union (EU), 125, 135; Poland and, 145–147

F

Faber, Peter, 99
Faud, king of Saudi Arabia, 205, 206, 207
Federal Bureau of Investigation (FBI), terrorism and, 101–103
Federation of American Scientists, 74
financial industry: in China versus India, 182; credit cards and, 24–27
financial system, of Mexico, and NAFTA, 114
"force caps," 98–99
foreign direct investment (FDI), economic growth in China and, 180–183
Foreign Intelligence Surveillance Act, 103
Fortuyn, Pim, 139, 140
Fox, Vincente, 118–119, 121
France, 88, 134–135, 136; armed forces of, 142; immigrants in, 139, 140
free trade: China and, 172–173; NAFTA and, 113–115
"freedom fighter" movements, 195
Friedman, Thomas, 4, 38
Fukuyama, Francis, 3

G

garage bombs, 70–72
Garcia, Alan, 120
gas centrifuges, 64, 65–66
genetically modified organisms (GMOs), biobased economy and, 36
geography, civil war and, 29
Germany, 88, 135, 136; armed forces of, 142; immigrants in, 139, 140, 141
Ghanians, 18–19
Gladstone, Ewart, 99
Global Fund to Fight AIDS, Tuberculosis, and Malaria, 212, 214
globalization, 3–7; biobased economy and, 38; illegal trade activities and, 224–229
Gottemoeller, Rose, 73
Greeks, 17
Gutiérrez, Lucio, 118, 121

H

Habiger, Eugene E., 69, 70, 71
Hale, Matt, 16
Halliburton, 208–209
Hamas, 14, 191
Hershey, John, 93
Hezbollah, 13, 14

highly-enriched-uranium (HEU) nuclear program, of North Korea, 48, 49, 50, 63, 64, 175–179
Hinton, Deane, 96
HIV, Africa and, 212, 213, 214
Hizb-ut-Tahrir (Party of Liberation), 164
Hong Kong, credit cards in, 26
Hu Jintao, 169, 171, 173–174
Huber, Albert, 16
human organs, illegal trafficking in, 226
human rights: Africa and, 214; China and, 172–173; North Korea and, 55; U.S. military trainers and, 95
Huntington, Samuel, 3
Hussein, Saddam, 14, 45

I

Ibrahim, Jawhara al-, 205, 206, 207
immigrants, in Europe, 139–141. *See also* emigrants
Inácio da Silva, Luiz "Lula," 116, 117
India, 18, 185; credit cards in, 25; economic growth in, versus China, 180–183
Indonesia, credit cards in, 25
information technology, war and, 9
Infosys, 181–182
intellectual property rights: AIDS and, 214; illegal trafficking and, 226–227
Intergovernmental Council (IGC), 125, 126
international aid organizations, 20–21
International Institute for Islamic Thought (IIIT), 210
International Islamic Front for Jihad Against the Jews and Crusaders (IIF), 13, 15
International Islamic Relief Organization (IIRO), 210
Internet: civil defense against bioterrorism and, 77–79; use of, by terrorists, 14–15
Iran, 87
Iraq, 71–72; nation-building and, 199–202; weapons of mass destruction program in, 45–47; U.S. war with, 83–90, 131, 136–137, 151, 153–154
Irish-Americans, 17
"iron majors," 94
Islam, anti-Americanism and, 189–918
Islamic Movement of Uzbekistan (IMU), 12–13, 157
Islamic Renaissance Party (IRP), of Uzbekistan, 157
Italians, in Argentina, 18
Italy, immigrants in, 140

J

Jackson-Vanik Amendment, 154
Jamaat ul-Fuqra, 15
Japan, North Korea and, 48, 178, 179
Jiang Zemin, 172, 173

Joppolo, Victor, 93

K

Kansi, Mir Aimal, 14
Kay, David A., 45–47
Kennard, William, 208
Khan, Asif Reza, 15
Khmer Rouge, 29
Khyber Rifles, 94
Kim Jong Il, 54, 176–177, 178
Kissinger, Henry, 3, 94
Korean Peninsula Energy Development Organization (KEDO), 179
Kosovo, 19

L

Lashkar-e-Taiba, 13, 15
Latin America, new political leaders of, 116–121
Latvia, 19, 21
leaderless resistance, terrorists and, 14
Lepper, Andrzej, 145
Levi, Michael A., 67
Levitt, Arthur, 208
Lindh, John Walker, 15
Linn, Brian McAllister, 97
Lithuania, 19
living modified organisms (LMOs), biobased economy and, 36
Lundestad, Geir, 134
Luongo, Kenneth, 70

M

Macedonia, 17
Mahler, Horst, 16
Major, John, 208
Mandela, Nelson, 212
Marquez, Gabriel Garciá, 117–118
Mbeki, Thabo, 212
McCarthy, Joseph, 220
Mexico, 17, 118–119, 121; credit cards in, 25–26; NAFTA and, 113–115
military. *See* armed forces
Millennium Challenge Account (MCA), 214
Miller, Leszek, 145
mines, terrorist training and, 159
minorities, in Europe, 139–141
Mir, Hamid, 14
missile technology, North Korea and, 50–51
missionaries, terrorist training and, 163–164
money laundering, illegal trafficking and, 226
Mongolia, 94
Mugabe, Robert, 20, 215
Muhammad, John Allen, 14
Murthy, Narayana, 181–182
Muslim World League, 210

Index

N

Nam Chon Gang, 64
nation building, Iraq and, 199–202
National Union for the Total Independence of Angola (UNITA), 29, 30
natural resources, civil war and, 29, 30, 31
neoconservatives, doctrine of preemption and, 84–85
Netherlands, immigrants in, 139, 140
New Partnership for African Development (NEPAD), 212
Nodong, 50
North American Free Trade Agreement (NAFTA), 113–115
North Atlantic Treaty Organization (NATO), 125, 126–128, 132, 134; Russia and, 126, 153
North Korea, 87; nuclear weapons and, 48–62, 63–66, 172, 175–179
nuclear power plants, terrorism and, 74–75
Nuclear Threat Initiative, 72–73
nuclear weapons: Iraq and, 46; North Korea and, 48–62, 63–66, 172, 175–179; terrorism and, 67–76
Nunn, Sam, 70, 72–73

O

Objective Force, 39
Omar Saeed Sheikh, Ahmed, 15
Optronic, 64–65
Ortega y Gasset, José, 99

P

Padilla, Jose, 15
Pakistan, 15, 185, 190, 192
PALS. See permissive action links
patents. See intellectual property rights
Patriot Act, 10
peacekeeping: civil war and, 31; UN, 221–222
Peloponnesian War, 87
permissive action links (PALS), nuclear weapons and, 70
Peru, 118–119, 225, 121
pharmaceutical industry, AIDS in Africa and, 214
Philippines, 98
Pinochet, Augusto, 96
Plant Protection and Quarantine (PPQ) Division, of APHIS, and bioterrorism, 37
plutonium production, North Korea and, 49, 177–178
poisons, terrorist training and, 159–160
Poland, 145–147
political correctness, UN and, 221
political globalization, 4
polymers, biobased economy and, 34
Potter, William C., 73

precision agriculture, 38
preemption: Bush and, 83–90, 220–221; UN Charter and, 220–221
preemptive wars, 87
preventive wars, 87
Putin, Vladimir, 69, 75, 152, 153

R

radiation bombs, 68, 73, 74, 75
Reid, Richard, 15
Revolt of the Masses, The (Ortega y Gasset), 99
Revolutionary Armed Forces of Colombia (FARC), 32, 91, 225
Revolutionary United Front (RUF), of Sierra Leone, 29, 30
Rice, Condoleezza, 208
Ridge, Tom, 68
Roach, J. S., 95
Rumsfeld, Donald, 85, 88, 129, 133
Russia, 21, 136; Afghanistan and, 153, 184–185; NATO and, 126, 153; North Korea and, 48, 55–56, 179; nuclear weapons of, and terrorism, 68–69, 72, 73, 74; U.S. and, 151–155

S

SAAR Foundation, 210
Safa Trust, 210
Sánchez de Lozada, Gonzalo, 119
Saudi Arabia, 203–211
Savage Wars of Peace, The (Boot), 98
Seelye, Talcott, 94
Shachnow, Sidney, 93
Shales, Robert, 39–40
Shikalov, Vladimir, 67
Shineski, Eric, 39
Sierra Leone, 29, 30
small arms, terrorist training and, 158–159
South Africa, 212
South Korea: credit cards in, 24, 25, 26; North Korea and, 48, 51–53, 177, 179
sovereignty: aid and, 20–21; UN and, 222
Spain, immigrants in, 140
Sri Lanka, 18
stolen art, illegal trafficking in, 226
Suez crisis, 134
surveillance warrants, terrorism and, 101, 103

T

Tablighi Jamaat (TJ), 15
Taepodong 1 (TD–1), 50
Taepodong 2 (TD–2), 50–51
Taiwan, China and, 169, 170, 171, 172–173, 174
Tamils, 18
targeting, terrorist training and, 159
Taylor, Theodore, 73–74

terrorism, 3, 4, 6, 12–16; bio-, 37–38, 77–79; doctrine of preemption and, 83–90; intelligence network and, 101–103; nuclear weapons and, 67–76; Saudi Arabian sponsorship of, 206–207, 210–211; Saudi Arabian oil fields and, 203–204; security issues of U.S. and Canada and, 104–108, 109; U.S.–China relations and, 170–171, 172; U.S.–Russian relations and, 152, 153; Uzbekistan and, 156–165
Thailand, credit cards in, 25, 26
Toledo, Alejandro, 118–119, 121
toxic waste, illegal trafficking in, 226
Trade-Related Aspects of Intellectual Property Rights (TRIPS) Agreement, 35
Treaty of Moscow, 153
Truppel, Hans Werner, 64–65
Turkey, 19
Turner, Ted, 72–73

U

Umar, Warith Deen, 15
unions, armed forces of Europe and, 143
United Nations (UN), 219–223; bureaucracy of, 222; peacekeeping and, 221–222; political correctness and, 221; relationship with U.S. and, 220; sovereignty and, 222
United Nation Charter, doctrine of preemption and, 220–221
United Nations Framework Convention of Biological Diversity, 35–36
United Nations Security Council, 9, 97, 219–220; enlarging, 92, 222; Iraq War and, 153–154
United States Coast Guard, 97
United States Department of Agriculture (USDA), bioterrorism and, 37
United States Information Agency, 99
United States, NAFTA and, 113–115
uranium. See highly-enriched-uranium (HEU) nuclear program
urbanization, biobased economy and, 36–37
Uribe, Álvaro, 96–97, 119–120
U.S. Army and Counterinsurgency in the Philippine War, The (Linn), 98
Uzbekistan, terrorism and, 12–13, 156–165

V

Valynkin, Igor, 69
van Bebber, Charles, 94
Venezuela, 116–118, 120–121
Ville de Virgo, 63, 64–65
virtual networks, terrorists and, 14
Volodin, Yuri G., 72
von Hippel, Frank, 72, 74

W

Wahhabi movement, 205, 206–207
war, 8–11; airpower and, 10–11; asymmetrical, 10; civil, 28–32; democracy and, 9; information technology and, 10; preemptive, 87; preventive, 87; United Nations and, 9
Warburton, Robert, 94
water, biobased economy and, 34, 38
weapons of mass destruction (WMDs), 3–4, 129, 153; doctrine of preemption and, 83–90; Internet civil defense against bioterrorism and, 77–79; Iraq and, 45–47; North Korea and, 48–62, 63–66, 175–179; terrorism and, 67–76, 77–79
white supremists, 16
Wilhelm, Tom, 94
Willrich, Mason, 73–74
Workers' Party (PT), of Brazil, 116, 117
World Trade Organization (WTO), 214; China and, 170, 173; Russia and, 152, 154

Z

Zawahiri, Ayman al-, 12, 13
Zimbabwe, 20, 215

Test Your Knowledge Form

We encourage you to photocopy and use this page as a tool to assess how the articles in *Annual Editions* expand on the information in your textbook. By reflecting on the articles you will gain enhanced text information. You can also access this useful form on a product's book support Web site at *http://www.dushkin.com/online/*.

NAME: DATE:

TITLE AND NUMBER OF ARTICLE:

BRIEFLY STATE THE MAIN IDEA OF THIS ARTICLE:

LIST THREE IMPORTANT FACTS THAT THE AUTHOR USES TO SUPPORT THE MAIN IDEA:

WHAT INFORMATION OR IDEAS DISCUSSED IN THIS ARTICLE ARE ALSO DISCUSSED IN YOUR TEXTBOOK OR OTHER READINGS THAT YOU HAVE DONE? LIST THE TEXTBOOK CHAPTERS AND PAGE NUMBERS:

LIST ANY EXAMPLES OF BIAS OR FAULTY REASONING THAT YOU FOUND IN THE ARTICLE:

LIST ANY NEW TERMS/CONCEPTS THAT WERE DISCUSSED IN THE ARTICLE, AND WRITE A SHORT DEFINITION:

We Want Your Advice

ANNUAL EDITIONS revisions depend on two major opinion sources: one is our Advisory Board, listed in the front of this volume, which works with us in scanning the thousands of articles published in the public press each year; the other is you—the person actually using the book. Please help us and the users of the next edition by completing the prepaid article rating form on this page and returning it to us. Thank you for your help!

ANNUAL EDITIONS: World Politics 04/05

ARTICLE RATING FORM

Here is an opportunity for you to have direct input into the next revision of this volume.
We would like you to rate each of the articles listed below, using the following scale:

1. **Excellent: should definitely be retained**
2. **Above average: should probably be retained**
3. **Below average: should probably be deleted**
4. **Poor: should definitely be deleted**

Your ratings will play a vital part in the next revision.
Please mail this prepaid form to us as soon as possible.
Thanks for your help!

RATING	ARTICLE
	1. Clash of Globalizations
	2. War
	3. The Protean Enemy
	4. A World of Exiles
	5. The People's Sovereignty
	6. Charging Ahead
	7. The Market for Civil War
	8. From Petro to Agro: Seeds of a New Economy
	9. Ex-Inspector Says C.I.A. Missed Disarray in Iraqi Arms Program
	10. The Nuclear Crisis on the Korean Peninsula: Avoiding the Road to Perdition
	11. N. Korea Shops Stealthily for Nuclear Arms Gear
	12. Nuclear Nightmares
	13. Towards an Internet Civil Defence Against Bioterrorism
	14. Bush's Revolution
	15. Supremacy by Stealth: Ten Rules for Managing the World
	16. The Watchful and the Wary
	17. Economic Crossroads on the Line
	18. Canada Links Arrest of 19 to Possible Terrorism Ties
	19. Free Trade on Trial
	20. Latin America's New Political Leaders: Walking on a Wire
	21. Europe Enlarged, America Detached?
	22. America as European Hegemon
	23. Forget Asylum-Seekers: It's the People Inside who Count
	24. How the Armies of Europe Let Their Guard Down
	25. A Nervous New Arrival on the European Union's Block
	26. US-Russian Relations: Between Realism and Reality
	27. The Terrorist Notebooks
	28. Changing Course on China
	29. How to Deal with North Korea
	30. Can India Overtake China?
	31. Dangerous Neighbours

RATING	ARTICLE
	32. "Why Do They Hate Us?"
	33. The Reluctant Nation Builders
	34. The Fall of the House of Saud
	35. America and Africa
	36. United Nations
	37. The Five Wars of Globalization

(Continued on next page)

ANNUAL EDITIONS: WORLD POLITICS 04/05

NO POSTAGE
NECESSARY
IF MAILED
IN THE
UNITED STATES

BUSINESS REPLY MAIL
FIRST CLASS MAIL PERMIT NO. 551 DUBUQUE IA

POSTAGE WILL BE PAID BY ADDRESEE

McGraw-Hill/Dushkin
2460 KERPER BLVD
DUBUQUE, IA 52001-9902

ABOUT YOU

Name　　　　　　　　　　　　　　　　　　　　　　　　　　　Date

Are you a teacher? ❐　A student? ❐
Your school's name

Department

Address　　　　　　　　City　　　　　　　　State　　　　Zip

School telephone #

YOUR COMMENTS ARE IMPORTANT TO US!

Please fill in the following information:
For which course did you use this book?

Did you use a text with this ANNUAL EDITION?　❐ yes　❐ no
What was the title of the text?

What are your general reactions to the *Annual Editions* concept?

Have you read any pertinent articles recently that you think should be included in the next edition? Explain.

Are there any articles that you feel should be replaced in the next edition? Why?

Are there any World Wide Web sites that you feel should be included in the next edition? Please annotate.

May we contact you for editorial input?　❐ yes　❐ no
May we quote your comments?　❐ yes　❐ no